# Understanding Statistics

# Understanding Statistics

**Bruce J. Chalmer**
*Montpelier, Vermont*

MARCEL DEKKER, INC.                    New York and Basel

Library of Congress Cataloguing-in-Publication Data

QA
276.12
.C446
1987

Chalmer, Bruce J.
    Understanding statistics.

    Bibliography : p.
    Includes index.
    1. Statistics. I. Title.
QA276.12.C446   1987   519.5   86-16614
ISBN 0-8247-7322-5

Marcel Dekker, Inc. and Statpal Associates make no warranty with regard to the *Statpal* software, its accuracy, or its suitability for any purpose other than specified in the manual. *Statpal* software is licensed solely on an "as is" basis. The only warranty made with respect to the *Statpal* diskette is that the diskette medium on which the software is recorded is free of defects. Marcel Dekker, Inc. will replace a diskette found to be defective if such defect is not attributable to misuse by the purchaser or his agent. The defective diskette must be returned within ten (10) days of receipt to:

Customer Service
Marcel Dekker, Inc.
Post Office Box 5005
Cimarron Road
Monticello, NY  12701

Comments, suggestions or bug reports concerning *Statpal* are welcome and should be sent to:

Statpal Associates
18 Summer Street
Montpelier, VT  05602

IBM PC is a registered trademark of International Business Machines Corporation.

MARCEL DEKKER, INC.,
270 Madison Avenue, New York, New York  10016

Current printing (last digit):
10 9 8 7 6 5 4 3 2 1

PRINTED IN THE UNITED STATES OF AMERICA

# Preface

This book is intended for a one-semester introductory course in statistical methods. It has two main distinguishing features. First, it stresses concepts rather than mathematical notation, while at the same time allowing the instructor to include as much or as little mathematical detail as is appropriate for the students. Mathematical notation has been included, but is not relied upon. Thus, this book can be used for a survey of statistical methods for students with as little mathematical background as high school algebra, or for a more traditional statistical methods course for students with greater mathematical sophistication. Either way, the text is intended to teach statistical concepts, not mere squiggle manipulation.

Second, both text and exercises assume that students have access to a computer—preferably a microcomputer. Many exercises are specifically written to use the Statpal statistical package for microcomputers (although they can be adapted to other software if necessary). Data sets are provided in both printed and diskette form.

Each chapter is divided into sections headed by a key sentence which expresses the main idea of the section. Interspersed throughout the text, following each section, are study questions (exercises). The study questions have two functions. First, they are intended to solidify understanding of the concepts already presented. Second, and equally important, they are designed to set the stage for concepts soon to be presented. For this reason, study questions should generally be done in their place—that is, before reading subsequent sections. Study questions labeled ''Statpal exercise'' are intended for use with Version 5 of Statpal for the IBM PC and compatibles (included with the text). However, instructors can adapt these exercises to other software if necessary.

Chapters 1 and 2 introduce some of the main concepts of statistical inference, using nothing more complicated than proportions and (unspecified) averages in the examples. Chapter 3 then introduces measures of central tendency and variability. The rationale behind this approach is to stress the generality of inferential concepts and to avoid some of the confusion that students often experience when inferential concepts (particularly the key concept of a sampling distribution) are presented in the context of unfamiliar methods.

Chapter 4 presents the normal and binomial distributions, laying the foundation for the inferential procedures to be covered in subsequent chapters. Chapters 5 and 6 then introduce the general techniques of interval estimation and hypothesis testing and Chapter 7 applies these techniques to inferences about a single population mean. Chapter 8 extends these techniques to cover proportions, and introduces nonparametric approaches.

Chapters 9 and 10 expand the methods to cover inferences about group differences, with Chapter 9 concentrating on the two-group case (again including both parametric and nonparametric approaches) and Chapter 10 covering one-way ANOVA and its nonparametric analog.

Chapter 11 turns to the problem of describing relationships between two variables. Of particular interest is the use of an intuitively defined measure here called "easy-r"—a modified phi coefficient—to motivate the discussion of the correlation coefficient. Spearman's rho and Kendall's tau are also presented.

Chapters 12 and 13 comprise a survey of basic regression methods, including simple and multiple regression. Among the topics covered is stepwise regression, with considerable emphasis given to the dangers of misapplied stepwise methods.

Chapter 14 introduces the analysis of variance of factorial designs, including a discussion of the concept of interaction. Finally, Chapter 15 introduces the analysis of categorical data using the chi-square test.

I am grateful to the many people who have, directly or indirectly, helped shape this book. Special thanks go to Paul Chalmer and Karen Chalmer, whose comments on an earlier form of the manuscript were tremendously helpful; to an anonymous reviewer, whose thoughtful, page-by-page remarks were a model of constructive criticism; to the many students at the University of Vermont who have previewed various sections of this book; and to Vickie Kearn, Executive Editor, Physical Sciences, at Marcel Dekker, Inc., whose patience, skill, and courtesy have greatly lightened the task of preparing this book.

To *Judith Chalmer* I owe more than I can say; here I will mention only her generosity in applying her professional editing skills to the task of clarifying and refining my prose. To her I dedicate this book.

*Bruce J. Chalmer*

# Contents

# Understanding Statistics

# 1

# Variables, Populations, and Samples

## 1.1 Statistics is concerned with describing and explaining how things vary.

### What is statistics?

Things vary. You do not need a course in statistics to tell you that. People, objects, words—all things that have attributes that can be categorized or measured—vary. *Statistics* is the discipline that deals with describing and explaining how things vary.

### Variables and data

We use the term *variables* for the attributes we are interested in. Variables used in studies of people, for example, include height, hair color, attitude toward nuclear power as indicated by response to a question on a particular survey, age at death, and so on. Individuals can be categorized or measured in terms of each variable, yielding *data*. We will use the terms *data* and *information* synonymously.

### Populations and samples

The term *population* refers to the entire set of individuals in whom we are interested. (A synonym is *universe*.) A bit more precisely, the term *population* refers to the data that could (at least theoretically) be obtained for these individuals on one or more variables. The important point is that the population is the whole set of interest. A population can be real, such as the population of weights of babies born in 1980, or theoretical, such as the population of weights of babies who have ever been born or will ever be born to mothers who smoke one pack of cigarettes a day.

The term *sample* refers to a subset of a population. Most statistical studies are based on samples, since data for the complete population are usually impossible to obtain. (A notable exception is a census, which seeks to obtain information on the entire population.) It is important to use the term ''sample'' correctly. If you select 50 individuals from a population, you have one sample of size 50. You do *not* have 50 samples. A sample is a group of individuals from a population.

## Inferences about populations: parameters versus statistics

We often wish to draw inferences about populations based on sample data. For example, we might wish to estimate the percentage of households in which the Super Bowl was viewed, based on data from a sample of households. Or we might wish to predict whether a new pneumonia vaccine will reduce the number of deaths among people at high risk, based on the experience of a sample of such people.

The term *parameter* refers to some feature of the population as a whole. In our examples, the percentage of households watching the Super Bowl and the number of deaths among high-risk people vaccinated with the new vaccine are parameters.

The term *statistic* refers to some feature of a sample. In our examples, the percentage of households watching the Super Bowl *in a particular sample* and the number of deaths *among a sample* of high-risk people vaccinated with the new vaccine are statistics.

If we were able to obtain complete data for the population, we could calculate the parameters in which we are interested, and there would be no need to use sample information. Usually, however, the best we can do is to use statistics to make inferences about parameters.

### STUDY QUESTIONS—*Section 1.1*

1.1.1 Distinguish between *population* and *sample*.

1.1.2 Distinguish between *parameter* and *statistic*.

1.1.3 Suppose that you wish to determine the current unemployment rate in your community. How would you do it? In planning your strategy, consider the points discussed below. Don't expect to be able to answer the questions in detail, but be sure you understand what is being asked and why it is important.

a. What is the population in this case? Does it include everyone who lives in your community, or everyone over a certain age, or only people who are willing and able to be employed? Do you wish to find out about total unemployment, or ''insured'' unemployment (people who are receiving unemployment benefits)? Note that the decisions you make on these points will greatly affect the result.

b.  Do you have access to information for the entire population, or will it be necessary to obtain a sample? If you must obtain a sample, how will you select the individuals to be included?

c.  If you need to obtain a sample, how many individuals should be included? Why does this matter? Think about this carefully!

d.  Once you have determined the individuals for whom you want to get data, how will you get the data? Will you have interviewers visit people, or call people on the telephone, or get information from other sources?

e.  Once you have obtained the necessary data, how will you calculate the unemployment rate? Will you be absolutely sure that your result is correct? On what will your degree of certainty depend?

1.1.4  In an effort to determine the concerns of its membership, an environmental organization in Vermont sent out surveys to about 2500 members. About 600 members responded. Of the members who responded, 97% said they read the organization's newsletter regularly.

a.  Would it be reasonable to infer that nearly all the members of the cooperative read the newsletter regularly? Why not?

b.  Suppose that instead of mailing our surveys, the organization had picked, say, every tenth member on the membership list, contacted each member individually, and asked if the member reads the newsletter regularly. Assuming that nearly all people asked were willing to answer the question truthfully, why would this method yield much more reliable results?

## 1.2  Inferences about populations based on sample information are subject to error.

Any time we make a statement about a population based on incomplete information, our statement might be wrong, for any of a variety of reasons.

### Blunders

One source of error might be called ''blunders''—errors in writing down the answers to questions, arithmetical errors, sloppy laboratory technique, faulty computer programs, and so on. Blunders can have an astonishing effect on the results of a study. There is a maxim in the computer field that applies equally well to statistical work: ''garbage in, garbage out.'' No matter how fancy the methodology, faulty data lead to faulty conclusions.

Blunders are a major concern in practical work, and we must guard against them. However, there is another source of error that we must be concerned about whenever we use sample information to draw inferences about a population. Our inferences about the population can be wrong even if we are careful about how we collect the sample data. The reason is that the sample may not be representative of the population. Thus sample statistics may be quite different from population parameters. This source of error—that is, error resulting from the fact that a sample may not give an accurate picture of the population as a whole—is known as *sampling error*.

## Systematic sampling error

*Systematic sampling error* is present when a sample is selected in a biased fashion. For example, suppose that you wish to estimate the percent of adult women in your community who are employed outside the home. A sample consisting of people encountered on a downtown street during the noon hour would almost certainly contain a higher proportion of employed women than in the population. A sample obtained by telephoning people at home during a weekday would almost certainly contain a lower proportion. Either way, the sample is biased and the resulting conclusion will be off in a predictable direction.

Other examples of systematic sampling error are more subtle. Consider the problems involved in mail surveys, in which some proportion of the people who receive questionnaires do not return them. If the people who do not respond are different in some important way from the people who do respond, the results of the survey can be quite misleading. The *Literary Digest* received millions of responses to its voter preference survey in 1936—and confidently predicted that Landon would defeat Roosevelt. In this case, the people who mailed back the surveys tended to have higher incomes and were less pleased with New Deal policies than people who did not. Thus the sample was seriously biased. This type of problem might be termed *self-selection bias*, since the bias resulted from the fact that the sample consisted of people who selected (or excluded) themselves.

Often, people try to eliminate systematic bias by carefully selecting individuals who seem to be typical of the population. This method is nearly always a failure. No matter how careful we are, our own biases will creep into our decisions. What we need is a way of selecting samples without systematic bias.

## Randomization

The best way to eliminate systematic bias is to *randomize*. The idea of randomizing is to leave the decision about which individuals get into the sample up to chance. For example, suppose that we need to select a sample of 100 state employees to interview for a study of job satisfaction. We would first obtain a list of all state employees (or more likely, gain access to such a list stored in a computer). Suppose that there are 10,000 names on the list. We might number the names from 1 to 10,000. Next, we obtain 100 numbers in the range between 1 and 10,000 by some random process. Our sample consists of the employees on the list with the selected numbers.

### Simple random sampling

The type of sampling procedure described above is called *simple random sampling*. The idea of simple random sampling is that every individual in the population has the same chance of being selected for the sample. In our state employees example, as long as the 100 numbers were chosen by a random process, everyone on the list had the same chance of being picked. Generally, the numbers would either be generated by a computer program or taken from specially prepared lists of random numbers. Entire books of random numbers have been published. (They make great bedtime reading.)

### Advantages of randomization

Randomization does not guarantee, of course, that the sample will look like the population. We could be unlucky and get a sample that gives a distorted picture of the population. Just by chance, our sample might include a disproportionate number of very satisfied employees, so that we would fail to detect some problems. Still, randomization has some very compelling advantages.

First, randomization (if performed properly) totally eliminates systematic bias. Although any particular random sample might mislead us one way or the other, there will be no general trend. Compared to the proportion in the entire population of employees, the proportion of dissatisfied employees in our sample is just as likely to be too high as too low.

Second, randomization allows us to use the mathematics of probability to help us draw conclusions. For example, suppose the state employees' union asserts that at least 50% of all state employees feel dissatisfied. Our sample reveals that 85 of the 100 employees feel dissatisfied. Is it possible that the union is wrong and that the sample was loaded with dissatisfied employees just by chance? Of course, it is theoretically possible. But if we apply some mathematical principles, we find that the chance of this occurring in a random sample is extremely small—so small that we conclude that the union is correct.

### STUDY QUESTIONS—*Section 1.2*

**1.2.1**  A part of a random number table appears as Figure 1-1. The process that generated the numbers was designed to make sure that each digit appears with roughly equal frequency and that there is no particular pattern in the digits. The numbers are printed in blocks purely for convenience—there is no meaning to the blocks.

| | | | | | | | |
|---|---|---|---|---|---|---|---|
| 14758 | 20898 | 32039 | 78196 | 85948 | 89663 | 93285 | 25742 |
| 08325 | 26425 | 31300 | 55022 | 69337 | 10656 | 26367 | 27952 |
| 49116 | 27272 | 38886 | 23254 | 50388 | 41643 | 22059 | 22935 |
| 68243 | 48713 | 33787 | 85013 | 38184 | 07778 | 65255 | 00014 |
| 75295 | 09423 | 04010 | 75371 | 33189 | 01479 | 46927 | 43455 |
| | | | | | | | |
| 52802 | 58433 | 36591 | 84317 | 37758 | 54433 | 99788 | 91679 |
| 20441 | 63445 | 73205 | 03906 | 70296 | 96444 | 78726 | 69223 |
| 86399 | 46773 | 13534 | 24850 | 29883 | 61564 | 27926 | 03588 |
| 90581 | 76778 | 54640 | 29996 | 45219 | 72250 | 12927 | 03645 |
| 22102 | 97491 | 58690 | 48111 | 79346 | 56150 | 33255 | 81460 |
| | | | | | | | |
| 40854 | 52964 | 69475 | 02032 | 25931 | 07371 | 15149 | 83187 |
| 93435 | 17635 | 57384 | 33739 | 60714 | 92600 | 96203 | 04774 |
| 71329 | 97557 | 97511 | 99314 | 54010 | 44791 | 78911 | 74272 |
| 68944 | 65531 | 38805 | 00835 | 14684 | 79348 | 50188 | 19827 |
| 40221 | 65850 | 80174 | 51756 | 53825 | 85919 | 22254 | 69627 |
| | | | | | | | |
| 60028 | 35646 | 27371 | 71274 | 42600 | 18805 | 22232 | 95961 |
| 17823 | 77021 | 28219 | 14845 | 59842 | 83831 | 97361 | 68612 |
| 69162 | 58851 | 68149 | 31872 | 04211 | 54603 | 16549 | 59699 |
| 25067 | 93987 | 82244 | 12429 | 10042 | 40931 | 17694 | 09874 |
| 48195 | 98907 | 91956 | 89220 | 54004 | 39590 | 84731 | 00205 |
| | | | | | | | |
| 06765 | 56961 | 46108 | 61742 | 48253 | 95125 | 38866 | 88944 |
| 74701 | 61809 | 66811 | 08306 | 69181 | 40256 | 05761 | 50862 |
| 33404 | 99779 | 92416 | 90025 | 45272 | 98217 | 65369 | 52904 |
| 81422 | 31318 | 04078 | 12623 | 15342 | 97072 | 22556 | 84449 |
| 88865 | 86800 | 87957 | 31618 | 23338 | 68830 | 19414 | 41415 |
| | | | | | | | |
| 28035 | 87160 | 05518 | 85412 | 87750 | 88459 | 72999 | 86851 |
| 70938 | 63113 | 21068 | 11043 | 40498 | 33145 | 69658 | 90120 |
| 60559 | 26312 | 28079 | 08779 | 14546 | 16039 | 25002 | 09180 |
| 43627 | 17505 | 30125 | 69234 | 98666 | 98441 | 45013 | 05228 |
| 37791 | 94556 | 33913 | 37806 | 75568 | 73532 | 93160 | 45142 |
| | | | | | | | |
| 66511 | 18995 | 96336 | 43968 | 20213 | 95013 | 52613 | 83094 |
| 00752 | 42593 | 97685 | 69903 | 94364 | 06270 | 61273 | 17422 |
| 59689 | 96799 | 28393 | 82697 | 64792 | 24273 | 76980 | 74916 |
| 30379 | 67525 | 76656 | 91186 | 28652 | 16819 | 13430 | 14078 |
| 72590 | 28833 | 26407 | 05741 | 60643 | 20667 | 51901 | 46219 |

**Figure 1.1** A table of random numbers.

a. Suppose that you have a list of 900 small businesses, numbered from 001 to 900, from which you wish to select a sample of size 9. Starting anywhere in the table, write down the next nine three-digit numbers. If you get a duplicate or a number greater than 900, skip it and use the next three-digit number. Consider these nine numbers to represent your sample.

b. Suppose that the first 300 businesses on the list are in manufacturing, the next 400 are in wholesaling, and the last 200 are in retailing. How accurately does your sample of nine businesses reflect these proportions?

c. Now select another sample, this time of 100 businesses. Are the proportions any more accurate than they were with nine? Generally, they will be. Why?

d. How could you select a random sample of 100 businesses so as to be certain that the proportions of the three types of businesses are exactly as they are in the population?

e. If you do succeed in matching the population proportions exactly, does that mean that your sample will be a perfect picture of the population in every respect? (No!) Why not?

f. Could you do better by picking out some businesses you think are "typical" for your sample?

1.2.2 Suppose someone claims that the random number table is defective because it underrepresents the digit "4." Of course, if the table was generated properly, 10% of the digits in the entire table should be 4's.

a. Select a sample of, say, 20 one-digit numbers from the table, and find the proportion of 4's.

b. Your proportion probably wasn't exactly 10%. Assuming that it was not, does this mean that the table is defective?

c. How far off should the percentage be before you would conclude that the table is defective? You can not really answer this without applying some mathematics. But what would be the basis for your decision?

## 1.3 Other forms of random sampling include stratified and cluster sampling.

### Stratified random sampling

*Stratified sampling* is a way of ensuring that a random sample will represent various groups in the same proportions as in the whole population. For example, suppose we wish to be sure that our sample of state employees will contain the same proportions of the two sexes as in the entire state work force—say, 55% male and 45% female. If we select our random sample of 100 employees as above, we might end up with too many employees of one sex or the other. A solution is simply to divide the population by sex and then select a simple random sample of the appropriate size for each sex—in this case, 55 males and 45 females. Note that the sampling is still random within each sex. The only difference is that we have made sure that we end up with the right number of individuals in each group. The groups are called *strata*, hence the name "strat-ified" random sampling.

If we had wanted to be even fussier, we could have stratified not only by sex but also by other factors, such as age, salary level, or agency. Of course, with a small sample size we are limited in how much we can stratify. If only, say, one-tenth of 1% of the population falls into a particular group—for example, 30- to 35-year-old males in the human services agency earning between $45,000 and $50,000—then a sample of 100 people should have one-tenth of a person in that group! One way around the problem is to *poststratify*, that is, include people in each group, but weight the responses so that the people in overrepresented groups are counted less heavily.

### Cluster sampling

*Cluster sampling* is a way of reducing the cost of gathering data. It is especially useful when gathering data involves travel time. The idea is to select clusters of individuals for the sample, rather than single individuals, so that less time is required to track down the individuals.

In cluster sampling, the population is divided into clusters on the basis of convenience. For example, in a survey of households, the entire population of households is divided into clusters such as city blocks. Then, instead of selecting individual households for the sample, we randomly select clusters and survey all the households in those clusters. Thus much less travel time is required than that needed to locate scattered households for a simple random sample.

Of course, individuals who live in the same neighborhood are likely to be similar to each other. We cannot be as certain that we have adequately repre-sented all segments of the population if our sample consists of clusters rather than individuals, so we are not quite as confident about our inferences as we would be with a simple random sample of the same size. The statistical methods used to deal with data from cluster samples take account of this fact.

Sometimes, the cost savings of cluster sampling are so great that we can afford to obtain a much larger sample—and thus greater certainty in our re-sults—than we could using a simple random sample. Formulas have been worked out to determine which type of sampling can give the most reliable information at the lowest cost. Methods for dealing with stratified or cluster samples are not covered in this book. See, for example, Cochran (1953) for a discussion of these methods.

### Subtle biases

Sometimes samples are assumed to be random. For example, many studies use rats or other animals to test the effects of a drug. The rats used in the study are simply assumed to be a random sample of all rats of the same type. As long as there was no systematic bias in how the rats were picked, this assumption will be valid.

However, it is important to watch out for subtle biases. For example, in studies using animals, *cage effects* often occur. If the cages in which the animals are housed are stacked against a wall (as in many animal care facilities), the cages near the ceiling could be substantially warmer than the cages near the floor; or the lighting levels could be different; or the animals in some cages might get less water than those in other cages because of the way the water is piped.

Generally, animal studies involve assigning the animals to groups, treating each group a certain way, and then comparing the groups. To test the toxicity of a drug, for example, some rats might be injected with the drug, while others (the "control" group) are injected with an equivalent volume of saline solution. The idea, of course, is to have the groups be equivalent in every way except for the treatment of interest. Differences in how the rats react to their injections, beyond what might happen just by chance, are then assumed to be due to the effect of the drug.

But suppose that the rats are assigned to the groups by cage, with the top few rows being given the drug while the bottom few rows are given the saline. If cage effects are present, we can no longer say for sure whether differences between the groups are due to the drug. Some of the differences could be due to cage effects. The solution is to randomize the assignment of rats to groups, usually with some restrictions similar to stratification to make sure that cage effects are balanced between the groups.

### STUDY QUESTIONS—*Section 1.3*

**1.3.1** Suppose that you wish to determine the proportion of adults in a large state who smoke at least one pack of cigarettes a day. How would you select the sample? Consider these points:

1. Cluster sampling will be *much* less expensive than simple random sampling.
2. You want to be sure to represent various areas of the state in the right proportions.
3. You are interested in the proportion of adults, but it is much easier to sample households. However, households differ in the number of adults who live in them. How do you decide whom to ask in each household? Should you simply ask the person who answers the door (or telephone)?

**1.3.2** Many psychological studies involve college students, since they are available. (In many places, participation in a research project, as a subject, has been a requirement of introductory psychology courses.) What must researchers assume if they wish to make inferences about people in general based on samples of college students? Do you see any problems?

**1.3.3**   [Statpal exercise—Coronary Care data set (see Appendix A)]

    a.   Read the description of the Coronary Care data set in Appendix A.

    b.   Read the chapter of the Statpal user's manual that covers the creation of Statpal system files.

    c.   Create a Statpal system file for the data listed in Appendix A. (Note that you will find it particularly easy to do this if your instructor provides a disk file with the raw data already entered!) In Statpal terminology, what is a "case" in this context? What are the variables?

    d.   Use Statpal's LIst facility to obtain a printed listing of the data, and check it against Appendix A.

**1.3.4**   [Statpal exercise—Electricity Consumer Questionnaire data set (see Appendix B)]

    a.   Read the description of the Electricity Consumer Questionnaire data set in Appendix B.

    b.   Create a Statpal system file for the data listed in Appendix B. (Again, this can be made easier if your instructor provides a disk file with the data already entered.)

    c.   Use the LIst facility to get a printed listing of the data, and check the results against Appendix B.

**1.3.5**   A radio advertisement for a smoker's tooth polish made the following claim: "We surveyed over 250,000 users of our product, and of those who returned reply cards, over 95% said they would buy the product again." What does the advertiser want you to infer from this information? What can you legitimately conclude?

# 2

# Basic Ideas of Statistical Inference

## 2.1 There are two basic tasks of statistical inference.

The first task is to make statements about parameters using statistics. The second task is to determine how certain we are that those statements are true. We discuss these tasks in this chapter.

There are many different types of parameters. We discuss a variety of parameters, together with the corresponding statistics, in later chapters. Each type of parameter describes a particular feature of the population. For example, one parameter often of interest is some sort of average. Or we might want to know the proportions of individuals that fall into various categories on some variable—say, the proportion of light bulbs that burn out within 10 hours, or between 10 and 20 hours, or between 20 and 30 hours, and so on. Other types of parameters involve more than one variable, such as the average loss in fuel economy for each additional kilogram of weight in a car.

But whatever parameters are of interest, the basic tasks of statistical inference are the same: to make statements about parameters and to determine how certain we are about those statements. We consider each task in turn.

### Making statements about parameters using statistics

The first task for which we use statistical inference is to make statements about parameters using statistics. Of course, if we have access to data for the entire population, no inference is needed; we can simply calculate the parameters in which we are interested. But usually, we must use sample statistics to help us make statements about population parameters. These statements can be of several types.

### Estimation and hypothesis testing

One type of statement about a parameter is simply to estimate its value. An estimate can be a *point estimate*, meaning a single value, or an *interval estimate*, meaning a range of values. For example, we might estimate that the number of people who will die in traffic accidents over a certain weekend will be 450 (a point estimate), or we might estimate the number to be between 430 and 470 (an interval estimate).

Another type of statement about a parameter derives from a procedure called *hypothesis testing*. Hypothesis testing is related to estimation, but the emphasis is different. For example, we might wish to determine whether there is any difference in average amount of pain reported after oral surgery between people taking a drug and people receiving a placebo (a substance that has no inherent effect on pain). We are looking for statistical evidence that the drug has some effect on pain beyond the placebo effect. (Curiously, many studies have shown that about a third of people who receive placebos derive considerable pain relief.) The emphasis is not on estimating the degree of difference between drug and placebo, but simply on testing whether there is a difference.

### Determining degree of certainty

The second task for which we use statistical inference is to determine how certain we are that our statements are true. It may seem odd to make statements if we are not sure they are true. (Actually, the odd part is admitting it!) But as we noted in Chapter 1, inferences about parameters based on statistics are always subject to error. We can never be completely sure that a sample gives a good picture of the whole population—unless, of course, our sample is the whole population. So there is always the possibility that our statements will be false. The beauty of statistical inference based on random samples is that we can state how worried we are about that occurring.

### The importance of statements about certainty

It is important to understand why a statement about our degree of certainty is so useful and necessary. If you are told that the price of a certain stock, based on past trends, is estimated to increase by $2 to $10 a share over the next 3 months, you may be inclined to invest heavily in the stock. If, however, you are also told that past trends are so capricious that we can be only 60% sure of the prediction, you might hesitate. Such a relatively low degree of confidence would suggest that you at least hedge your bet.

How do statistical methods allow us to evaluate degree of certainty? As we have already noted, the basic idea is to apply the mathematics of probability. We discuss the general principle first and then consider some specifics.

## 2.2   To evaluate degree of certainty, first consider all possible samples.

### All possible samples

Think about the entire population from which you selected your random sample. Let's assume that the population is large. Then there are an enormous number of possible samples you could have selected. The sample you actually selected was simply a random pick from that vast number of possible samples.

Picture it for a moment. For example, suppose that your sample consists of 25 trees selected at random from a large forest of, say, 10,000 trees. The number of possible samples of 25 trees from a forest of 10,000 trees is *much* greater than 10,000/25. (Do you see why?) In fact, the number is somewhere in the neighborhood of 1,000,000,000,000,000,000,000,000,000,000,000,000,000,000, 000,000,000,000,000,000,000,000,000. Each of those possible samples of 25 trees had the same chance of being selected if you chose your sample in a random fashion. You have selected one of them. (The set of all possible samples is known as the *sample space.*)

Now suppose that the parameter you are interested in is the average diameter of the trees in the forest. You would find the 25 trees in your sample, measure each tree using calipers at a standard height above the ground, and then calculate the sample average. (We'll have more to say about averages in Chapter 3. For now, we could be talking about *any* parameter and its corresponding statistic.)

Your one sample of 25 trees, then, is simply a random pick from a vast number of possible samples. Its average is a random pick from the same vast number of possible sample averages. How close will your sample average be to the population average? Of course, you have no way of knowing for sure. If you make a statement about the population average based on your sample average, you might be wrong. How can you tell what the chance is that a given statement will be wrong?

### The reasoning of determining degree of certainty

The key is to think about all those possible samples, from which you have selected one. If you knew what proportion of those samples are "good," in the sense that their statistics are close enough to the parameter of interest, you would have the answer. If half of the possible samples are good, you are 50% sure that the sample you selected is good. If 90% are good, you are 90% sure.

But how could we ever know what proportion of all the possible samples are good? Wouldn't we have to look at all of them—an impossible task? Wouldn't we also need to know the value of the parameter, so that we could tell if each sample's statistic were close enough to call the sample good? If that were true, we might just as well use the entire population and forget about sampling altogether.

The answer is that in many situations, we *can* know what proportion of all possible samples are good, even without looking at all of them and even without knowing the value of the parameter in advance, by applying the laws of probability. How this works is the subject of a later section. The important point to understand now is that knowing this makes it possible to say how certain we are about our statements. Remember the reasoning: If, say, 90% of all the possible samples are good, we are 90% sure that the particular sample we have selected at random is good.

### The trade-off between certainty and precision

We have seen that knowing the proportion of good samples among all possible samples allows us to say how certain we are about our inferences. A sample is good if it yields a statistic that is close enough to the parameter of interest. We do not know for sure if the particular sample we have is one of the good ones, but we can at least say how confident we are that it is.

Of course, which samples are good depends on how close you want the statistic to be to the parameter. For example, if we want to use the sample average to estimate the population average, and we want our estimate to be within, say, 3 centimeters (cm) of the actual value, then a sample is good if its average is within 3 cm of the population average. If we insist on being within 1 cm, fewer samples will be good. If we were to insist on hitting the population mean exactly, there would probably be no good samples at all, since it is unlikely that a sample average would be *exactly* the same as the population average—carried out to enough decimal places.

So if we are willing to allow for a greater margin for error—that is, less precision—in a statistic and still call it good, then more of the possible samples will be good and we can be more sure that the sample we actually picked is one of the good ones. If, on the other hand, we insist that a statistic be very accurate before we call it good, very few of the possible samples will be good and we will not have much confidence that we got a good one. This simply means that we can be more sure of ourselves if we are willing to be less precise in our statements about parameters.

STUDY QUESTIONS—*Sections 2.1 and 2.2*

2.2.1   Suppose that you are interested in the average height of adult males. You obtain a random sample of five adult males and find that their average height is 175 cm. How certain would you be about each of the following statements? Do not worry about putting an exact percentage on it, but make an educated guess based on your general knowledge and the information given. (*Note*: 175 cm is about 5 ft 9 in.)

   a.   "The average height of adult males is between 0 and 1000 cm." (1000 cm is about 33 ft.)
   b.   "The average height of adult males is between 125 and 225 cm."
   c.   "The average height of adult males is between 170 and 180 cm."
   d.   "The average height of adult males is exactly 175 cm."
   e.   "The average height of adult males is greater than 150 cm."
   f.   "The average height of adult males is greater than 175 cm."

2.2.2   Suppose that your interest is not only in the average height, but in a whole distribution of heights—say, the proportion of adult males less than 100 cm, between 100 and 120 cm, between 120 and 140 cm, and so on. How useful is your sample of five adult males for estimating this distribution in the population?

2.2.3   Of course, you are pretty familiar with the heights of adult males, so what you find in one particular sample of five adult males probably will not make much difference in your opinions about either the average height or the entire distribution. If you ended up with a sample of five males all over 190 cm (about 6 ft 3 in.), you would certainly not conclude that all adult males are that tall, nor would you conclude that the average height of adult males is that great. But what might you conclude about your sampling procedure?

2.2.4   Suppose that the population from which you drew your sample was not adult males in general, but male professional basketball players. Suppose further that your sampling procedure was strictly random and unbiased. Now suppose again that you ended up with a sample of five males all over 190 cm tall. This is, of course, a lot less surprising with basketball players than it was with adult males in general.

   a.   Can you conclude from your sample that the entire population of male professional basketball players are over 190 cm tall? (Of course not!) Why not?
   b.   Suppose that the average height of the five players was 205 cm. In the absence of other information, your best estimate of the average for the population is 205 cm. Again, this does not necessarily mean that the population average is that tall—you might simply have chosen an unusually tall sample. A commentator claims that the average height of basketball players is only 200 cm. Your best guess—a point estimate—is higher than that. How certain are you that the population average is really greater than 200 cm? What type of information do you need to answer the question?
   c.   Would it make a difference in your answer if the players in the sample still had an average of 205 cm but ranged in height from 160 to 240 cm?

## 2.3   To find degree of certainty, find the sampling distribution of the statistic.

Whenever we use a statistic to make a statement about a parameter, our statement will be either right or wrong. (Of course, when we make the statement we hope it is right, but we cannot know for sure.) As we have seen, our statement will be right if we are lucky enough to get a random sample that gives us a statistic close to the parameter of interest. We have also seen that our degree of certainty is simply the proportion of all possible samples that are good in just that sense. But how can that proportion be calculated?

### Sampling distributions

Let's go back to our tree example: You wish to use the average diameter of a random sample of 25 trees to estimate the average diameter of the all the trees in the population. Each one of the stupendous number of possible samples of 25 trees has a sample average. Picture them for a moment—a vast array of sample averages waiting for you to choose one.

The problem of figuring out the proportion of good samples comes down to that of figuring out the proportion of good statistics—in this case, good sample averages. Since "good" depends on how close we want to be, what we really want is to know what proportion of all the possible sample averages out there are within any given distance of the population average. That is, we want to know what proportion of the possible sample averages are within, say, 10 cm of the population average, or within 5 cm, or within 1 cm, or within any margin for error we choose.

This set of information—the proportion of all possible sample averages that lie within any given distance of the population average—is called the *sampling distribution* of the sample average. Be careful not to confuse "sampling distribution" with "sample distribution," which has an entirely different meaning. (Sample distribution refers to the distribution of individuals in a particular sample—*not* the distribution of some statistic in all possible samples.) If we can calculate the sampling distribution of a statistic, we can state how certain we are that our particular sample gave us a statistic within any given distance of the parameter of interest.

### Sampling distributions in statistical inference

*The concept of the sampling distribution of a statistic is the most important idea in statistical inference.* Virtually every inferential procedure involves knowing, approximating, or at least assuming knowledge of the sampling distribution of some statistic.

Whenever you see the words "sampling distribution" or use a procedure that involves the use of the concept, you should form an image in your mind's eye of all possible samples, each yielding its own value of the statistic. Use whatever imagery you like. For example, you might think of the sampling distribution of some statistic as an anthill teeming with ants, each ant representing a sample of whatever size you are using. Each ant is wearing a T-shirt bearing a number. What number? The value of the statistic! When you pick a random sample, you are reaching in (blindfolded, of course) with a tweezers and removing one ant. If you are lucky, the ant you pick is wearing a T-shirt with a number close to the parameter of interest. What is the chance you are lucky? It is simply the proportion of ants wearing lucky T-shirts, that is, T-shirts with numbers close to the parameter on them.

Notice that a sampling distribution is itself a population—but instead of being a population of individuals, it is a population of sample statistics. When you select a random sample of, say, 25 individuals from a population, you can look at the process as making 25 selections from the population of individuals or as making one selection from the population of possible samples. The thing to keep in mind is that the population of individuals is not at all the same as the population of possible samples (unless, of course, you are dealing with the trivial case of ''samples'' of size 1).

## Normal sampling distributions

It turns out that many statistics have a *normal sampling distribution*. The term ''normal'' has a specific meaning, referring to a particular mathematical function. (It does not mean that there is something wrong with statistics that do not have a normal sampling distribution!) The term *Gaussian distribution* is also used for the same mathematical function (after the mathematician Gauss).

A basic feature of a normal distribution is that most individuals are fairly close to the average, with relatively few individuals far away from the average. In the case of a sampling distribution of some statistic, this means that most of the possible samples give a statistic that is fairly close to the parameter, with relatively few samples giving a statistic that is far from the parameter.

It should not be surprising that statistics such as a sample average tend to have a normal distribution. After all, we would expect most random samples to give an average somewhere near the population average, particularly if we are talking about large random samples (i.e., samples with many individuals in them). Even if there are many extreme scores in the population, you would expect most sample averages to be less extreme, because it would be very unusual for a sample to contain mostly extreme scores.

## The central limit theorem

As we have seen, knowing the sampling distribution of a statistic is the key to using the statistic to make inferences about the parameter of interest. So the fact that many statistics are known to have a particular type of distribution—the normal distribution—is of great value. But how do we know this? After all, the normal distribution is a very specific mathematical function.

The answer is that the mathematics of probability can be used to prove it. Specifically, a theorem called the *central limit theorem* (CLT) says that *any statistic that is calculated by summing or averaging will tend to have a normal distribution*. Since many statistics are averages of one type or another, the CLT applies to them. We do not need to bother with a formal statement or proof of the CLT (which requires some advanced mathematics), but at least we know that our intuitive explanation is backed up by hard-nosed mathematical proof.

STUDY QUESTIONS—*Section 2.3*

2.3.1   Suppose that you wish to estimate the proportion of circuit boards from a particular production line that operate properly for at least 300 hours. You select a random sample of 40 boards from the line (from a very large population) and plan on testing each of the 40 boards for 300 hours, so that you can see how many fail.
   a.   What is the parameter of interest here?
   b.   What statistic will you use to estimate it?
   c.   What do we mean by the ''sampling distribution'' of the statistic? Be specific!
   d.   Suppose you knew that 90% of possible samples of 40 boards would give a proportion of failing boards that is within .08 (i.e., 8%) of the population proportion. In doing your tests, you find that 4 of the 40 boards (10%) fail. How certain are you that your observed proportion (10%) is within 8% of the population proportion?
   e.   Based on your answer to part d, give an interval estimate of the population proportion in which you have 90% confidence.

2.3.2   We noted in Section 2.3 that statistics that are calculated by some type of averaging process tend to have sampling distributions that are normal. In fact, the normal distribution shows up in many contexts, precisely because many things in nature arise as the result of the action of many factors working more or less independently. For example, scores on the Scholastic Aptitude Test tend to have a normal distribution, probably because a person's score on the SAT depends on so many different facets of ability. To get an extremely high or extremely low score is very rare, because whereas many people are extremely good or extremely bad at *some* types of problem, very few people who take the SAT are extremely good or extremely bad at *all* types of problem. Thus people tend toward the center of the distribution.
   a.   Suppose that you were to sort a large set of SAT answer sheets into piles according to total score, scaled from 200 to 800. Each pile contains the answer sheets for a range of, say, 10 points, so that you would have a total of 60 piles. Make a sketch of what the piles might look like if they were ordered by test score.
   b.   Now suppose that instead of sorting individual answer sheets, you are sorting

score summaries for individual high schools according to the average score in the school. Make a sketch of what the piles might look like. In what way is the distribution of high school averages different from the distribution of individual scores?

## 2.4  A handy way of representing a distribution is to draw a histogram.

### Histograms

A *histogram* is a graph that shows the proportion of individuals that have each possible value on some variable or statistic. It is as if all the individuals were piled up on a number line according to their score on the variable. In statistical terminology, a histogram is not simply a bar graph. Rather, a histogram is a graph that represents the frequency of occurrence of the scores on a variable. Let's look at some examples of histograms.

Figure 2.1 shows a histogram of the weights (in kilograms) of a sample of 2150 four-year-old girls in Basel, Switzerland, in 1956–1957 (Heimendinger, 1964). (The histogram is based on extrapolations from summary data. The figures in this study, incidentally, are still often used to evaluate growth for children all over the western world, in spite of the narrow geographic range covered by the sample.) As we might expect, the histogram shows most children near the middle of the weight distribution, with few children at the extremes. This sounds like our description of the normal distribution in the preceding section, for good reason: As is true of many other naturally occurring variables, children's weights do tend to have an approximately normal distribution.

Note that Figure 2.1 does not show an *exactly* normal distribution, which is a theoretical abstraction. Figure 2.2 shows an idealized histogram of a normal distribution. We discuss this in more detail in Chapter 3, but for now, notice the general bell-like shape of a normal distribution, indicating the preponderance of scores toward the center.

### Skewed distributions

Not all variables have distributions that are close to normal. A common example is household income level. Figure 2.3 shows a histogram of household income levels for the state of Vermont as reported in the 1980 census. Of course, since few households have incomes at the high end, the histogram is concentrated toward the lower and moderate incomes.

A nonsymmetrical distribution such as the one in Figure 2.3 is referred to as *skewed*. If the tail points to the right, as in Figure 2.3, the distribution is said to be skewed to the right, and if the tail points to the left, the distribution is said to be skewed to the left.

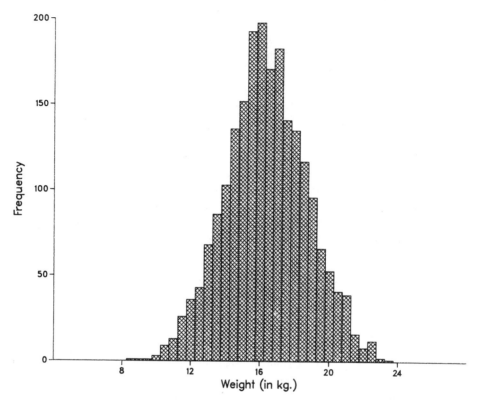

**Figure 2.1**  Weights of 2150 four-year-old girls. (From Heimendinger, 1964.)

## Histogram of a sampling distribution

The histograms in Figures 2.1 and 2.3 display information about individual scores; that is, we can think of the histograms as representing the individuals piled up on the number line, with each individual placed on the pile over its score on the variable. In Figure 2.1 each "individual" is a child's weight, and in Figure 2.3, a household's income. Putting all the individuals together produces the histogram.

In the case of a sampling distribution, the "individuals" are the statistics associated with *all possible samples from the population*. For example, if we wanted to make a histogram of the sampling distribution of a statistic—say, the sample average—in samples of size 30 from the population of weights of 4-year-old girls, we would need to calculate and record the sample average from every possible sample of 30 weights of 4-year-old girls. Having written down this (unthinkably long) list of sample averages, we could then prepare a histogram of

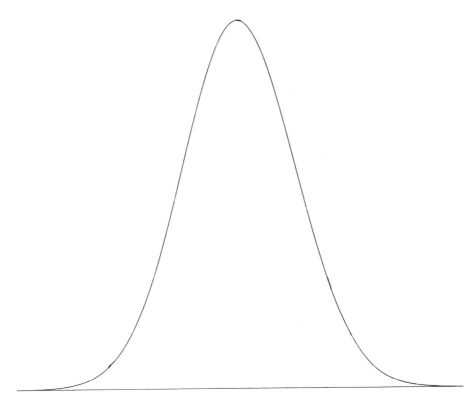

**Figure 2.2**   Histogram of normal distribution.

the averages just as we did for the individual scores. Of course, this is not feasible. But it is possible, and instructive, to simulate the process by computer.

### Simulated sampling distributions

Figure 2.4 shows an actual histogram of a large number of sample averages using samples of 30 individuals. Each sample of 30 individuals was selected from a population of randomly generated "weights" via computer. The computer then calculated and stored the sample average. This process was repeated thousands of times, resulting in thousands of sample averages. The histogram shows the distribution of these averages.

The resulting distribution is called a *simulated sampling distribution* because it was generated by computer simulation rather than by mathematical theory. Notice how closely the simulated sampling distribution approximates the idealized normal distribution shown in Figure 2.2. This serves as a nice illustration

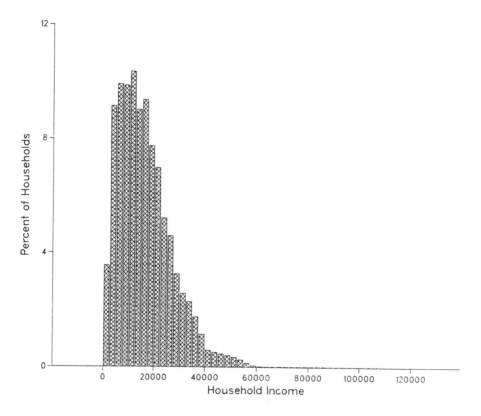

**Figure 2.3**   Household incomes in Vermont.

of the CLT: Sample averages using samples of size 30 really do seem to have a normal distribution.

## 2.5   The larger the sample, the less spread out is the sampling distribution.

*Sample size and the spread of a sampling distribution*

The larger the sample, the closer a statistic will tend to be to the parameter of interest. So the sampling distribution will be more concentrated toward the center. This will be true for virtually any statistic we ever want to use.

   To illustrate how this works, let's crank up our computer again to make some simulated sampling distributions. As before, we will use the sample average as

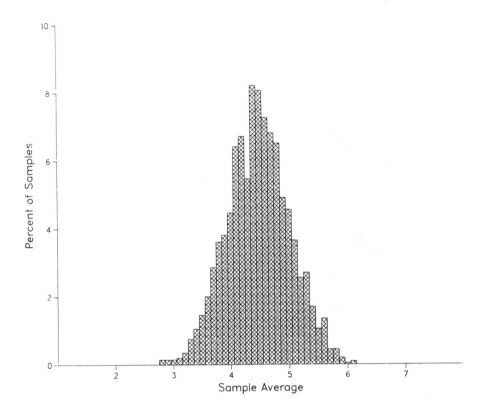

**Figure 2.4**  Histogram of sample averages using samples of size 30.

our statistic; but this time, we have the computer pick samples randomly from the digits 0 through 9, much as if we were selecting digits from a random number table.

## *Averages from a population with a uniform height distribution*

First, let's consider the rather trivial case of samples of size 1. That is, each "sample" is simply a single individual, and each sample "average" is simply that individual's score—in this case a random digit from 0 through 9. As before, the computer randomly generated thousands of such samples (each sample being of size 1 in this case). The sampling distribution of the sample average is simply the distribution of the digits in the population. Figure 2.5 shows the histogram.

Notice that the histogram is more or less flat. (Theoretically, of course, it would be exactly flat, but we are using a finite number of individuals, so there is

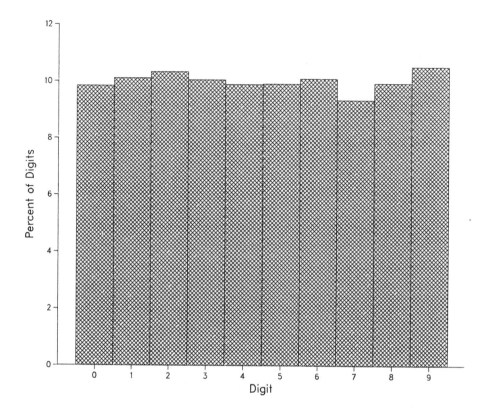

**Figure 2.5**   Random one-digit numbers.

a bit of random fluctuation.) It is not normal. This distribution is sometimes called the *uniform height* or *discrete uniform distribution*.

Now let's consider averages from samples of size 2. Figure 2.6 shows the histogram. It is still not normal, but notice that it is more concentrated toward the center than is Figure 2.5. This is because the average of two randomly picked scores is more likely to be near the center of the distribution than is either of the scores. Notice how seldom, relatively speaking, an average of 0 to 9 occurs.

Figure 2.7 shows what happens with samples of size 10. Now the distribution is starting to look approximately normal, with very few sample averages occurring in the tails of the distribution. By the time we get to Figure 2.8, generated using samples of size 100, the extreme averages almost never occur; the histogram is much less spread out.

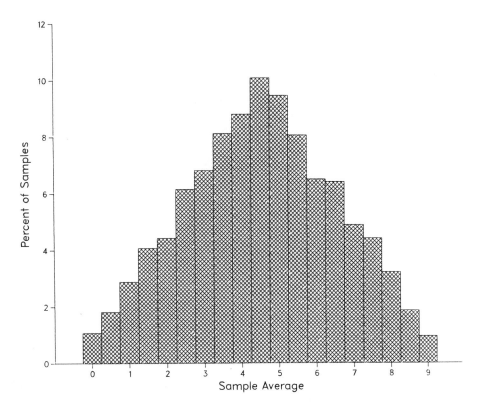

**Figure 2.6**   Random sample averages using samples of size 2.

## Sampling distributions from skewed populations

Since we were sampling random digits, the sampling distributions in Figures 2.5 through 2.8 came from a population with a uniform height distribution. In actual statistical work, most populations of interest have different distributions. As we have noted, many are at least approximately normal. This will be especially true when the variable we are talking about is itself some type of sum or average.

But many populations have distributions that are not only nonnormal but also nonsymmetrical; that is, they are skewed, as in the sample of household income. What happens to the sampling distribution of a statistic when the population has a skewed distribution? It depends on the statistic. Remember, not all statistics are covered by the CLT. But for those that are, such as a sample average, the CLT says that despite the skewness in the population, the statistic will have a normal sampling distribution as long as the sample size is large enough.

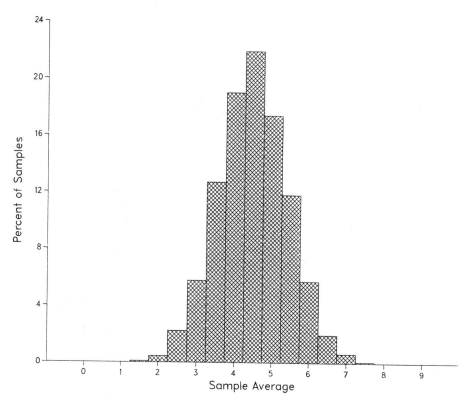

**Figure 2.7**  Random sample averages using samples of size 10.

Figures 2.9 through 2.12 illustrate this. Figure 2.9 shows a skewed distribution similar to the household income distribution shown in Figure 2.3. Figure 2.10 shows the sampling distribution of the sample average using samples of size 2 from the same skewed population. Notice that it is still skewed, but less so. In Figure 2.11 the sample size is 10, and the skewness is even less. In Figure 2.12, with a sample size of 100, it is virtually undetectable.

## STUDY QUESTIONS—*Sections 2.4 and 2.5*

2.5.1  What would you guess the distribution of the ages of students in introductory statistics courses would look like? Sketch a histogram showing what you think. Be sure to label the axes.

2.5.2  Suppose that you selected a random sample of, say, 100 introductory statistics students. What would the distribution of their ages look like? Does the CLT say anything at all about the distribution of a particular sample? (No!)

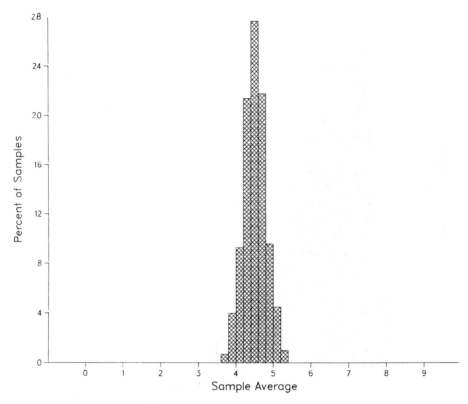

**Figure 2.8**   Random sample averages using samples of size 100.

**2.5.3**   Now consider the *average* age of introductory statistics students. Your sample of 100 students provides an estimate of the population average. So would any of the vast number of possible samples of 100 students that you might have selected. If you were able to draw a histogram showing the distribution of all those possible sample averages, what would it look like? How do you know?

**2.5.4**   Suppose that your histogram of all possible sample averages indicated that 95% of them were within 1 year of the population average. (Knowing that the distribution is of a particular form is what makes it possible to make such a statement.) Your sample of 100 students has an average age of 22.2 years.

Suppose that the average age of students in all courses is 20.6 years. Are introductory statistics students older on the average? Of course, you can not say for sure, because you only have a random sample. It is theoretically possible that the average age of the entire population of introductory statistics students is really the same as for all courses, and your sample was, by chance, older.

Possible, yes. But how plausible is it? What can you say about the chance of

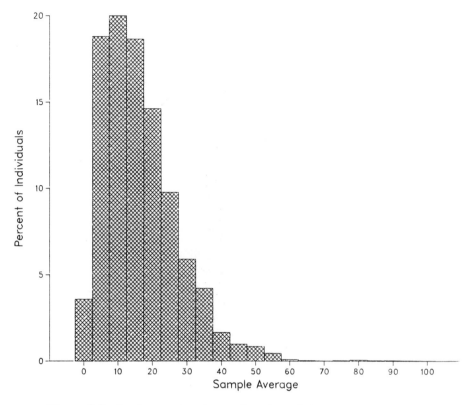

**Figure 2.9**   Random numbers from a skewed population.

getting a sample average as large as the one you got, if in fact the average for the population were only 20.6? What would you conclude?

Do you see how useful it is to know the sampling distribution? (If your answer is no, reread the last few sections!)

## 2.6   So what?

Let's recap what we have discussed about inference, so that we can see the importance of sampling distributions and the CLT. The tasks of inference are to use statistics to make statements about parameters and to say how certain we are that those statements are true. To determine how certain we are about a given statement, we need to consider all the samples we might have obtained. Our statement will be true if we were fortunate enough to have chosen a sample that

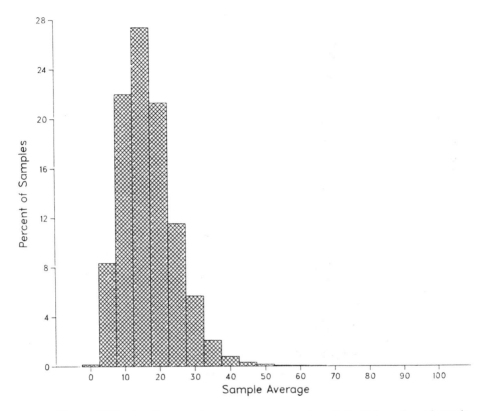

**Figure 2.10** Random sample averages using samples of size 2 from skewed population.

gives a statistic that is close enough to the parameter of interest. What is close enough depends on how precise we insist on being in our statement.

Our degree of certainty will be the proportion of all possible samples for which the statistic of interest is close enough to the parameter of interest. The more precision we demand from our statistic, the smaller this proportion will be. To find out what the proportion is for any given degree of precision, we need to know what proportion of all possible samples yield a statistic within any given distance of the parameter of interest. This information is called the *sampling distribution of the statistic*.

The central limit theorem (CLT) says that many types of statistic have a particular type of sampling distribution, called the normal distribution. This will be true if the statistic comes from a large enough sample, even if the population from which the sample is taken has a skewed distribution.

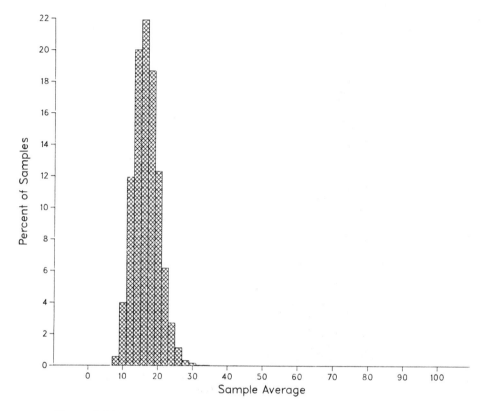

**Figure 2.11**   Random sample averages using samples of size 10 from skewed population.

## The importance of the CLT

The reason the CLT is so important is that it lets us use the normal distribution to figure out our degree of certainty even when we have little knowledge of the population from which we are sampling. It does not matter whether the population is normal or not—if we are using a statistic to which the CLT applies and have a large enough sample, we can assume that the sampling distribution of the statistic is normal.

The fact that the normal distribution shows up so often in nature is essentially a manifestation of the CLT. It is nevertheless an awe-inspiring fact that this sort of statistical regularity occurs. Some people have pointed to this regularity as a demonstration of the existence of a supreme intelligence. However you look at it, it is important to appreciate at least the utility, if not the cosmic significance of the CLT. It provides a degree of predictability in the midst of randomness.

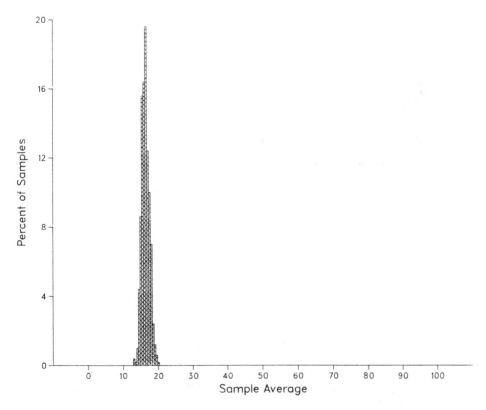

**Figure 2.12**    Random sample averages using samples of size 100 from skewed population.

# 3

# Describing Data for a
# Single Variable

## 3.1 There are many ways of summarizing
##     a set of data.

In Chapter 1 we discussed the distinctions between a sample and a population
and between a statistic and a parameter. We considered the ideas of systematic
bias and randomization and mentioned some techniques used in selecting random
samples. In Chapter 2 we discussed how we can use information from samples
(statistics) to draw inferences about population characteristics (parameters). We
saw how knowledge of the sampling distribution of our statistic allows us to say
how certain we are about our conclusions.

In our examples up to now, our parameter of interest has been either a
proportion (indicating how prevalent some characteristic is in the population) or
an average (indicating how large the scores on some variable tend to be in the
population). But as we have noted previously, the concepts we have been
discussing can be used to draw inferences about any type of parameter. In this
chapter we begin our discussion of various types of parameters and the statistical
methods used to draw inferences about them.

There are many types of parameter, with each type describing a particular
characteristic of the population. Sometimes we can summarize a great deal of
information about a population in just a few parameters, at least approximately.
This capability is what makes statistical analysis so useful. In this chapter we
consider techniques for describing a population in terms of a single variable.

Later chapters deal with situations involving several variables or several populations.

## 3.2   Often we are interested in the entire population distribution.

### The population distribution

The population distribution for some variable is simply the proportion of individuals with each possible score on the variable. As we have already seen, a histogram is a picture of a distribution. Everything we ever want to know about a variable in a population is contained in the population distribution.

Even if we are interested primarily in an average or some other summary measure, we often need to consider the entire distribution. For example, if we are concerned about the effect of certain government policies on family income, we might look at changes in average family income (perhaps using several types of average). But an average might not reflect other important considerations. Some policies could tend to polarize family incomes—the rich getting richer and the poor getting poorer—a tendency that would not necessarily be reflected in an average. A picture of the distribution would make such an effect easier to spot.

### The population distribution in statistical inference

Another reason we need to consider the population distribution relates to the problem of statistical inference. Procedures for statistical inference generally depend on knowing the sampling distribution of the statistic being used. In many cases we can be sure that the sampling distribution is normal even if the population distribution is not because of the CLT. But there are other situations, also very common, in which our knowing the sampling distribution depends on our knowing at least the general shape of the population distribution. In these cases we need to get information on the population distribution.

Of course, there is a bit of a paradox here. The time we need to use statistical inference is when we cannot look at the entire population but only at a sample. But we just noted that some inferential procedures require that we know something about the shape of the population distribution. So technically, we cannot do inference in those situations unless we do not need to!

Fortunately, in most cases we can get out of this mess, because the requirement that the population distribution have a certain shape is usually not too critical. This means that we can look at a histogram of the sample—which, after all, should look something like the population distribution—and as long as the shape of the histogram is not too far from what it is supposed to be, we can proceed with our inference about the parameter.

## 3.3 Some distributions can be approximated by mathematical functions.

### Distributions specified by mathematical functions

An example is the normal distribution, which is a specific mathematical function. As we have discussed, many variables have distributions that are approximately normal. Other important distributions that often occur in real data include the Poisson, binomial, lognormal, and Weibull distributions. Three other distributions that are often referred to include the chi-square, Student's t, and F distributions. These are used mostly in statistical inference, because many kinds of statistics have sampling distributions that can be approximated by one of the three.

### Parameters of mathematical distributions

The reason mathematical distributions are so useful is that they allow us to convey a great deal of information in very compact form. If we know, for example, that the distribution of scores on some test is Normal, then all we need to know is two parameters—the mean and the standard deviation—and we have completely specified the distribution. That is, just knowing two numbers (and the fact that the distribution is normal) would allow us to figure out what proportion of scores is in any interval we care to name.

Moreover, the parameters needed to specify a particular distribution often have meaningful interpretations in their own right. This is particularly true of the parameters of the normal distribution: The mean is an indicator of the general magnitude of the scores, and the standard deviation is an indicator of how variable the scores are.

### The normal distribution specified by mean and standard deviation

Figures 3.1 through 3.4 show how the normal distribution changes according to the values of the mean and standard deviation. All four figures are drawn to the same scale. Notice how the mean affects the general location of the distribution on the number line, whereas the standard deviation affects how spread out the distribution is. We have more to say about the mean and standard deviation, together with other important parameters, later in this chapter.

### STUDY QUESTIONS—*Sections 3.1 through 3.3*

**3.3.1**   Following is a sample of 40 scores on a test. Each score is the number of items answered correctly by a student on a 15-item "fill-in-the-blanks" test.

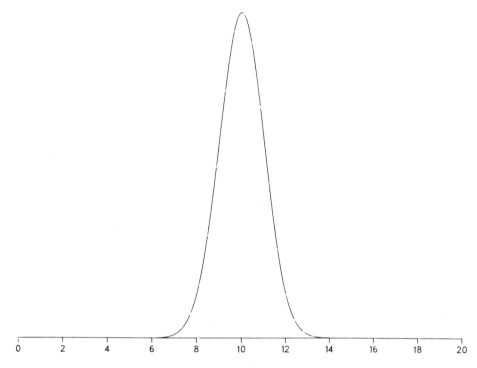

**Figure 3.1**   Normal distribution with mean = 10, standard deviation = 1.

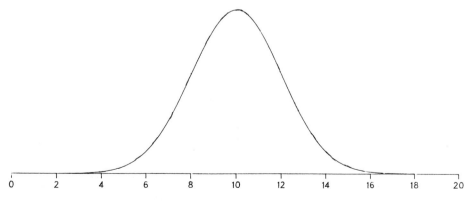

**Figure 3.2**   Normal distribution with mean = 10, standard deviation = 2.

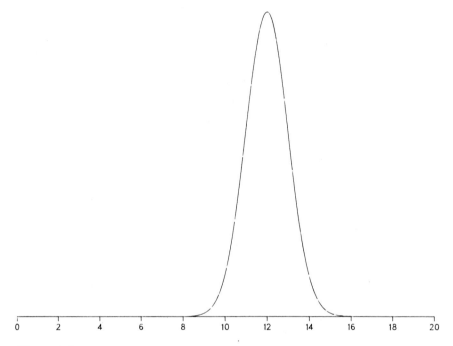

**Figure 3.3**   Normal distribution with mean = 12, standard deviation = 1.

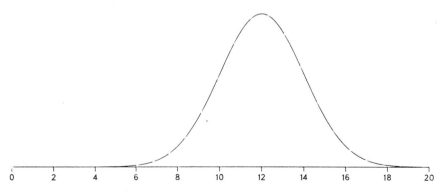

**Figure 3.4**   Normal distribution with mean = 12, standard deviation = 2.

```
12  11  13  14  14  12  15   2
11   0  11  14  12  13  10   3
13  14   5  15  13  14  14  13
15  12  10  13  12  12  12  14
14  13  14  13  12   9  15  13
```

How would you describe the performance of the students on the test? Consider each of the following points:

a  How well did an "average" student do? What might we mean by an average student? There are several possible answers—try to think of at least three.

b. How consistent were the students' scores on the test? Can you think of a way of summarizing the degree of variability in the scores? *Suggestion*: How far away from the average were the scores, on the average?

c. What does the distribution of students in the sample look like? You might wish to group the scores in, say, 3-point intervals before you draw a histogram.

d. What does the presence of a few very low scores do to your description of the "average" student?

e. What is the 75th percentile of the sample? (Recall that the 75th percentile is the score that divides the upper 25% from the lower 75%.)

f. If the test had been a multiple-choice test instead of fill-in-the-blanks, with, say, four choices per item, how would that change your description of the data? (*Hint*: What would you think about a student who got a score of 0 on a 15-item multiple-choice test?)

**3.3.2**  Suppose that you have test scores (for the same 15-item test as above) for two different samples of students. Students in sample 1 used computer-aided instruction (CAI) to study the material covered on the test, while students in sample 2 were taught by traditional lectures. Here are the scores for the 20 students in each sample:

Sample 1 (CAI):

```
Student:  1  2  3  4  5  6  7  8  9  10
Score:    7  6  8  8  7  7  9  8  6   8

Student:  11  12  13  14  15  16  17  18  19  20
Score:     7   9   8   6   8   9   7   7   5   8
```

Sample 2 (traditional):

```
Student:   1   2  3  4   5  6  7  8   9  10
Score:    14   8  5  0  11  4  7  3  12   6

Student:  11  12  13  14  15  16  17  18  19  20
Score:     8  11  10  12   6   9  13  10   6   2
```

a. How would you characterize the differences between the two samples? Compare the samples in terms of both the performance of the "average" student and the degree of variability.

b. You will notice that the sample averages (however you define it) are not exactly the same. Can you conclude that CAI and traditional lectures differ in their effectiveness for the material covered on the test? Specifically, what information do you need to decide if such a conclusion is warranted?

## 3.4  A histogram and frequency distribution are very useful descriptive tools.

We have already encountered histograms in Chapter 2. As we noted, a histogram is a graph that shows the proportion of individuals in a sample or population with each possible score (or range of scores) on a variable. To draw a histogram, you need to figure out how many individuals in the sample or population have each score. This set of information is called a *frequency distribution* for the variable. A histogram is a picture of a frequency distribution.

### Frequency distributions for ungrouped data

When there are relatively few distinct scores in the data, a frequency distribution can include information for each score. Let's look at an example. In 1976, the U.S. Department of Health, Education, and Welfare questioned a total of 7697 people who had entered college 4 years previously regarding their current status. The frequency distribution of the respondents' current status was as follows:

Status of 1972 College Entrants as of October 1976
*Source*: Eckland and Wisenbaker (1979).

| Status in 1976 | Frequency | Relative frequency (%) |
|---|---|---|
| Dropped out | 2694 | 35.0 |
| No degree, but still enrolled | 2001 | 26.0 |
| Graduated | 3002 | 39.0 |
| Total | 7697 | 100.0 |

The column headed "relative frequency" gives the percent in each category and makes interpreting the numbers somewhat easier. The percents add up, of course, to 100. Figure 3.5 shows a histogram for this distribution.

### Qualitative versus quantitative variables

Note that Figure 3.5 concerns a variable for which the "scores" being represented in the histogram are simply categories rather than numbers representing some measurable quantity. Such variables are often called *qualitative variables*. Sometimes the categories being represented can be meaningfully ordered, as in

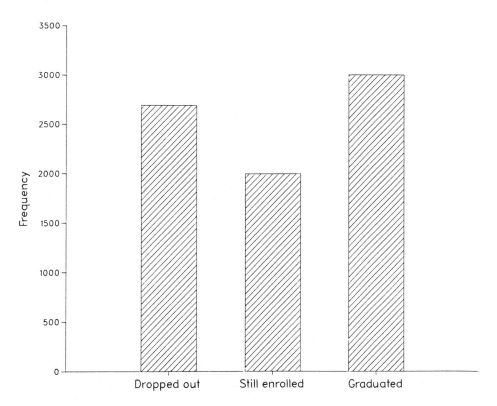

**Figure 3.5**  Status of 1972 college entrants in 1976.

Figure 3.5. In many situations, however, the order in which the categories are presented is entirely arbitrary. Consider, for example, the following distribution:

Ethnic Distribution of Residents of Buffalo, New York
*Source*: 1980 Census.

| Ethnic group | Frequency | Relative frequency (%) |
|---|---|---|
| Asian | 1,322 | .36 |
| Black | 95,116 | 25.89 |
| Hispanic | 9,499 | 2.59 |
| Native American | 2,383 | .65 |
| White | 252,365 | 68.69 |
| Other | 6,684 | 1.82 |
| Total | 367,369 | 100.00 |

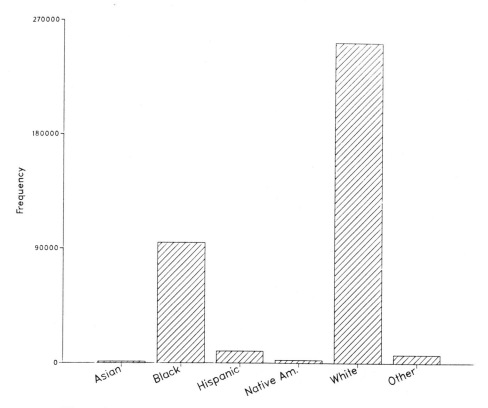

**Figure 3.6**   Ethnic distribution of Buffalo, New York.

Figures 3.6 and 3.7 both represent this set of data. Why are the shapes of the two histograms so radically different? Think about it for a moment. (What does the shape of a histogram depend on? Would it be meaningful to characterize Figure 3.7 as skewed?)

## Frequency distributions for grouped data

If there are many distinct scores in the data, it is necessary to group the data into categories before making a frequency distribution. This will generally be true when the variable being considered is "continuous" as opposed to "discrete."

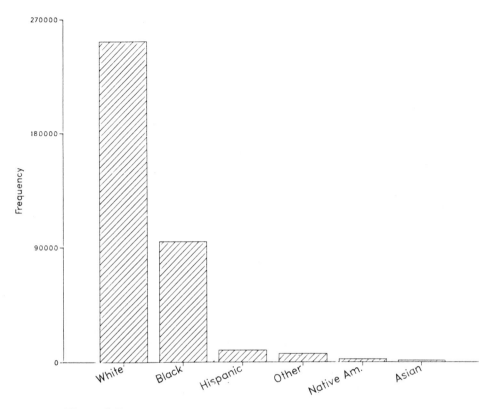

**Figure 3.7**   Ethnic distribution of Buffalo, New York.

*Continuous variables* are variables that can take on any value (or, for practical purposes, any of a large number of possible values) in some range—for example, height. It would be silly to report the frequency of individuals with a height of 140 cm, and 141 cm, and 142 cm, and so on (although that is what some computer programs will do if you do not tell them to group the data first!).

Figure 3.8 shows a histogram for the distribution of scores on the Scholastic Aptitude Test mathematics section for students in an introductory statistics course at the University of Vermont in 1982. Here is the frequency distribution for these data:

Scores on the Scholastic Aptitude Test Mathematics Section for Students in an Introductory Statistics Course at the University of Vermont, Spring 1982

| Score range | Frequency | Relative frequency (%) | Cumulative relative frequency (%) |
|---|---|---|---|
| 350–399 | 5 | 1.9 | 1.9 |
| 400–449 | 12 | 4.5 | 6.4 |
| 450–499 | 32 | 11.8 | 18.2 |
| 500–549 | 70 | 25.9 | 44.1 |
| 550–599 | 72 | 26.7 | 70.8 |
| 600–649 | 44 | 16.2 | 87.0 |
| 650–699 | 23 | 8.5 | 95.5 |
| 700–749 | 9 | 3.4 | 98.9 |
| 750–799 | 3 | 1.1 | 100.0 |
| Total | 270 | 100.0 | |

The column labeled "cumulative relative frequency" shows, for each score range, the percent of students with those scores or below. This column makes it easy to find which score range contains any particular percentile. For example, we can see that the 75th percentile occurs somewhere between 600 and 649, while the 50th percentile occurs somewhere between 550 and 599.

## Stem-and-leaf displays

Another way of representing a distribution that has become popular in recent years is called a *stem-and-leaf display* (Tukey, 1977). A stem-and-leaf display presents essentially the same information as a histogram but supplies additional detail in the same space.

For example, consider the following set of scores on a statistics exam given to 30 students (the scores have already been arranged in order for convenience):

```
47  55  62  63  68  72  72  73  74  76  77  78  78  78  80
80  81  82  82  82  83  85  86  86  86  88  89  89  90  93
```

A frequency distribution and histogram for this set of data (as generated by Statpal) is shown below, with the data grouped into 10-point ranges. Note that the histogram is presented sideways (i.e., with increasing frequency moving to the right, rather than up). Each asterisk on the histogram represents one individual's score.

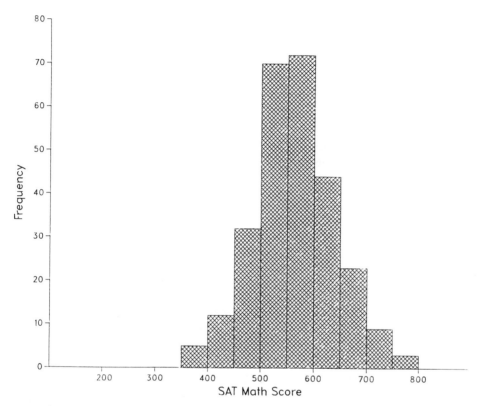

**Figure 3.8** Scores on the SAT math section for students in an introductory statistics course: histogram.

| Lower bound | Upper bound | Frequency | | |
|---|---|---|---|---|
| | | N | Prop. | |
| 40.0000 | 50.0000 | 1 | .033 | * |
| 50.0000 | 60.0000 | 1 | .033 | * |
| 60.0000 | 70.0000 | 3 | .100 | *** |
| 70.0000 | 80.0000 | 9 | .300 | ********* |
| 80.0000 | 90.0000 | 14 | .467 | ************** |
| 90.0000 | 100.0000 | 2 | .067 | ** |

Here is a stem-and-leaf display for the same data:

| Stem | Leaf |
|------|------|
| 4 | 7 |
| 5 | 5 |
| 6 | 238 |
| 7 | 223467888 |
| 8 | 00122235666899 |
| 9 | 03 |

In the "stem" column, we list the digits from 4 through 9. Each digit represents a 10-point interval, with 4 representing the 40s, 5 the 50s, and so on. In the leaf column, we put the trailing digit of each score in the appropriate row. Thus we can see at a glance that, for example, there were three scores in the 60s: 62, 63, and 68. Note that the shape formed by the leaves is precisely the same as that formed by the asterisks in the histogram. But the stem-and-leaf display gives us more information in the same space, because we can recover the original data from the display.

Sometimes, of course, this degree of precision is not necessary or even desirable. For the SAT score distribution discussed above, a stem-and-leaf display using 10-point intervals would be too detailed. In that case we might wish to use the hundreds digit for the stem, plot the tens digit for the leaves, and ignore the units digit altogether.

## STUDY QUESTIONS—*Section 3.4*

**3.4.1** [Statpal exercise—Coronary Care data set (see Appendix A)]
   a. Produce a histogram for the variable called AGE. Note that Statpal's HIstogram and frequency distribution procedure allows you to group the data into specified categories or to report frequencies for each possible age. What is the advantage of grouping the data into categories in this case?
   b. How would you characterize the shape of the age distribution (i.e., generally symmetric, or skewed left or right)? Note that it is difficult to judge skewness if you have too many or too few categories.

**3.4.2** [Statpal exercise—Electricity Consumer Questionnaire data set (see Appendix B)]
   a. Produce a histogram for the variable called BESTMNTH, which indicates which month the respondent feels is the best for the electric cooperative's annual meeting.
   b. Now redisplay the histogram specifying four categories. What does each category represent?

## 3.5   In a histogram, area represents relative frequency.

This property of histograms turns out to be very useful. If we call the entire area of the histogram 100%, *the area of the histogram above any range of scores is simply the relative frequency of those scores.* In fact, this correspondence between area and relative frequency is a defining characteristic of histograms, and it is important to maintain the correspondence whenever a histogram is drawn. If, for example, we had combined the two highest categories of our SAT data into one, giving a combined relative frequency of 4.5% for scores in the range 700 through 799, the height of the bar for the last category should *not* be drawn at 4.5% but at 2.25%, since the last category is twice as wide as the others. This keeps the total area of the histogram over the category the same as it was in Figure 3.8.

### Frequency polygons

If we draw a line connecting the tops of the bars of the histogram, we get a *frequency polygon.* Figure 3.9 shows the frequency polygon for the SAT data superimposed over the histogram, and Figure 3.10 shows just the frequency polygon. A little inspection should convince you that the correspondence between area and relative frequency is preserved in a frequency polygon: If the total area under the line is 100%, the area under the frequency polygon above a range of scores represents the relative frequency of those scores.

### Chunky versus creamy-smooth histograms

One way to think of a histogram is to think of each individual score turning into a little block, which is then piled on the horizontal axis above its score. The more blocks with certain scores, the bigger the pileup at those scores. When you group the data into categories, you are simply piling blocks with similar scores over the same point on the line.

As we noted in the preceding section, grouping individuals is simply a way of making the frequency distribution (and histogram) more useful, since a histogram that showed only one or two individuals for each of a great many possible scores would not be very informative. When we have relatively few individuals, then, we tend to group the data and the histogram looks "chunky." The corresponding frequency polygon consists of straight lines, as in Figure 3.10.

But suppose that we wished to plot a histogram that showed the SAT-math scores of not just 270 students but of all students who took the test in a particular year. The number of students who take the SAT in any particular year is well over 1 million. Since reported scores on the SAT are whole numbers between 200 and 800 inclusive, there are 601 possible scores on the SAT. With 1 million students taking the test, we would certainly have plenty of students at each

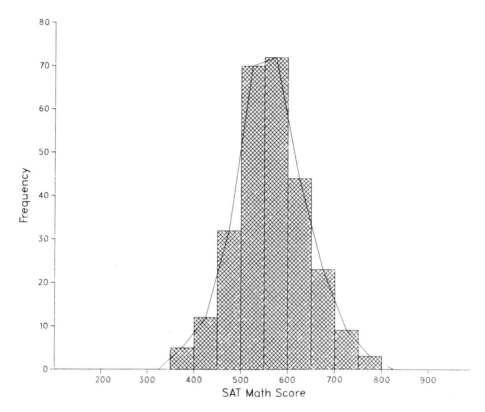

**Figure 3.9**  Scores on the SAT math section for students in an introductory statistics course: frequency polygon superimposed over histogram.

possible score. Moreover, we would expect that similar scores would have similar relative frequencies; that is, the percent of students with a score of 584 would be about the same as the percent with a score of 583 or 585. The effect of this is that the histogram should look "creamy smooth" instead of chunky. Figure 3.11 shows a hypothetical frequency polygon of 1 million students distributed over 601 SAT scores. Note that the frequency polygon is still not perfectly smooth, because the 601 separate lines are still (barely) discernible— but from a distance it is starting to look smooth.

Of course, the correspondence between area and relative frequency is still preserved no matter how many categories we use for our histogram. As the number of individuals represented in the histogram grows, the amount of area that represents a single individual decreases; nevertheless, the area above any interval gives the relative frequency of scores in that interval.

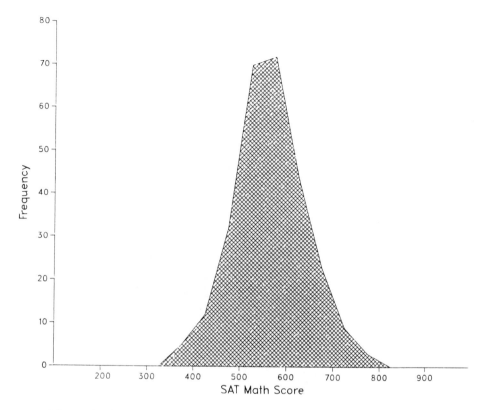

**Figure 3.10**  Scores on the SAT math section for students in an introductory statistics course: frequency polygon.

If you have been exposed to calculus, you can probably guess what is coming next. (If not, read on anyway!) As the number of possible scores represented in the histogram increases without bound, the histogram gets smoother and smoother (provided, of course, that we have an inexhaustible supply of individuals so that someone can have each possible score). In the limit (as the mathematicians say), we have a smooth curve for a frequency polygon. Each individual gets an infinitesimal bit of area in the histogram, but if you aggregate enough of them, they define a smooth curve. If you put together enough grains of sand, you can make a mountain.

Now the frequency polygon is not a set of straight lines but a smooth curve, although the area under the curve still has the same meaning. Finding the area over a particular interval is a bit more complicated with a smooth curve than with a set of chunks, but the principle is the same. Fortunately, numerical tables are

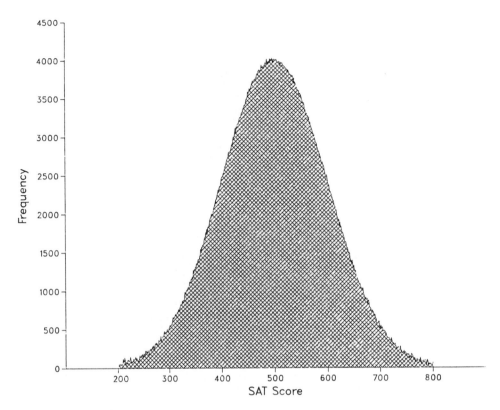

**Figure 3.11**   Hypothetical SAT scores of 1 million students.

available for most commonly used smooth distributions, so that it is easy to find any areas we might need.

As we have already seen, a very important case of a smooth distribution is the normal distribution. We discuss how to find areas under the normal distribution in Chapter 4. First, we need to consider some special parameters (and their corresponding statistics) in detail.

## 3.6   The mean, median, and mode are useful for describing central tendency.

Each of these three parameters (or statistics, when used on sample data) might be called an *average*; which you use in a particular situation depends on what you want to convey.

## The mean

The *mean* is simply the arithmetic average—the type of average you probably first learned to calculate when you were trying to predict what your grade would be in school. To get a mean, you add up all the scores and divide by their number.

The mean is very widely used, largely because it is easy to compute and has simple mathematical properties. In formulas, the mean of a *population*—the mean as a parameter—is usually designated by the Greek lowercase letter mu, $\mu$. The mean of a *sample*—the mean as a statistic—is usually designated by a Latin letter with a bar on top, such as $\overline{X}$. In mathematical notation, the population mean can be denoted as

$$\mu \equiv \frac{\Sigma X}{N}$$

where the symbol $\equiv$ means "is defined as," $N$ represents the number of individuals in the population, and the symbol $\Sigma$ means that all scores are to be added up.

The mean corresponds to what physicists call the "center of mass." If we think of the horizontal axis of a histogram as a bar on which we have piled up all the scores, with each score having an equal weight, the mean is the point at which the bar will balance.

## The median

The *median* is the middle score when the scores are in order. In a histogram, the median is the score that divides the area of the histogram (and therefore the relative frequency) in half. So the median is simply the 50th percentile. If there are an even number of scores, there is no single middle score, so the convention is to split the difference (if any) between the two middle scores and call that the median.

Notice that the score that divides the area in half is not necessarily the score that "balances" the distribution. Suppose that five children, all the same size, are sitting on a seesaw. The median position on the seesaw will always be the position of the middle child, wherever that happens to be, because that point is the middle position. But the mean position—the balance point—will depend on how far out each child is sitting. If the child on one end suddenly moves outward, that child's end will move toward the ground. To make the seesaw balance again, we would have to move the fulcrum toward the child that moved outward.

## The average of a symmetrical distribution

When will the two points (the median and the balance point) coincide? They will coincide if the children are positioned *symmetrically* about the center of the seesaw. Or, getting back to distributions, the mean and median will coincide if the distribution is symmetrical. (Symmetry is not the only way for the mean and median to coincide, but it is the most interesting.) The point of symmetry is, of course, the point where the mean and median occur.

So if we are dealing with a symmetric distribution—as, for example, when we are dealing with the normal distribution—it makes no difference whether we talk about the mean or the median. Since the normal distribution is symmetric, we could talk about either one as simply the average of the distribution. (Usually, though, we refer to it as the mean rather than the median.) That is why there is usually no reason to be concerned about which average to use when we are discussing IQ scores, or errors in the size of machined parts, or sampling distributions of statistics covered by the CLT, all of which tend to have normal distributions.

## Averages of skewed distributions

But which average should be used for skewed distributions? Consider, for example, salaries paid by a large corporation. As you can imagine, the distribution of salaries would be highly skewed to the right. There are relatively few employees who are paid very large salaries, with most employees getting the lower and middle salaries. Figure 3.12 shows an example of such a distribution, with the mean and median indicated.

So which average is more appropriate to report? It depends on your point of view. The management of the company might want to stress the mean, not only because it is higher, but also because it reflects the overall economic impact the corporation has on the community in which it is located (at least in terms of dollars spent on salaries). The union representing employees of the company might be more concerned with the median, because it is representative of the experience of more individuals.

## The effect of extreme scores

Note that the median is not affected by changes in the extremes. If our hypothetical company raises the salary of its president by $1 million, there is no difference in the salary that divides the lower from the upper half of the distribution. However, the mean salary is affected by the extremes, since the calculation of the mean involves adding up all the scores in the distribution. The fact that the mean is affected by extremes has a curious effect: In some cases, it is not possible to calculate the mean even though it is possible to calculate the median! How can this be?

Consider a study of the effectiveness of a certain treatment—say, a new chemotherapeutic agent—on the survival of patients with a usually fatal form of cancer. A sample of 35 patients is followed from the time of treatment (all at

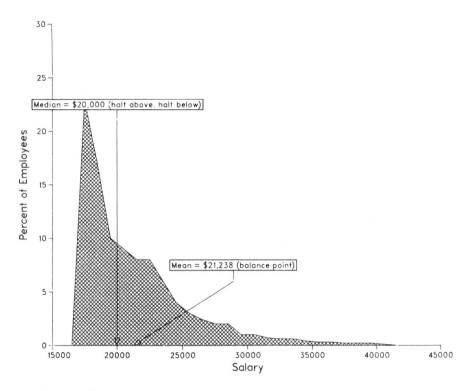

**Figure 3.12**   Salaries paid by a large corporation.

around the same time), and the length of time they survive is recorded. Of the 35 patients, 25 have died of their disease at the time the data are analyzed, while the other nine are still alive. All we know about the nine fortunate patients is that they have survived at *least* a certain length of time. We do not know what their ultimate survival time will be (aside from the fact that it is finite!). So we cannot calculate the mean survival time for the 35 patients.

But we *do* know that the nine surviving patients survived longer than any of the other patients. We also know how long each of the other 26 patients survived. Thus we can figure out the median survival time—it is simply the length of time survived by the patient with the middle survival time (the eighteenth, in this case).

## The mode

How about the mode? The *mode* is the score obtained by the greatest number of individuals. On a histogram, the mode is the score that has the highest point above it. Of course, there may not be a unique mode if two or more scores are represented by the same number of individuals.

The mode is particularly useful when the data are in categories rather than scores on a continuum. Indeed, when there is no inherent order in the scores, the mode is the only kind of average that makes sense. For example, it makes no sense to talk about the mean ethnic group among people in the United States—although many computer programs will happily calculate it for you! But it does make sense to say that a particular ethnic group comprises the largest number of individuals.

## STUDY QUESTIONS—*Sections 3.5 and 3.6*

**3.6.1** [Statpal exercise—Coronary Care data set (see Appendix A)]
   a. Compute the mean, median, and mode of the variable called AGE. Use the DEscriptive statistics procedure to get the mean and the HIstogram and frequency distribution procedure to get the median and mode. (Save your output for a future exercise.)
   b. Mark the mean and median on the frequency distribution you generated in Exercise 3.4.1. What is the relationship of mean and median in a skewed distribution?
   c. How did you determine the mode? Note that the mode is not very useful if you have too many or too few categories.
**3.6.2** [Statpal exercise—Electricity Consumer Questionnaire data set (see Appendix B)]
   a. Compute the mean, median, and mode of the variable BESTMNTH, the respondent's preference for the month in which to hold the annual meeting. Which of the three measures of central tendency is the most meaningful in this instance? (Save your output for a future exercise.)
   b. Now do the same for the variable WDATTEND, which asks whether the respondent would attend the annual meeting if it were held at a convenient time. A response coded 1 means yes and 2 means no. Which of the three measures of central tendency are meaningful in this case?
   c. What does the mean indicate? (Think about it!)
**3.6.3** Calculate the mean and median of each set of numbers. For each pair of sets notice what happens to the mean and median between the first and second sets.

   a.  1   1   1   1   1
       1   1   1   1   21
   b.  −2   −1   +1   +2   0   0
       −2   −1   +1   +2   0   36
   c.  0   0   0   1   1   1   1   1   2   3
       0   0   0   1   1   1   1   1   2   3   0   0   0

**3.6.4** Indicate which symbol ($\overline{X}$ or $\mu$) would be used in reporting each of the following means.
   a. Mean height of the population of maple trees

    b.  Mean height of 30 randomly selected maple trees
    c.  Mean survival time of a sample of mice in a study of a new drug
    d.  Mean survival time of all mice that might theoretically receive a new drug
    e.  Mean response on a questionnaire scale as calculated from a sample
    f.  Mean response on a questionnaire scale as estimated from a sample

## 3.7 The range and interquartile range are measures of variability.

### Variability

In Chapter 1 we noted that statistics is concerned with describing and explaining how things vary. In particular, many statistical procedures involve measuring the *degree* to which things vary.

Parameters and statistics that measure variability are important for two reasons. First, we are often interested in the variability of some distribution in its own right. For example, consider the manufacture of machined parts. Often, the tolerance for error is very small. It is important that the parts be of the correct average size—but it is just as important that they vary as little as possible about that average.

Second, measures of variability are important in drawing inferences about other parameters, such as the population mean. Why should variability matter when you want to draw inferences about another parameter? Because our degree of certainty about inferences that we draw based on a *statistic* will depend on how much that statistic varies from sample to sample. The more variable the statistic, the less precise we can be in our statements for a given level of certainty. Determining how certain we are about our inferences thus requires that we have some measure of the variability of the sampling distribution of whatever statistic we are using to draw inferences.

### The range

How can we measure variability? One very simple way is to consider the *range* of the scores in the distribution: that is, the highest score minus the lowest score. The range is very crude, in that it considers only the two extremes and ignores the intermediate scores altogether. Nevertheless, the range does give a rough indication of variability.

The problem with the range as a measure of the overall variability of a distribution is that one unusually high or low score can obscure the fact that the distribution as a whole is relatively homogeneous. One way around the problem is to consider not the entire range, but the range of some subset of the scores. By excluding the extremes of the distribution, we can get a range that is more representative of the distribution as a whole.

## The interquartile range

One commonly used example of this type of range is the *interquartile range*, which is defined as the distance between the 25th and 75th percentiles of the distribution. (The name derives from the fact that the 25th and 75th percentiles can also be called the first and third "quartiles," respectively.) Figure 3.13 shows an example of a distribution, with the 25th and 75th percentiles indicated. Notice that the interquartile range is the range of the middle 50% of the scores in the distribution. Thus, in a histogram, the interquartile range is the range spanned by the middle 50% of the area in the histogram, with 25% of the area below and 25% of the area above. It makes no difference how far below and above the extreme areas fall, because the calculation of the interquartile range is not affected by the extremes.

## Calculating quartiles

Incidentally, there are at least two ways you can handle the problem of calculating the 25th and 75th percentiles if you have a relatively small sample where the number of individuals is not divisible by 4. For example, suppose that you want to compute the interquartile range of the following 11 scores:

    37   15   92   86   14   25   28   72   53   46   28

The first thing to do, of course, is to arrange the scores in order:

    14   15   25   28   28   37   46   53   72   86   92

Notice that with 11 scores, there are no scores that exactly divide the distribution into quarters. Now to calculate the 25th and 75th percentiles, we could do one of two things.

One thing we could do is to interpolate painstakingly so as to find the point just the right distance between two scores. For example, we could reason that we want the interquartile range to span the middle 50%—that is, 5.5—scores. To do this, we should include the middle five scores, plus .25 of the way to the scores on either side (to give a total of 5.5). So we would define the 25th percentile as three quarters of the way between 25 and 28, or 27.25, and the 75th percentile as a quarter of the way between 53 and 72, or 57.75. The interquartile range would therefore be equal to 57.75 minus 27.75, or 30.

The other thing we could do is adopt a simpler convention: Use the scores that come as close as possible to encompassing the middle 50%. In this case we would use 28 and 53 as our 25th and 75th percentiles. This way is certainly easier, and since the purpose of calculating the interquartile range is to give a simple, intuitively reasonable measure of variability, there is usually no practical value in going through the arithmetic involved in interpolating.

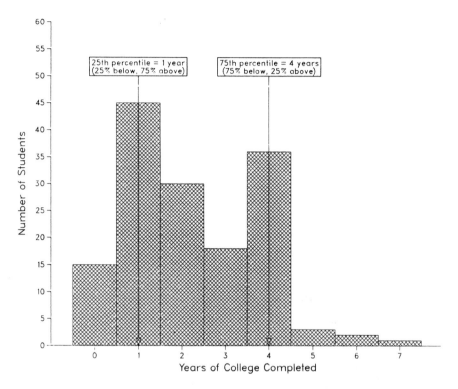

**Figure 3.13**   Years of college completed by a sample of statistics students.

## *Boxplots*

A handy graphical device that makes use of quartiles is called a *boxplot* or a *box-and-whisker plot* (Tukey, 1977). We shall present a simplified version of box-plots here. The idea of a boxplot is to show on a numerical scale both the range of the data (thus showing the spread of the entire distribution) and the interquartile range (thus showing the spread of the middle 50% of the distribution).

To make a boxplot, we first draw a numerical scale appropriate to the situation. Next, we locate the three quartiles (i.e., 25th, 50th, and 75th percentiles) of the distribution and draw a rectangular box extending from the first to the third quartile. In the box we draw a vertical line showing the 50th percentile. Finally, we draw "whiskers" from the two ends of the box extending to the lowest and highest scores, thus showing the range. Figure 3.14 shows a boxplot for the same distribution as that shown in Figure 3.13.

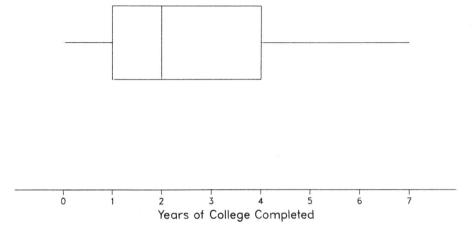

**Figure 3.14**  Boxplot of years of college completed by a sample of statistics students.

### Other interquantile ranges

It is possible, of course, to chop off other proportions from the extremes besides 25%. For example, if you chop off 10% from each end and look at the range, you get the *interdecile range*—the difference between the first and ninth deciles. If your knowledge of Latin permits, you can come up with impressive terms for lots of similarly derived ranges. (Interquintile? Interoctile?)

## 3.8  The standard deviation is the most commonly used measure of variability.

The prevalence of the standard deviation as a measure of variability stems largely from its importance in statistical inference, as we will see in Chapter 4. But it is also useful in its own right as a descriptive measure of variability: The larger the standard deviation, the more variable the distribution.

### The mean absolute deviation

To understand the standard deviation, it is useful to talk first about a different measure called the *mean absolute deviation* (MAD). The MAD itself is not used very often in statistical work because it is difficult to work with mathematically, but it is very appealing intuitively as a measure of variability.

What is the MAD? The MAD is simply the mean distance from the mean. The word ''deviation'' refers to how far away each score is from the mean. The word ''absolute'' indicates that we are talking about the absolute distance (i.e., without worrying about whether the sign of the deviation is positive or negative). The word "mean" we have already discussed—it is simply the arithmetic average. So to calculate the MAD, you first find the mean of the scores, then find out how far each score is from the mean (yielding "deviation" scores), and then find the mean of the deviation scores. In mathematical notation (for a population MAD):

$$\text{MAD} \equiv \frac{\Sigma \ |X - \mu|}{N}$$

where the symbol $|\cdot|$ indicates the absolute value of the expression.

For example, the MAD of the numbers 12, 8, 17, 16, and 7 is 3.6. (Why? Because the mean of the five numbers is 12, giving deviation scores of 0, 4, 5, 4, and 5, which have a mean of 3.6.) The MAD of the numbers 2, 27, 22, 1, and 8 is 10. (Verify this!) Notice that the two sets of numbers have the same mean, but the MAD of the second set is much bigger than the MAD of the first set because the second set of numbers is less homogeneous than the first. Make sure that you understand what the numbers 3.6 and 10 mean: the first set of numbers deviate from their mean (of 12), on the average, by 3.6, while the second set of numbers deviate from their mean (also 12), on the average, by 10.

Notice that the MAD is measured in the same units as the original scores. If the numbers in the sets above represent the ages of people in years, the mean age for each set was 12 years and the MAD for the first set was 3.6 years. We can say that the people in the first set differed from the mean by an average of 3.6 years, while the people in the second set differed from the mean by an average of 10 years.

### The standard deviation

The standard deviation is very similar to the MAD and can be interpreted in almost the same way. The difference is that instead of taking the mean of *absolute* deviations, we take the mean of *squared* deviations and then take the square root of the result to get us back to our original units. The mean squared deviation—before we take the square root—has its own name: it is called the *variance*. Thus the standard deviation is simply the square root of the variance.

As in the case of the mean, the standard deviation of a population is usually denoted with a Greek letter, while the standard deviation of a sample is usually denoted with a Latin letter. The letters usually used are $\sigma$ for the parameter—that is the lowercase sigma—and $s$ for the statistic. The notation used for the variance is, sensibly enough, $\sigma^2$ for the parameter and $s^2$ for the statistic. Thus we can write the definition of the population standard deviation as

$$\sigma = \sqrt{\frac{\Sigma \ (X - \mu)^2}{N}}$$

Let's use the same sets of numbers to illustrate the calculation of the standard deviation. For the first set, the deviation scores were 0, 4, 5, 4, and 5. So the squared deviations are 0, 16, 25, 16, and 25. The mean squared deviation—or variance—is therefore 82 divided by 5, or 16.4. For the second set, the variance is 112.4. (Do not let this paragraph get away without verifying each of these steps!) These variances, of course, are in *squared* units. So if our numbers are ages in years, we can say that the variance of the first set is 16.4 squared years, and the variance of the second set is 112.4 squared years. "Squared years" is not a very useful concept—so we will take the square root of our variances, giving us the standard deviation for each set: 4.05 for the first set and 10.60 for the second set. These are back in the original units.

How can we interpret these standard deviations? Well, notice that they are not all that far from the MADs. On the other hand, they are not exactly the same either. (Does it surprise you that they are not the same? Notice that taking the mean of the squared deviations does *not* give you the square of the mean of the unsquared deviations!) Nevertheless, it is appropriate to think of the standard deviation, like the MAD, as a measure of the average amount of deviation from the mean—as long as we are a bit loose about what we mean by "average." In the case of the standard deviation, we took a roundabout route to get the average deviation from the mean.

### STUDY QUESTIONS—*Sections 3.7 and 3.8*

3.8.1   Calculate the MAD, variance, and standard deviation of each set of numbers. In stating the results, include the correct units.
   a.   Five temperatures, in degrees Celsius: 10, 10, 12, 13, 15
   b.   Four interest rates, in percents: 11, 8.5, 9, 12
   c.   Six sample average weights, in grams: 26, 13, 9, 30, 100, 12
   d.   Five consumer preference ratings, in scale points: 1, 1, 1, 1, 1
   e.   Four heights, in centimeters: 88, 90, 90, 92
   f.   Four heights, in meters: .88, .90, .90, .92
   g.   Four heights, in inches: 223.52, 228.60, 228.60, 233.68
   h.   Four heights as differences from 80, in centimeters: 8, 10, 10, 12

## 3.9   The standard deviation is computed differently for samples versus populations.

### N versus n − 1

There is a slight difference between the formula for the population standard deviation and that for the sample standard deviation. The difference is in how the mean of the squared deviations (i.e., the variance) is calculated. As you know, to calculate a mean of some squared deviations you simply add up the squared deviations and divide by the number of them—and that is precisely how it is done in the case of the *population* variance. But for a *sample* variance, you add up the squared deviations and divide not by the number of them, but by one less than the number of them. Since the number of scores is usually denoted by $N$ for a population and $n$ for a sample, this means that in calculating a population variance you divide by $N$, while in calculating a sample variance you divide by $n - 1$. Thus the formula for the *sample* variance is

$$s^2 = \frac{\Sigma (X - \overline{X})^2}{n - 1}$$

and, of course, the formula for the *sample* standard deviation is

$$s = \sqrt{\frac{\Sigma (X - \overline{X})^2}{n - 1}}$$

### Bias

Notice that dividing by $n - 1$ rather than $n$ has the effect of making the sample variance a bit larger than it would have been. The reason this is done is that random samples tend to have variances (if computed using $n$) that are a bit smaller than the population variance. This is not *always* true, but it is true on the average. This tendency is known as *bias*: The sample variance would be negatively biased if it were computed using $n$. It turns out that dividing by $n - 1$ instead of $n$ removes the bias.

In other words, if we always compute sample variances using $n - 1$, there will not be any tendency for us to estimate the population variance too high or too low. Any *particular* sample variance will be different from the population variance one way or another, but the mean of *all possible* sample variances will be equal to the population variance.

Curiously, even if we calculate the sample variance with the "unbiased" method (using $n - 1$), when we take the square root to obtain the standard deviation, we again introduce a bias! This may seem paradoxical: The sample variance (using $n - 1$) does not tend to underestimate the population variance, but the sample standard deviation (still using $n - 1$) *does* tend to underestimate the population standard deviation—even though the standard deviation is computed directly from the variance. This type of thing is possible for the same reason that the standard deviation and the MAD are different: When you take the square root of some numbers, their average is not simply the square root of the average of the numbers.

Unfortunately, there is no quick fix available for the standard deviation, such as dividing by something else to remove the bias. But using $n - 1$ at least tends to make the bias less than it would have been in most practical situations.

## Calculating the standard deviation

These days, people rarely calculate standard deviations without the aid of a computer or calculator. There is a shortcut formula for calculating the standard deviation which saves a few keystrokes but is not very intuitively appealing. But in order to *understand* the standard deviation it is best to stick with the method that we outlined above. The method can be summed up in the phrase *root-mean-squared deviation* (RMSD). (The short-cut formula will be given a bit later.) Following is an example of how the standard deviation is calculated.

**Example 3.1**   Six rats are to be used in an experiment. Their weights, in grams (g), are 317, 290, 285, 330, 322, and 276. Find the standard deviation of their weights.

**Step 1**:   *Find the mean of the six weights*:

$$\frac{317 + 290 + 285 + 330 + 322 + 276}{6} = \frac{1820}{6} = 303.3333$$

**Step 2**:   *Calculate the deviation scores*:

$$317 - 303.3333 = \quad 13.6667$$
$$290 - 303.3333 = -13.3333$$
$$285 - 303.3333 = -18.3333$$
$$330 - 303.3333 = \quad 26.6667$$
$$322 - 303.3333 = \quad 18.6667$$
$$276 - 303.3333 = \quad 27.3333$$

**Step 3**:   *Square each deviation score*:

$$(13.6667)^2 = 186.7787$$
$$(-13.3333)^2 = 177.7769$$
$$(-18.3333)^2 = 336.1099$$
$$(26.6667)^2 = 711.1129$$
$$(18.6667)^2 = 348.4457$$
$$(-27.3333)^2 = 747.1093$$

**Step 4**:   *Take the mean of the squared deviation scores (dividing, however, by 5 instead of 6), giving the sample variance*:

$$\frac{186.7787 + 177.7769 + 336.1099 + 711.1129 + 348.4457 + 747.1093}{5}$$

$$= \frac{2507.3334}{5}$$

$$= 501.4667$$

***Step 5***: *Take the square root of the variance, giving the sample standard deviation*:

$$\sqrt{501.4667} = 22.3935$$

Thus the weights of the six rats have a mean of 303.33 g, with a standard deviation of 22.39 g.

Make sure that you understand what is happening: We take the square root of the mean of the squared deviation scores. Even though we rarely find it necessary to calculate standard deviations by hand, it is *vital* that you understand what the standard deviation indicates and how it is defined. *Do not proceed until you are sure you understand Example 3.1!*

## Note on computational methods

Nearly all computer programs assume that you are working with sample data—since you nearly always are—and therefore use $n - 1$ in calculating the standard deviation. Some computer programs, and some scientific calculators, offer you a choice. Note that the larger the number of scores, the less difference it makes whether you divide by $n$ or by $n - 1$. (For example, compare the numbers $20/2$ and $20/3$; then compare the numbers $20/150$ and $20/151$. The larger the denominator, the less difference it makes if you change it by 1.) So if your sample is large enough, it makes little practical difference whether you use $n$ or $n - 1$.

As we noted above, the RMSD method is not the only way to calculate the standard deviation—indeed, nearly all computer programs use one or another mathematically equivalent but computationally more efficient method. The various methods, even though they are mathematically equivalent, may produce different results depending on how numbers are rounded off in the midst of calculations. When there are many numbers involved, different methods can produce strikingly different results. A well-designed computer program will use a method that takes this problem into account and produces accurate results even when there is a large amount of data. (One such method is known as the *provisional means algorithm*.)

A method that is not appropriate with large amounts of data, but that works fine with relatively small amounts, is to use the shortcut formula we mentioned earlier. It is algebraically equivalent to the RMSD method, but requires that the sum of the squared numbers be accumulated (and can thus require a potentially large sum to be stored.) Here it is (assuming $n - 1$ is to be used):

$$s = \sqrt{\frac{\Sigma X^2 - (\Sigma X)^2/n}{n - 1}}$$

STUDY QUESTIONS—*Section 3.9*

**3.9.1**  a.  Write down the ages, in years, of the people in your immediate family. Then calculate the mean, median, range, MAD, and standard deviation of those ages.

b.  Now write down the (approximate) heights, in inches, of the people in your immediate family, and calculate the mean, median, range, MAD, and standard deviation of those heights.

c.  Which is more variable among the members of your family, their ages or their heights? Does this question make sense? Can you think of a way of making the two distributions comparable in terms of variability?

d.  Now change all the height figures from inches to centimeters by multiplying each height by 2.5 (close enough for our purposes). Now figure out the mean and standard deviation. (*Hint*: Before you do any calculating, think about it!) When you multiply all the scores by a constant, what happens to the mean and standard deviation?

e.  This time, change all the height figures from inches to inches under 8 ft (i.e., subtract each height from 96 in. and write down the result). Now what happened to the mean and standard deviation? When you add a constant to (or subtract a constant from) all scores, what happens to the mean and standard deviation?

**3.9.2**  [Continuation of Exercise 3.9.1] Let's consider another statistic, called the *coefficient of variation* (CV). The CV is defined as the standard deviation divided by the mean—or to put it another way, the standard deviation expressed as a proportion of the mean.

a.  What is the unit of the CV for the age distribution? The height distribution? Any distribution?

b.  Calculate the CV for the age and height distributions of your family. Which CV is bigger? What does that indicate?

c.  Here is another example in which the CV is useful. Which is more variable, the salaries of auto workers who work for General Motors or the salaries of members of the board of directors for the 500 largest companies in the United States? Suppose that the mean and standard deviation of the two distributions are as follows:

|  | Auto workers | Directors |
|---|---|---|
| Mean | $22,000 | $240,000 |
| Standard deviation | 4,000 | 30,000 |

Which distribution shows greater variability? In terms of absolute dollar amount of variability about the mean, the directors have a more variable distribution. What do the CV figures say?

**3.9.3**  Calculate the standard deviation of each set of numbers.
   a.  A sample of six randomly selected leaf lengths, in millimeters: 60, 63, 52, 45, 50, 38
   b.  A sample of three ages of randomly selected trees, in years: 37, 50, 100
   c.  Lengths of the four books written by a particular author, in pages: 380, 500, 450, 285 (What determines which kind of standard deviation to calculate in this case?)

**3.9.4**  [Statpal exercise—Coronary Care data set (see Appendix A)]
   a.  Find the standard deviation of the variable AGE for the entire data set and write a sentence interpreting it. (Consult your output from Exercise 3.6.1.)
   b.  Now use the BReakdown procedure to find the mean and standard deviation of AGE for each of the two groups specified by the variable SEX. Which group is more variable?
   c.  Use BReakdown again to find the mean and standard deviation of AGE for the groups specified by the variable RISK. Which of the three groups has the most variable age?

**3.9.5**  [Statpal exercise—Coronary Care data set (see Appendix A)]
   a.  Using the TRansformations utility, make a new variable called AGEDEV containing deviation scores for the variable AGE. To do this, use the transformation statement

$$AGEDEV = AGE - k$$

where, for k, you substitute the mean AGE (obtained in Exercise 3.6.1). (In other words, your new variable will be equal to AGE minus the mean AGE.)
   b.  Now go back to the Main Menu and use the DEscriptive statistics procedure to get the mean and standard deviation for AGEDEV. Are you surprised by the results?
   c.  Next, go back to the TRansformations utility and make a new variable called AGEDEVSQ, defined as AGEDEV squared. The statement to do this is

$$AGEDEVSQ = AGEDEV \wedge 2$$

(Note that the symbol ''$\wedge$'' is Statpal's way of indicating that the expression following is an exponent.) Then get the mean of the new variable AGEDEVSQ.
   d.  What is the mean of AGEDEVSQ the same as?
   e.  Now square the standard deviation you obtained for AGE previously (either by hand or using a calculator). This gives the variance of AGE. Although this number is not exactly equal to the result of part d, it should be pretty close. Why are the two numbers not the same, and how are they algebraically related?

# 4

# Some Distributions Used in
# Statistical Inference

## 4.1 Knowing the sampling distribution of a statistic allows us to draw inferences from sample data.

### Distributions in statistical inference

We noted in Chapter 2 that knowing the sampling distribution of a statistic is the key to using the statistic to draw inferences about the parameter of interest. We noted also that the central limit theorem assures us that statistics calculated by summing or averaging have (at least approximately) normal sampling distributions.

In this chapter we discuss the details of how to use the normal distribution to find the proportion of individual scores in any given interval. This will lay the groundwork for the inferential procedures of estimation and hypothesis testing which we cover in Chapter 5. We also discuss the binomial distribution, which is another important distribution used in statistical inference. We conclude the chapter with a description of the relationship between the binomial and normal distributions.

## 4.2 The standard normal distribution is used to find areas under any normal curve.

### Characteristics of normal distributions

In Chapter 3 we noted that the mean and standard deviation completely specify a normal distribution. That is, once you know the mean and standard deviation of a distribution known to be normal in shape, you can say exactly what proportion of scores in the distribution are in any given range. Let's consider how this is done.

First, it is handy to consider some general characteristics. (In fact, you will find it convenient to *memorize* these characteristics of a normal distribution, since you will be using them very frequently.) Refer to Figure 4.1.

1. A normal distribution is symmetric; therefore, it is centered about its mean (and, of course, its mean and median are equal).
2. About 68%—a little over two-thirds—of the scores are within 1 standard deviation of the mean.
3. About 95% of the scores are within 2 standard deviations of the mean.
4. Nearly all the scores in a normal distribution are within 3 standard deviations of the mean.

Item 3 is especially handy: the mean ±2 standard deviations includes about 95% of the scores in a normal distribution. For example, if a normal distribution

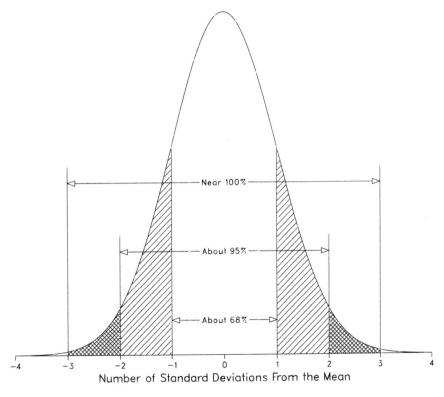

**Figure 4.1**   Areas in a normal distribution.

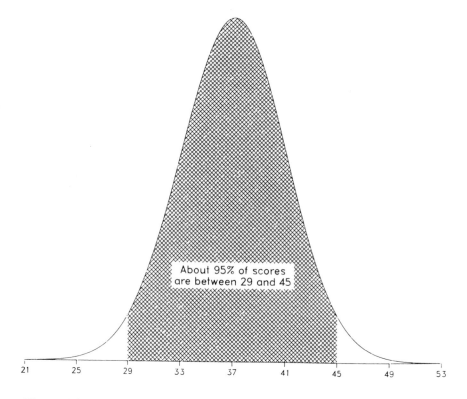

**Figure 4.2** Normal distribution with mean = 37, standard deviation = 4.

has a mean of 37 and a standard deviation of 4, we can say that 95% of the scores are between 29 and 45 (see Figure 4.2).

## Drawing a picture

Now, let's get more specific. In our normal distribution with mean 37 and standard deviation 4, what proportion of scores are between 37 and 39? Or between 38 and 42.86? How do we figure that out? There are two ways to do it. One way is to let a computer figure it out for you; the other is to use a table of the standard normal distribution. Since it is vital to understand how the normal distribution works even if a computer does carry out the calculation, we will cover the second method.

There are three rules for using a table of the standard normal distribution: (1) draw a picture, (2) draw a picture, and (3) draw a picture. What picture should you draw? A histogram of a normal distribution, of course. As we have already

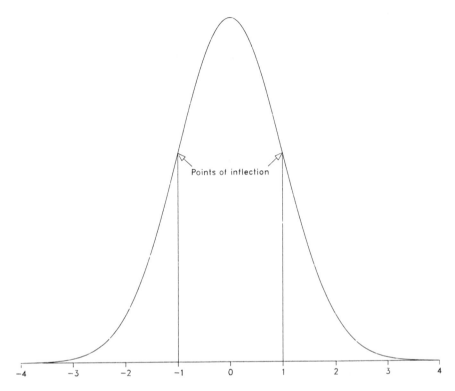

**Figure 4.3**   Points of inflection in a normal distribution.

seen, the proportion of scores in any particular range is represented by the area in the histogram above that range. So finding proportions in the distribution is the same as finding areas in the histogram. The first thing to do when you want to find a proportion in some region of a normal distribution is *draw a picture of the distribution and shade in the region in which you are interested.*

In drawing your picture, it is handy to take advantage of another feature of a normal distribution. This feature relates to the *points of inflection* of the curve. Look at Figure 4.3. If you start at the top of the curve and move to the right, the curve moves downward, first more steeply, then less steeply. The point at which the steepness stops increasing and starts decreasing is the point of inflection. Of course, because of the symmetry of the curve, there is also a point of inflection on the left side. As noted in the figure, the points of inflection are located exactly 1 standard deviation on either side of the mean. You can scale your sketches accordingly.

## The standard normal distribution and z scores

Table D.1 in Appendix D gives areas in various pieces of the standard normal distribution. The term *standard normal distribution* refers to the particular normal distribution that has a mean of zero and a standard deviation of 1. Why choose these particular values of mean and standard deviation to make a table? Because by a very simple algebraic trick, any normal distribution can be turned into the standard normal distribution.

The trick is to express the scores not in their original units but rather as *number of standard deviations away from the mean*. For example, suppose that the heights of cornstalks of a certain variety have a normal distribution with mean 220 cm and standard deviation 25 cm. Consider a cornstalk that is 190 cm in height. We can express this height not merely as 190 cm, but also as 1.2 standard deviations below the mean. (Do you see where 1.2 comes from?)

This trick of expressing original scores as number of standard deviations away from the mean is so useful that it has a special name: It is called *standardizing* the scores, and scores so expressed are called *standard scores*. Another term for standard score is *z score*, because a standard score is usually denoted in formulas with the letter *z*. The sign of a *z* score indicates whether the score is above $(+)$ or below $(-)$ the mean.

So our cornstalk with a height of 190 cm would have a *z* score of $-1.2$, which indicates that it is 1.2 standard deviations *below* the mean. A cornstalk with a height of 230 cm would have a *z* score of $+.4$. There is a formula for *z* scores, but you do not need it—just remember that *a z score is the number of standard deviations above $(+)$ or below $(-)$ the mean*.

Notice that a cornstalk with a height of 220 cm (right on the mean) has a *z* score of zero, while a cornstalk exactly 25 cm (1 standard deviation) above the mean has a *z* score of $+1.0$. This illustrates the fact that *z* scores have mean zero and standard deviation 1, simply as a consequence of how they are defined. So changing the original, "raw" scores to *z* scores gives you scores that have mean zero and standard deviation 1. If your original scores have a normal distribution, the corresponding *z* scores have the standard normal distribution.

## Finding normal curve areas

Since you can always express any score as a *z* score (as long as you know the mean and standard deviation), you can use the table of the standard normal distribution to find areas for *any* normal distribution.

First, let's look at Table D.1 and observe what it tells us. For a given *z* score, *the table tells us what proportion of the area under the normal curve lies between the mean and that z score*. Notice that the picture above the table shows what area is tabulated. To find the entry for a particular *z* score—say, 1.37—you first find the row that gives you the first decimal place (1.3) and then find the column for the second decimal place (7). The number in that row and column (.4147) is the proportion of the area under the normal curve between the mean ($z$ score $= 0$) and the $z$ score you are looking up.

Now let's consider some examples. We'll use the normal distribution with mean 37 and standard deviation 4.

**Example 4.1**  Find the proportion of scores between 37 and 41 in the normal distribution with mean 37 and standard deviation 4.

*Step 1*:  *Draw a picture.* (See Figure 4.4.)

*Step 2*:  *Figure out how to use the table to get the area we want.* In this case, it is easy, since we are looking for an area between the mean and some other score—which is precisely what the table gives.

*Step 3*:  *Calculate the z score for the scores we need to look up.* In this case all we need to do is calculate the $z$ score for 41. Since 41 is 1.00 standard deviation above 37, the $z$ score for 41 is $+1.00$.

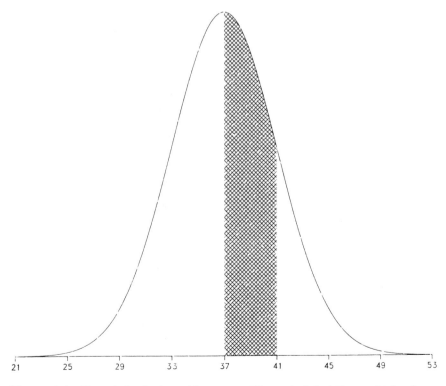

**Figure 4.4**  Normal distribution with mean $= 37$, standard deviation $= 4$, showing scores between 37 and 41.

***Step 4***:  *Find the required areas in the table*. The area between the mean and a z score of +1.00, from Table D.1, is .3413.

***Step 5***:  *Finish up*. In this case the table told us just what we needed. The proportion of scores between 37 and 41 is .3413, or 34.13%—a little over a third.

**Example 4.2**    Find the proportion of scores between 36 and 41.

***Step 1***:  *Draw a picture*. (See Figure 4.5.)

***Step 2***:  *Figure out how to use the table to get the area we want*. Note that the table will give us the area between the mean and 41—indeed, we just found that in Example 4.1. So all we have to do is get the area between the mean and 36 and add it to .3413 (the area between the mean and 41).

***Step 3***:  *Calculate the z score for the scores we need to look up*. The z score for 36 is −1/4, or −.25. The table does not include negative z scores, but that is not a problem, thanks to the symmetry of the normal distribution: The area

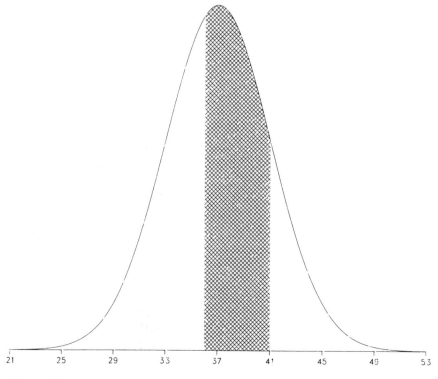

**Figure 4.5**  Normal distribution with mean = 37, standard deviation = 4, showing scores between 36 and 41.

between the mean and a *z* score of −.25 is the same as the area between the mean and a *z* score of +.25. So the *z* score that we need is .25.

**Step 4**: *Find the required areas in the table*. The area between the mean and a *z* score of .25 is .0987.

**Step 5**: *Finish up*. So the area that we seek is .0987 + .3413, or .4400. That is, 44% of the scores in this distribution are between 36 and 41.

**Example 4.3**   Find the proportion of scores between 38.4 and 41.

**Step 1**: *Draw a picture*. (See Figure 4.6.)

**Step 2**: *Figure out how to use the table to get the area we want*. Again, we already have the area between 37 and 41, from Example 4.1. But this time we want to subtract the area between 37 and 38.4, leaving only the area shaded in the figure. So all we need to do is find the area between 37 (the mean) and 38.4.

**Step 3**: *Calculate the z score for the scores we need to look up*. The *z* score for 38.4 is 1.4/4, or .35.

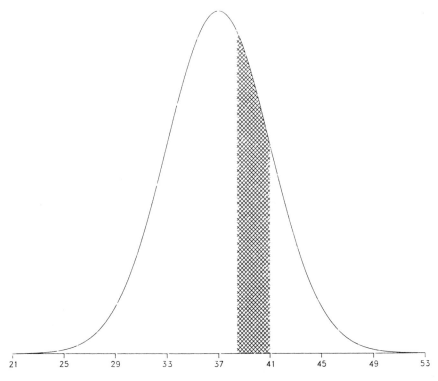

**Figure 4.6**   Normal distribution with mean = 37, standard deviation = 4, showing scores between 38.4 and 41.

**Step 4**: *Find the required areas in the table.* The area between the mean and a *z* score of .35 is .1368.

**Step 5**: *Finish up.* We subtract .1368 from .3413 and get .2045. So 20.45% of the scores in the distribution are between 38.4 and 41.

**Example 4.4**   Find the proportion of scores that are greater than 41.

**Step 1**: *Draw a picture.* (See Figure 4.7.)

**Steps 2 through 5**:   All we have to do is notice that the area we want is the entire right half, minus the area between the mean and 41. Since we already know that the area between the mean and 41 is .3413, we calculate our desired area as .5000 − .3413, or .1587. Thus 15.87% of the scores are in the tail of the distribution above a score of 41.

Every problem involving normal curve areas fits into one of the four examples we have just considered (or a simple modification).

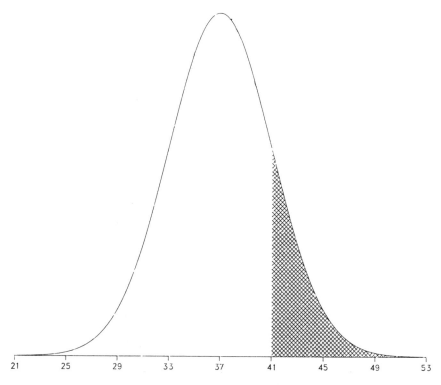

**Figure 4.7**   Normal distribution with mean = 37, standard deviation = 4, showing scores greater than 41.

## STUDY QUESTIONS—*Sections 4.1 and 4.2*

**4.2.1** Consider a normal distribution with mean 100 and standard deviation 5. Calculate the proportion of scores
    a.  Between 100 and 110
    b.  Between 90 and 110
    c.  Between 105 and 107
    d.  Above 100
    e.  Above 110
    f.  Below 92
    g.  Above 93

**4.2.2** Repeat Exercise 4.2.1 for the normal distribution with mean 100 and standard deviation 10. Compare the results of each part between the two distributions. Does doubling the standard deviation result in doubling the proportion of scores in a given area?

**4.2.3** Using the table of the standard normal distribution, verify the general characteristics (2, 3, and 4) listed in Section 4.2. Specifically, verify the following.
    a.  About two-thirds of the scores in a normal distribution are within 1 standard deviation of the mean. (*Hint*: The scores within 1 standard deviation of the mean are the scores between 1 standard deviation *below* the mean and 1 standard deviation *above* the mean. What is the $z$ score for 1 standard deviation above the mean?)
    b.  About 95% of the scores in a normal distribution are within 2 standard deviations of the mean.
    c.  Nearly all the scores in a normal distribution are within 3 standard deviations of the mean. (Using Table D.1, what do we mean by "nearly all"?)

**4.2.4** Now let's use Table D.1 a slightly different way. Instead of finding the area between the mean and a given $z$ score, let's find the $z$ score that yields a certain area between itself and the mean. Specifically:
    a.  What $z$ score is the 75th percentile of the normal distribution? (*Hint*: The 75th percentile is the score that separates the upper 25% of the scores from the lower 75% of the scores. Because the normal distribution is symmetrical, the mean is also the median, or 50th percentile. Thus the area between the mean and the 75th percentile is 25%, or .2500. Get as close as you can, but do not bother interpolating.)
    b.  Based on your results of part a, what is the interquartile range of the standard normal distribution?
    c.  What $z$ score cuts off the upper 5% (i.e., .0500) of the scores?
    d.  What $z$ scores bracket the middle 90% of the scores? (Draw a picture.)
    e.  What $z$ scores bracket the middle 95% of the scores?

**4.2.5** The basal metabolic rate (BMR) for 30-year-old men, measured in kilocalories per square meter per hour (kcal/m$^2$ per hour), has a mean of 37.6, with a standard deviation of 2.6. The distribution of BMR scores is approximately normal.
    a.  What proportion of 30-year-old men have BMR scores lower than 31 kcal/m$^2$ per hour?

    b.   A 30-year-old man has a BMR measured at 40 kcal/m² per hour. What percentile is his BMR?

    c.   What is the interquartile range of BMR scores for 30-year-old men?

    d.   What BMR scores encompass the middle 95% of the distribution?

## 4.3  The binomial distribution is used for variables that count the number of yeses.

### Some examples

To motivate our discussion of the binomial distribution, let's consider some examples. In any barrel of apples, as the saying goes, there will be some rotten ones. Suppose that in a certain large warehouse full of apples, 5% of the apples are rotten, with the rotten apples randomly dispersed among the apples in the warehouse. Apples are sold in dozen-apple bags, with the apples for each dozen randomly selected from the warehouse. What proportion of dozen-apple bags will contain no rotten apples? What proportion will contain exactly one rotten apple? What proportion will contain three or more rotten apples?

Here's another example: On a multiple-choice test with five choices for each question, someone with absolutely no clue as to the right answer on any question—say, someone who is unable to read the questions—would be expected to get about 20% of the questions right by chance. On a 15-item test, what proportion of such people would get six or more questions right?

One more example: When a certain type of thumbtack is tossed, it lands point up 30% of the time. A certain game involves tossing four thumbtacks. If exactly three tacks (not more, not fewer) land point up, you win. What proportion of the time will you win?

### Variables that count the number of yeses in a set of trials

What is the common theme of these three examples? In each example we have a situation in which something either does or does not occur—that is, a "yes-no" situation. An apple either is or is not rotten; a question either is answered correctly or is not; a tack lands point up or it does not. Moreover, the variable of interest is the number of times "yes" occurs in a set of trials. In the first example, we were interested in the *number* of rotten apples out of a dozen; in the second example, the *number* of questions answered correctly out of 15; and in the third example, the *number* of tacks landing point up out of four.

### Independence

We note also that in each situation, whether or not "yes" occurs in any one trial has no effect on whether or not "yes" occurs in any other trial. The rotten apples were randomly dispersed throughout the warehouse; thus the fact that a particular apple was rotten did not mean that the next apple was more or less likely to be rotten; getting one question right by chance does not mean that one is more or less likely to get the next question right by chance; and the result of one toss of a tack has no bearing on the result of another toss. This feature of these examples—the fact that the occurrence (or nonoccurrence) of a "yes" on one trial does not affect the chance of a "yes" on any other trial—is called *statistical independence* or simply *independence*. We can say that the result of each toss of a tack, for example, is independent of the result of any other toss. Thus in each case we are interested in the *distribution of a variable that counts the number of times "yes" occurs in a set of independent trials*. The distribution of such a variable is called the *binomial distribution*.

### A note on probability

How does the binomial distribution work? We shall present an intuitive derivation. More mathematically rigorous presentations are available in many standard probability texts (see, e.g., Schaeffer and Mendenhall, 1975). We shall use the term *probability* synonomously with *long-run proportion*, an intuitively appealing rather than a mathematically precise usage of the term. It is, nevertheless, a usage that will serve us well for the purposes of understanding the role of probability in statistical inference.

## 4.4   To calculate binomial probabilities, we need to find the probability of each possible outcome.

### Finding the probability of a particular outcome

Let's start with the tack-tossing example. Suppose that the game involved only one tack, and we win if the tack lands point up. Then we would expect to win 30% of the time, in the long run. Another way to say this is that the probability of winning this game would be 30%, or .3. And, of course, the probability of not winning would be 100% minus 30%, or 70%.

Now consider a two-toss game, involving, say, a red tack and a green tack, in which we win if *both* tacks land point up. Now what is the probability of winning? To figure this out, we note that there are four possible outcomes of the game: (1) both point up; (2) red up, green down; (3) red down, green up; and (4) both down. The outcome of interest to us is the first (both up). What proportion of the time, in the long run, will this outcome occur? Note that 30% of first tosses would land point up; and of those, 30% would be followed by a second toss that would land point up. Thus, in the long run we would win this game 30% of 30%, or 9%, of the time. Or, to use the terminology of probability, we can say that the probability of winning this game on any given occasion is 9%, or .09.

## Multiplying probabilities of independent events

Note that we obtained the probability that *both* tacks would land point up by simply multiplying the probability that the first tack would land point up (.3) by the probability that the second would land point up (also .3). It is the independence of the two tosses that allows us to do this, because if the tosses were not independent, the probability of the second tack landing point up would have changed, depending on the outcome of the first.

Now let's change the game slightly. Suppose that we win not if both tacks land point up, but rather if *exactly* one of the tacks lands point up. If both are up, or both are down, we lose; otherwise, we win. What is the probability of winning this game?

There are two distinct outcomes that lead to a win in this game: (1) red up, green down, or (2) red down, green up. To find the proportion of the time we would win this game, then, we need to find the proportion of the time that *either* of these two outcomes would occur. Since the two outcomes do not overlap, we can find the proportion of times *either* will occur by summing the proportion of the time each occurs.

By the same logic as above, we can calculate the probability of each outcome as follows:

probability of "red up, green down"
    = probability of "red up" × probability of "green down"
    =          .3     ×        .7
    = .21, or 21%

probability of "red down, green up"
    = probability of "red down" × probability of "green up"
    =          .7     ×        .3
    = .21, or 21%

So in the long run, 21% of the time we would get "red up, green down," and 21% of the time we would get "red down, green up." Since either of these outcomes is a win, we would win 21% plus 21%, or 42% of the time.

Having performed this calculation, we note that both of the "winning" outcomes (red up/green down and red down/green up) had the same probability (21%). This, of course, is not merely coincidental; it is a direct consequence of the fact that the winning outcomes here are precisely those outcomes that have exactly one tack down and one up. *Any* such outcome will have the same probability. You should go over the reasoning up to this point until you see why this is so.

4.5 We next find the number of relevant
outcomes and multiply it by the
probability of each relevant outcome.

*Finding the number of relevant outcomes*

Now we are ready to consider our original example with four tosses. (For clarity, we will assume that four tacks are tossed, with one red, one green, one blue, and one yellow.) What is the probability that exactly three of the tacks will land point up?

Before we consider the details, note the general approach. Of all of the possible outcomes, we are interested in those outcomes in which exactly three of the tacks land point up. Each such outcome has the same probability, since the four probabilities being multiplied will give the same product regardless of what order they are multiplied. Specifically, each "winning" outcome has probability .3 × .3 × .3 × .7, which works out to .0189, or 1.89%. So to find the probability that we will get *any* of the winning outcomes, we need only figure out the number of winning outcomes—let's call it W—there are in all possible outcomes. Each willing outcome will have probability 1.89%; thus the probability of getting any of them will be W times 1.89%.

For the four-tack case, here are all the possible outcomes (U means "up" and D means "down," and the number of tacks up is indicated for each outcome):

| Red | Green | Blue | Yellow | Number up | |
|-----|-------|------|--------|-----------|---|
| U | U | U | U | 4 | |
| U | U | U | D | 3 | • |
| U | U | D | U | 3 | • |
| U | U | D | D | 2 | |
| U | D | U | U | 3 | • |
| U | D | U | D | 2 | |
| U | D | D | U | 2 | |
| U | D | D | D | 1 | |
| D | U | U | U | 3 | • |
| D | U | U | D | 2 | |
| D | U | D | U | 2 | |
| D | U | D | D | 1 | |
| D | D | U | U | 2 | |
| D | D | U | D | 1 | |
| D | D | D | U | 1 | |
| D | D | D | D | 0 | |

The winning outcomes are indicated by a •. Since there are four of them, we can calculate the probability of winning this game as 4 × 1.89%, or 7.56%.

## A quicker way to find the number of relevant outcomes

The "brute force" method for finding the number of relevant outcomes works, as we saw above, but it is cumbersome, and it becomes more so very quickly if the number of "tacks"—that is, independent trials—involved in the game increases. In fact, the number of outcomes to search through is 2 to the power $n$, where $n$ is the number of trials. Thus for our rotten apple example, with 12 apples in a bag, we would have to search through 4096 outcomes, and for the multiple-choice test example, with 15 questions on the test, we would have to search through 32,768 outcomes! This is obviously impractical.

Fortunately, we can apply some mathematical counting principles to make the task easier. In particular, a general formula is available (and is derived in most probability texts) for the number of outcomes containing any particular number of yeses out of $n$ trials. The formula is

$$\binom{n}{k} \equiv \frac{n!}{k! \, (n - k)!}$$

where the symbol $\binom{n}{k}$ is read "$n$ choose $k$"; the symbol $n!$ is read "$n$ factorial" and is defined as $n \times (n - 1) \times (n - 2) \times \cdots \times 2 \times 1$. The special case of zero factorial is defined to be 1; thus "$n$ choose 0" is equal to $n!/n!$, or 1. Thus for our four-tack example, the number of outcomes containing exactly three U's is

$$\binom{4}{3} \equiv \frac{4!}{3! \, (4 - 3)!} = \frac{24}{6} = 4$$

Note, incidentally, that "$n$ choose $k$" is always equal to "$n$ choose $n - k$" (and you might want to verify this yourself from the definition of "$n$ choose $k$"). So we could have calculated "4 choose 1" instead of "4 choose 3" and gotten the same answer. (What is "$n$ choose 1" always equal to?)

## 4.6  Binomial probabilities can be computed from a general formula and are also available in tables.

### General formula for the binomial distribution

We are now ready to put together the general formula for the binomial distribution. Make sure that you understand the principle:

probability of getting $k$ yeses out of $n$ independent trials
= number of possible outcomes having exactly $k$ yeses (i.e., "$n$ choose $k$") $\times$ probability of each such outcome

If the probability of a "yes" on a single trial is called $p$, then, as we illustrated above:

probability of each possible outcome that has exactly $k$ yeses
$= p \times p \times \cdots \times p$     (a total of $k$ $p$'s)
$\times$
$(1 - p) \times (1 - p) \times \cdots \times (1 - p)$     [a total of $n - k$ $(1 - p)$'s]

which is easily seen to be equal to $p^k(1 - p)^{n-k}$.

So we arrive at the following general formula: The probability of observing exactly $k$ yeses is $n$ independent trials, where the probability of "yes" on a single trial is $p$, is equal to

$$\binom{n}{k} p^k(1 - p)^{n-k}$$

Note that the binomial distribution is determined by two parameters: $n$ (the number of trials) and $p$ (the probability of "yes" on a single trial). Let's apply this formula to the first two examples we mentioned at the beginning of our discussion of the binomial distribution in Section 4.3.

**Example 4.5**    Recall that we have bags of 12 randomly picked apples, with the probability of each apple being rotten equal to 5%. Thus we have $n = 12$ and $p = .05$. What proportion of such bags would have no rotten apples? Applying the binomial formula, the probability of getting 0 yeses is

$$\binom{12}{0}(.05)^0(.95)^{12} = (.95)^{12} = .5404$$

Thus slightly more than half (about 54%) of such bags (in the long run) will have no rotten apples; the remaining 46% will have at least 1.

We also asked what proportion of bags would have 3 or more rotten apples— that is, exactly 3 or exactly 4 or exactly 5 or . . . (etc.) up to 12. To find this, we could calculate each of the probabilities for 3, 4, 5, and so on up to 12, and then sum them. However, a simpler way is to realize that the proportion of bags with 3 or more rotten apples is simply 1 minus the proportion of bags with fewer than 3 rotten apples—that is, 1 minus the proportion of bags with 0, 1, or 2 rotten apples. To get this requires only three applications of the binomial formula rather than 10. The necessary numbers are:

probability of getting 0 rotten apples (from above) = .5404

$$\text{probability of getting 1 rotten apple } = \binom{12}{1}(.05)^1(.95)^{11}$$

$$= 12(.05)(.5688)$$

$$= .3412$$

$$\text{probability of getting 2 rotten apples } = \binom{12}{2}(.05)^2(.95)^{10}$$

$$= 66(.0025)(.5987)$$

$$= .0099$$

Thus the probability of getting a bag containing 0, 1, or 2 rotten apples is .5404 + .3413 + .0099, or .8916. Therefore, the probability of getting a bag containing 3 or more rotten apples is 1 − .8916, or about 11%. So about 11% of the bags would contain 3 or more rotten apples.

**Example 4.6**  On a 15-item multiple-choice test with five choices per question, what proportion of people would get six or more items right if they guessed randomly on every question? Here we have $n = 15$ and $p = .20$ (i.e., 1/5). As in Example 4.5, we could save a bit of calculation by adding up the binomial probabilities for 0, 1, 2, 3, 4, and 5 and then subtracting this total from 1. Thus we would have six probabilities to compute using the binomial formula.

Even more conveniently, however, we can make use of the table of binomial probabilities (Table D.2). As indicated in the table, the probability of observing each possible number of yeses in $n$ trials, for various values of $p$ is presented. Entering the table with $n = 12$ and $p = .20$, we find the following values:

probability of observing 0 = .069
probability of observing 1 = .206
probability of observing 2 = .283
probability of observing 3 = .236
probability of observing 4 = .133
probability of observing 5 = .053

The sum of these probabilities is .980, which is therefore the probability of getting fewer than six items correct by guessing. Thus 1 − .980, or 2%, of people taking the test could get six or more items right by pure chance.

## STUDY QUESTIONS—*Sections 4.3 through 4.6*

4.6.1   Eggs in supermarkets are generally sold in cartons, with each carton containing a dozen eggs.
   a.   Suppose the probability that a randomly picked egg is broken is 8%. What proportion of egg cartons would contain at least one broken egg?

b.  In using the binomial distribution to answer part a, what assumption were you making about the incidence of broken eggs? Is this a reasonable assumption?

**4.6.2**  Of all voters in a certain Vermont town in the 1980 U.S. presidential election, 20% voted for John Anderson, who was running as an independent. A committee of 10 voters is to be selected at random from the town to work on procedures for the next election.

a.  What is the probability that none of the voters selected voted for Anderson?

b.  What is the probability that half or more of the committee voted for Anderson?

c.  In the same town, 40% of the voters voted for Reagan. What is the probability that half or more of the selected committee will consist of Reagan voters?

**4.6.3**  In computer networks, information is sent in "packets" from one computer to another. Due to imperfections in the physical connection between computers, some packets are "damaged"—that is, information is lost in transit. The system includes a way of checking for damage, so that if a packet arrives damaged, a request for a retransmission is sent to the source computer and the packet is retransmitted. For purposes of this question, we will assume that retransmissions always get through. (This is possible if, for example, retransmissions are sent over a more secure, but more expensive line.) Suppose that in a certain network, the probability that a packet will be damaged in a single transmission is 10%, with the probability remaining constant in any sequence of transmissions (i.e., transmissions are independent of each other).

a.  If a message requires that five packets be transmitted, what is the probability that no retransmissions will be required?

b.  If a message requires that 10 packets be transmitted, what is the probability that fewer than two retransmissions will be required?

c.  In certain communications protocols, the system will give up and break the connection after, say, five retransmissions. If a message contains 20 packets, what is the chance the message will get through (i.e., that fewer than five retransmissions will be required)?

**4.6.4**  A production line turns out electrical resistors. Due to an equipment malfunction, 15% of the resistors in the production run from a particular week were defective, with the defective resistors randomly dispersed among all the resistors produced that week.

a.  Customers order shipments of 25 resistors. What proportion of such shipments have at least three defective resistors?

b.  If a shipment goes out with more than two defective resistors, the net cost to the company is $500 (including the replacement costs of the resistors and the additional time of the customer-service staff needed to remedy the problem). The cost of testing an entire shipment of 25 resistors is $50. If the company sends out 400 shipments of resistors during the week in question, should it test each shipment before shipping it?

**4.6.5**  [Continuation of Exercise 4.6.4] As we discussed in Chapter 2, the central limit theorem assures us that statistics that are calculated by summing or averaging tend to have a normal sampling distribution. With that in mind, consider the following variable defined for each resistor:

$$\text{DEFECTIVE} = \begin{cases} 0 & \text{if the resistor is not defective} \\ 1 & \text{if the resistor is defective} \end{cases}$$

a.  Suppose that a case of 100 resistors contains eight defective resistors. What is the sum of the variable DEFECTIVE for that shipment?

b.  Consider a few thousand cases of 100 resistors. If you determined the sum of the variable DEFECTIVE for each case, you could find the mean and standard deviation of all those sums. If, say, 10% of the individual resistors were defective, what would you expect the mean of all the thousands of sums to be? In other words, what would you expect to be the average number of defective resistors in a case of 100?

c.  If you determined all of those thousands of sums, you could display them in a histogram, indicating the proportion of cases with 0, 1, 2, 3, and so on up to 100, defective resistors. What would the shape of the histogram be? How do you know?

d.  Given that the distribution of sums would be approximately normal, determine the proportion of cases of 100 resistors having five or fewer defective resistors, given that 10% of all resistors in the relevant population are defective. To do this, you need the mean of the distribution, which you guessed in part b, and the standard deviation, which (as we will discuss in Section 4.7) equals 3.

## 4.7   A binomial distribution with large $n$ and moderate $p$ is approximately normal.

### Binomial variables with large n

A binomial variable, as we saw in the previous sections, is a variable that counts the number of yeses in $n$ independent trials. When $n$ is fairly small, it is not too difficult to apply the general binomial formula to get whatever probability we need. If $n$ and $p$ happen to be one of the combinations included in the binomial table, we can bypass the formula altogether and get our desired probabilities from the table.

But what if $n$ is large—perhaps too large to compute results using the binomial formula? Let's consider an example. Suppose that 350 people are independently asked to sample two colas (let's call them CC and PC) and state their preference. The makers of PC claim that 70% of people prefer their product. If that claim is true, what is the probability that 220 or fewer people would prefer PC?

A "yes" here is defined as preferring PC. If the claim is true, $p$ equals .70. We have $n$ equal to 350. If you are not convinced of the difficulty of using the binomial formula to find the probability of observing 220 or fewer yeses with $n = 350$ and $p = .70$, try it. (Even if you have the mathematical sophistication needed to approximate the factorials involved, you are still faced with the necessity of calculating potentially dozens of probabilities.)

### Binomial variables as sums

How, then, can the probability be found? The key to finding a solution is to use the trick we illustrated in Exercise 4.6.5. That is, we note the fact that a binomial variable is also a sum. How is a binomial variable also a sum? The trick is to assign to each individual trial a score of 1 if the trial results in a ''yes,'' 0 otherwise. (Scores defined in this way are called *dummy variables*.) The binomial variable, which counts the yeses, can then be defined as the sum of the scores for the *n* trials.

This shows *how* we can view a binomial variable as a sum. But *why* should we

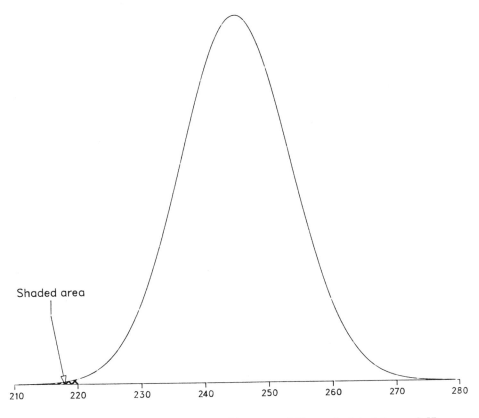

**Figure 4.8**   Normal distribution with mean = 245, standard deviation = 8.57.

view a binomial variable as a sum? The answer relates to what we noted in Chapter 2: The central limit theorem guarantees that a variable that is itself a sum tends to have a normal distribution, provided that *n* is large enough. Thus by viewing a binomial variable as a sum, we can see that the binomial distribution with large *n* becomes approximately the same as the normal distribution with the same mean and standard deviation. Thus *we can use the normal distribution to find binomial probabilities for large n.*

### The mean and standard deviation of a binomial variable

As we learned in Section 4.2, to find the proportion of scores in a normal distribution in any particular range, we need to know two parameters: the mean and the standard deviation. Thus to use the normal distribution to approximate binomial probabilities, we need to know the mean and standard deviation of a binomial variable.

The mean of a binomial variable is just what you would expect it to be

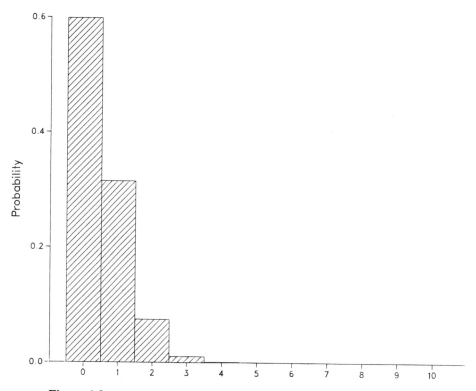

**Figure 4.9**  Binomial distribution with $n = 10$, $p = .05$.

intuitively. If (as in our example) 70% of individuals prefer PC, then if you independently poll 350 individuals, you would expect *on the average* that 70% of them, or 245, would prefer PC. Thus the mean of a binomial variable is simply $n$ times $p$. (This, of course, is not a mathematical proof, merely an intuitive explanation; fortunately, it is true!) In mathematical notation: For a binomial variable with parameters $n$ and $p$,

$$\mu = np$$

To get the standard deviation requires a bit of mathematical theory we have not developed in this book, so we will simply present the result: For a binomial variable with parameters $n$ and $p$,

$$\sigma = \sqrt{np(1 - p)}$$

Thus for our example with $n = 350$ and $p = .70$, we have $\sigma = \sqrt{350 \times .7 \times .3}$, or $\sqrt{73.5}$, which is approximately 8.57.

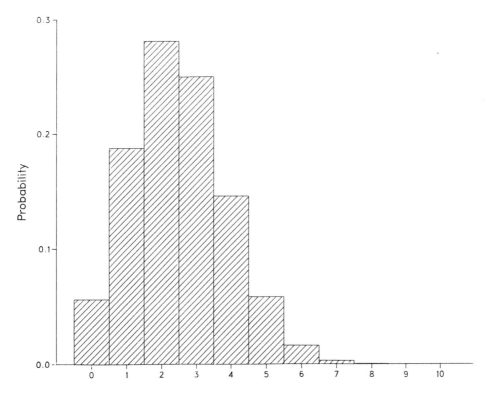

**Figure 4.10**   Binomial distribution with $n = 10$, $p = .25$.

## Using the normal distribution to find binomial probabilities

Having found the mean and standard deviation of our binomial variable, we are now able to use the normal distribution to find the desired probability for our example. Recall the question: What is the probability that 220 or fewer people will prefer PC cola out of a sample of 350 people, assuming that 70% of the individuals in the population prefer PC?

We first draw a picture of the appropriate normal distribution (see Figure 4.8), with $\mu = 245$ and $\sigma = 8.57$. The area of interest has been shaded in. Applying the methods we covered in Section 4.2, we observe that 220 is 2.92 standard deviations below the mean. Consulting the normal table (Table D.1), we find that .4982, or 49.82%, of the area under the curve is between the mean and a $z$ score of 2.92; thus the shaded area is 50% minus 49.82%, or .18%. Thus the probability is very small—less than two-tenths of 1%—that 220 or fewer people would prefer PC cola, assuming that the claim is true.

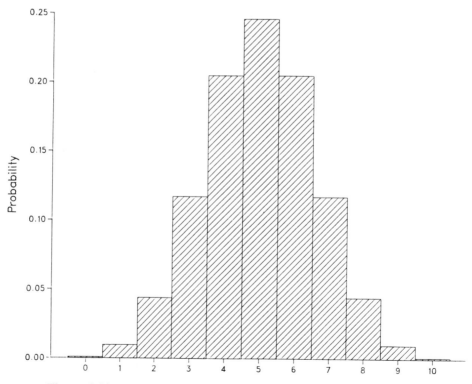

**Figure 4.11**   Binomial distribution with $n = 10$, $p = .5$.

### How large must n be?

We noted above that a binomial distribution "with large $n$" can be approximated by a normal distribution with the same mean and variance. How large must $n$ be for this to apply? The answer depends on $p$. To see why, look at Figures 4.9 through 4.14, which display histograms for binomial distributions with $n = 10$ and $n = 50$, each for $p = .05$, $.25$, and $.50$.

Consider first $n = 10$ (Figures 4.9 through 4.11). With $p = .05$ (Figure 4.9), the binomial distribution is markedly skewed to the right. A normal distribution with the same mean and standard deviation (.5 and .6892, respectively) would clearly be a poor approximation. With $p = .25$ (Figure 4.10), the approximation is somewhat improved, although some skew is still apparent in the binomial distribution. But with $p = .50$ (Figure 4.11), a normal distribution would come quite close.

Now consider $n = 50$ (Figures 4.12 through 4.14). Even with $p = .05$ (Figure

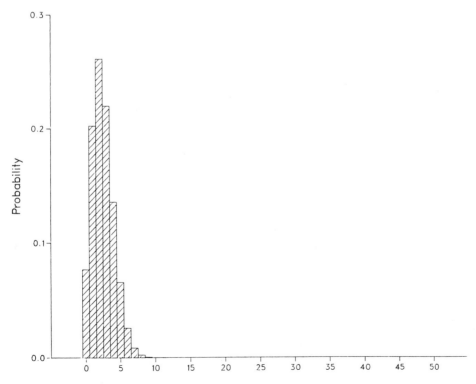

**Figure 4.12**   Binomial distribution with $n = 50$, $p = .05$.

4.12), the approximation is not too bad, and with $p = .25$ (Figure 4.13) and $p = .50$ (Figure 4.14), the approximation is very close.

So how large should $n$ be? A handy rule of thumb is to require that both $np$ and $n(1 - p)$ be at least 5. By this rule, if $p$ is .50, then $n = 10$ is sufficient. However, if $p$ is, say, .80, we would need $n = 25$, and if $p$ is .95, we would require $n = 100$ to be on the safe side.

### The continuity correction

Let's reconsider the case of $n = 10$ and $p = .50$. We have redrawn the binomial distribution, this time with a superimposed normal distribution, in Figure 4.15. As we noted above, the normal distribution does a good job of approximating the binomial in this situation.

Suppose that we wish to find out the probability of observing six or more yeses in 10 independent trials. The exact probability, from the binomial distribu-

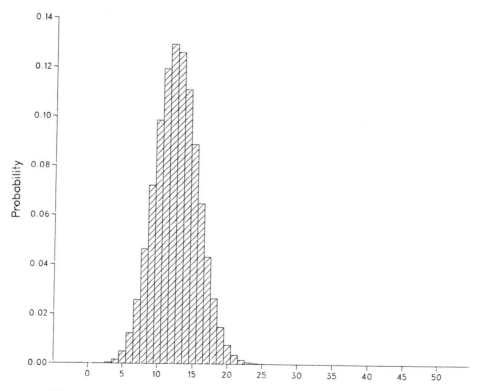

**Figure 4.13** Binomial distribution with $n = 50$, $p = .25$.

tion, would be the area shaded in the figure. This works out (from the binomial table) to be equal to .377.

Now let's use the normal distribution to figure out an approximation of the same probability, and compare it to the exact probability. Following the method outlined above, we find that

$$\mu = np = 5,$$

and

$$\sigma = \sqrt{np(1 - p)} = \sqrt{2.5} = 1.58$$

Now we need to find the $z$ score to look up in the normal table—and this is where our picture becomes essential. Note that if we simply found the $z$ score for 6, we would be *underestimating* the probability we are after. Why? Because, as is evident from the picture, the normal distribution is a continuous and smooth, whereas the binomial distribution is discrete and blocky. To best approximate the

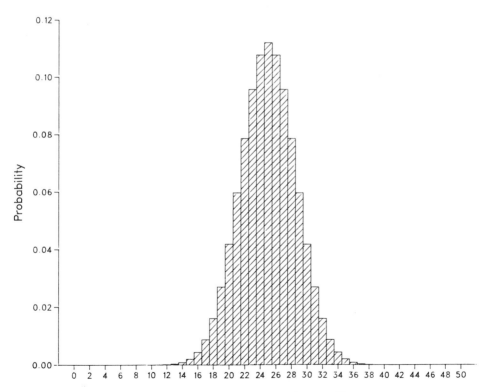

**Figure 4.14**  Binomial distribution with $n = 50$, $p = .50$.

area in the binomial blocks for 6 through 10, we need to find the area under the normal curve starting at 5.5, not 6.

To confirm this, note the results we get using the normal approximation: If we started at 6, which would be .63 standard deviation above the mean of 5, we would arrive at an area of .2676 (verify this!). Starting at 5.5, which is .32 standard deviation above the mean, we arrive at an area of .3745, which is a much better approximation of the exact probability found from the binomial table (i.e., .377). This correction—that is, the necessity of moving over half an interval when using the normal distribution to approximate a binomial—is called a *continuity correction*, since it compensates for the fact that the normal distribution is continuous.

A very handy way of remembering how to use the continuity correction (Wonnacott and Wonnacott, 1985) is to consider that there are two ways of asking the question in our example. One way is the way we asked it: What is the probability of getting *six* or more yeses? The other way is to ask: What is the

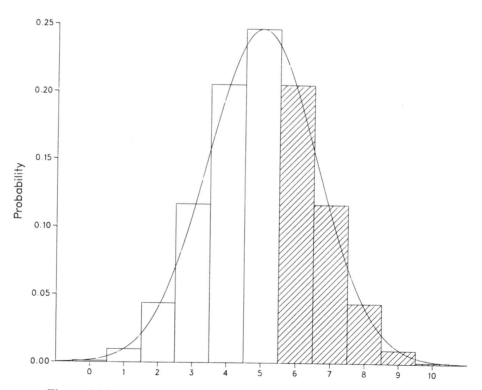

**Figure 4.15**   Binomial distribution with $n = 10$, $p = .5$.

probability of getting *more than five* yeses? Since the two ways are equivalent (because the number of yeses is a discrete variable, not a continuous one), we simply split the difference in using the normal approximation. Thus we ask: What is the probability of getting 5.5 or more yeses?

STUDY QUESTIONS—*Section 4.7*

**4.7.1**   Theoretically, the continuity correction will always improve the approximation in using a normal distribution to approximate a binomial. But as a practical matter, the continuity correction will not make much difference with large $n$, because the bigger $n$ gets, the narrower each individual block on the binomial distribution becomes. Thus with large $n$, moving over half the width of a block changes the result very little. For each of the following binomial probabilities, approximate the probability using the normal distribution (1) with and (2) without the continuity correction. Then compare the results and comment on how much of a difference the continuity correction makes.

a.   Probability of 6 or more yeses with $n = 14$ and $p = .4$
b.   Probability of 24 or more yeses with $n = 56$ and $p = .4$
c.   Probability of 48 or more yeses with $n = 112$ and $p = .4$
d.   Probability of 4 or more yeses with $n = 6$ and $p = .6$ (For this one, also find the exact answer and comment on the appropriateness of using the normal approximation.)

**4.7.2**   a.   A teacher gives a true-false test containing 50 items. A score of 30 items or more answered correctly is a passing grade. What is the probability of passing the test by "flipping a coin" (literally or figuratively) to answer each item?

b.   A teacher gives a true-false test containing 50 items. A score of 60% or more of the items answered correctly is a passing grade. What is the probability of passing the test by flipping a coin to answer each question?

c.   As you no doubt noticed, part b asks the same question as part a. What does this tell you about the distribution of the *proportion* of yeses in $n$ independent trials?

# 5

## Interval Estimation

### 5.1 The standard error of a statistic is the standard deviation of its sampling distribution.

#### The standard error

Recall from Chapter 2 what a sampling distribution is: the distribution of *all possible* values of a statistic. If you were able to select every possible sample (of the size you are using) from the population, you would be able to write down on an enormous list all the possible values of some statistic—say, the sample mean—that could be observed in samples from the population. A histogram of the numbers on this list would simply be a picture of the sampling distribution of the statistic. Make sure that you have this idea clearly in mind before you proceed.

Now in practice, of course, we cannot possibly look at all possible samples from the population (or if we can, we don't need statistical inference!). We generally select a single sample (containing $n$ randomly selected individuals) and would like to use that sample to draw inferences about the population. Suppose that we would like to use the mean of our sample to estimate the population mean. We know before we start that the sample mean is unlikely to be *exactly* the same as the population mean. But how far off are we likely to be? To put the question another way, if we use the sample mean to estimate the population mean, what will be our average error?

First, let's consider whether we are likely to be off systematically one way or the other. It turns out that the *mean* of all possible sample means is simply the population mean—in other words, the sample mean is an *unbiased* estimator of the population mean. This can (fortunately) be proven mathematically; we do not have to have access to the actual sampling distribution, or even an empirically derived estimate of the sampling distribution.

So there is no tendency to be off one way or the other, at least on the average. But in any particular case, we are likely to be off. The question is: how far off on the average? In other words, how variable are sample means about their average (the population mean)?

Well, if we could look at the entire sampling distribution, all we would have to do would be to calculate the average deviation of all possible sample means from the population mean, and we would have the answer. Thus the MAD of the sampling distribution of sample means would serve as a measure of the average error we would experience using the sample mean to estimate the population mean.

Similarly, the standard deviation of the sampling distribution of sample means would also serve as a measure of "average" error. To recognize this fact, the standard deviation of the sampling distribution of a statistic has a special name: the *standard error* of the statistic. The larger the standard error, the more spread out the sampling distribution—and hence the less certain we will be that a sample will give us a statistic close to the parameter of interest.

In practice, the standard error (rather than the MAD of the sampling distribution) is nearly always used to measure "average" error. This is because of the role of the standard deviation (and hence standard error, which is merely a particular kind of standard deviation) in the normal distribution. Since the sampling distribution of many statistics is at least approximately normal, by the CLT, the standard error is of great importance in drawing inferences using those statistics.

## Notation used for the standard deviation and standard error

Make sure that you understand the distinction between the standard deviation of the original population from which you are sampling and the standard error of a statistic. The former indicates how much the individuals in the population vary about the mean of the population. The latter indicates how much the possible values of a statistic vary about their mean.

We have already mentioned the notation used for the standard deviation of a population—the Greek lowercase letter sigma ($\sigma$). We also use sigma for the standard error, but with a subscript to indicate that we are talking about the variability of the sampling distribution of a statistic, not the variability of the original population. Thus when the statistic is the sample mean, itself denoted $\overline{X}$, we denote the standard error of the sample mean by $\sigma_{\overline{X}}$.

## The case of n = 1

In the trivial case of samples of size 1—in which each "sample" consists of a single individual—the set of all possible sample means is precisely the set of all possible individuals, that is, the population itself. So for this trivial case, $\sigma_{\bar{X}}$ is the same as $\sigma$, by definition.

The case may be trivial—but you should not try to read further unless you feel that you clearly understand two points: (1) why $\sigma_{\bar{X}}$ equals $\sigma$ for the trivial case of $n = 1$, and (2) why $\sigma_{\bar{X}}$ does not equal $\sigma$ for $n$ greater than 1. (Which one should be smaller? In other words, which would tend to give you a better estimate of the population mean: a single individual's score or the mean of a sample of individuals?)

## The standard error of the sample mean

It turns out that the relationship between $\sigma_{\bar{X}}$ and $\sigma$ is quite simple and appealing: $\sigma_{\bar{X}}$ is equal to $\sigma$ divided by the square root of the sample size. The mathematical proof of this is beyond the scope of this book, so we shall simply accept it. To put it as a formula,

$$\sigma_{\bar{X}} = \frac{\sigma}{\sqrt{n}}$$

What does the relationship between $\sigma_{\bar{X}}$ and $\sigma$ mean? First, notice that the bigger the sample size, the smaller the standard error of the sample mean. The bigger the sample, the less the degree of variability among all possible sample means. Sample means from big samples tend to be closer to the population mean than do sample means from small samples. These are all ways of saying the same thing: You can be surer of your inferences from big samples than from small samples (all other things equal).

Second, notice what you gain for a given increase on sample size. If you want to cut the standard error in half, you have to quadruple (not double) the sample size because of the square root in the denominator. This means that improvements in our degree of certainty (which we get by decreasing the standard error) can get very expensive in terms of sample size.

## STUDY QUESTIONS—Section 5.1

**5.1.1** Calculate the standard error of the sample mean, $\sigma_{\bar{X}}$, for each of the following situations.
   a. $\sigma = 100$, $n = 4$
   b. $\sigma = 100$, $n = 16$
   c. $\sigma = 100$, $n = 64$
   d. $\sigma = 50$, $n = 4$

**5.1.2** [Statpal exercise—Coronary Care data set (see Appendix A)]

a.  Consider the patients in the data set to be a randomly picked sample from a large population of similar patients. Using DEscriptive statistics, get the standard deviation and standard error of the two variables AGE and NDAYS (the number of days the patient was hospitalized).

b.  Verify that the computed standard error is correct in terms of the standard deviation and sample size.

c.  What does the standard error indicate for AGE? For NDAYS?

**5.1.3**  [Statpal exercise—Electricity Consumer Questionnaire data set (see Appendix B)]

a.  Consider the respondents in the data set to represent a random sample of the large population of people served by the particular company (a rural electric cooperative). Using DEscriptive statistics, obtain the mean, standard deviation, and standard error of the variable YRSMEMBR, which indicates how long the respondent has been a member of the electric cooperative.

b.  Verify that the computed standard error is correct in terms of the standard deviation and sample size.

c.  If you were to use the mean of this sample as an estimate of the population mean, how far off would you expect to be on the average?

d.  Now use the SElect utility to tell Statpal to include only those cases with a score on the variable LIKENUKE equal to 1. (The variable LIKENUKE has the value 1 for those approving of the cooperative's investments in nuclear power, 2 for those not approving.) Then use DEscriptive statistics to get the standard deviation and standard error for YRSMEMBR.

e.  What happens to the standard deviation and standard error when the sample size is reduced?

## 5.2   The CLT can be applied to draw inferences about the population mean.

### Sampling distributions and the CLT

Recall what the central limit theorem (CLT) says: Statistics that are calculated by summing or averaging have sampling distributions that are (at least approximately) normal in shape, as long as the sample size is large enough. What is "large enough"? It depends on the shape of the distribution of the population being sampled, but in practice a sample size of 25 is nearly always large enough for the CLT to apply. Of course, the sample mean is a statistic calculated by averaging, so the CLT applies. Thus the sampling distribution of the sample mean is normal.

Do not read past this paragraph until you have a clear picture in your mind of a normal sampling distribution of sample means. Imagine that you are about to select a random sample of, say, 30 individuals from a population. Now try to imagine *all possible* samples of 30 individuals from the population. Each possible sample has a mean. If you took all those means and piled them on a number line—that is, made a histogram of all possible sample means—the pile would be mound-shaped. Very few means would be at the extremes.

So you are about to pick one particular sample mean, at random, from a normal distribution of all possible sample means. As we discussed in Chapter 2, you can draw inferences about the population mean by considering what your chances are of getting a sample mean within any particular distance of the population mean. Recall the logic: If, say, 95% of the possible sample means are within some distance $k$ units of the population mean, we are 95% sure that the sample mean we select will be one of those. Thus we can say that we are 95% sure that the population mean is within $k$ units of the sample mean we actually select.

Since (by the CLT) the sample mean has a normal sampling distribution, all we need to know is how to find what $k$ is for a given level of certainty—or, equivalently, what level of certainty we can have for a given $k$. We discussed how to do this in Chapter 4. For example, we recall that 95% of the scores in a normal distribution are within (approximately) 2 standard deviations of the mean. Thus we know that 95% of the possible sample means are within 2 standard errors of the population mean.

So once we know the standard error of the sample mean, we are in a position to make statements about the population mean, with our level of confidence based on the normal distribution. In the next few sections we discuss how this works in the context of two related types of inferential procedures: interval estimation and hypothesis testing.

## STUDY QUESTIONS—*Section 5.2*

**5.2.1**  [Continuation of Exercise 4.2.5] Does regular exercise increase basal metabolic rate (BMR)? Suppose that you select a random sample of 25 men, all 30 years of age, and put them on a program of daily exercise for 3 months. You wish to estimate the mean BMR for 30-year-old men on the exercise program—which, of course, may differ from the mean BMR for *all* 30-year-old men, known to be equal to 37.6 kcal/m$^2$ per hour. The standard deviation is known to be 2.6. Let's assume (for simplicity's sake) that even though the mean BMR may be affected by exercise, the standard deviation will not change. So your sample of 25 men is a random sample from a population with an unknown mean but a known standard deviation (equal to 2.6).

    a.  When you have finished measuring the BMR of the 25 men in your sample, you will calculate the mean of the sample. As we have noted previously, you can think of your sample mean as a random pick from all possible sample means that could be obtained using samples of 25 men from this population— that is, from the sampling distribution of the sample mean. What is the shape of the sampling distribution? How do you know that it is normal?

    b.  We have noted that the sample mean is an *unbiased* estimator of the population mean. Therefore, what is the mean of all possible sample means equal to?

     c.  What is the standard deviation of all possible sample means—also known as the standard error—equal to in this case? (*Answer*: .52. Make sure that you see why!)

     d.  Recall that $z$ scores of $\pm 2$ bracket about 95% of the scores in a normal distribution. Thus about 95% of all possible sample means are within 2 standard errors of the population mean. Given that the standard error is equal to .52 kcal/m$^2$ per hour, 2 standard errors is equal to 1.04 kcal/m$^2$ per hour. So we can say that we have a 95% chance of getting a sample mean within 1.04 kcal/m$^2$ per hour of the population mean. Suppose that the sample mean you actually get is 39 kcal/m$^2$ per hour. Calculate an interval estimate for the population mean in which you have 95% confidence.

     e.  Following the same logic, calculate a 90% confidence interval for the population mean.

**5.2.2**  [Continuation of Exercise 5.2.1] Now let's consider the question of the effect of exercise on BMR in slightly different terms. Suppose that we are simply interested in knowing whether exercise increases BMR or not. Specifically, we seek proof for the hypothesis that exercise increases BMR.

     a.  Suppose that, in reality, exercise does not affect BMR. This would indicate that the population mean for men in the exercise program is 37.6 (i.e., the same as for the general population). If this were the case, how likely is it that you would observe a sample mean that is 39 or larger? (*Hint*: Note that 39 is over 4 standard errors above 37.6.)

     b.  But you *did* observe a sample mean of 39. What would you conclude?

## 5.3  If we know σ, we can use z scores to form a confidence interval for the mean.

### Confidence intervals for a population mean when σ is known

By now the logic of forming a confidence interval should be familiar; we have discussed it in several sections and study questions. Let's recap the logic and then pull all the details together. The logic of forming a confidence interval rests on the idea of a sampling distribution. In obtaining a random sample and calculating its mean, we are, in effect, picking a sample mean at random from the distribution of all possible sample means—that is, from the sampling distribution.

    Before we even look at the sample, we know that the sampling distribution is normal, by the CLT (as long as our sample size is large enough). Moreover, if we happen to know the standard deviation of the population from which we are sampling ($\sigma$), we can calculate the standard error of the sample mean ($\sigma_{\overline{X}}$), using the formula $\sigma_{\overline{X}} = \sigma$ divided by the square root of the sample size. So before we look at the sample, we know that we have a 95% chance of getting a sample mean that is within 2 standard errors of the population mean. How do we know this? Because of the fact that the sampling distribution is normal, and normal distributions have 95% of scores within 2 standard deviations of the mean.

Now we look at the sample mean we actually did observe. Since we had a 95% chance of getting a sample mean within 2 standard errors of the population mean, we can be 95% sure that the sample mean we *did* get is within 2 standard errors of the mean. Therefore, a 95% confidence interval for the population mean is the interval 2 standard errors on either side of the sample mean. In other words, *we are 95% confident that the population mean is between $\overline{X} - 2\sigma_{\overline{X}}$ and $\overline{X} + 2\sigma_{\overline{X}}$.* By the same reasoning, we can easily obtain other confidence levels. For example, a 90% confidence interval for $\mu$ is the sample mean $\pm 1.65$ standard errors. Why 1.65? Because in a normal distribution, 90% of the scores are within 1.65 standard deviations of the mean.

For 99% confidence, the interval is bounded by 2.58 standard errors on either side of the sample mean. Notice that as we discussed in Chapter 2, there is a trade-off between level of confidence and precision. If we want to be more sure that $\mu$ is really in our interval, we must make the interval wider. If we are willing to accept less certainty, we can have a smaller interval. Let's go over the procedure step by step, by example.

**Example 5.1**    Electrical resistors marked "500 kilohms" manufactured by a certain company vary somewhat in their actual resistance. The standard deviation of the resistances is known to be 15 kilohms (k$\Omega$). A sample of a hundred 500-k$\Omega$ resistors from this company yields a mean of 487 k$\Omega$. Give a 95% confidence interval for the mean resistance of the population of all 500-k$\Omega$ resistors manufactured by the company.

*Step 1*:    *Find the standard error of the statistic.* The standard error of the sample mean in this example is 15 divided by the square root of 100, which is 15/10, or 1.5.

*Step 2*:    *Determine how many standard errors on either side of the mean the interval needs to be.* For 95% confidence, we need 2 standard errors on either side of the sample mean. (If you want to be picky, you can look in the normal table and find that the exact $z$ score you need to encompass 95% of the area is 1.96, not 2. But for our purposes, 2 is close enough to 1.96.)

*Step 3*:    *Finish up.* Two standard errors equals $2 \times 1.5$, or 3 k$\Omega$. So we are 95% confident that 500-k$\Omega$ resistors manufactured by this company have a mean resistance between 484 and 490 k$\Omega$.

**Example 5.2**    Continuing with the resistor example, find a 90% confidence interval for the population mean.

*Step 1*:    *Find the standard error of the statistic.* We have already determined that the standard error is 1.5 k$\Omega$.

*Step 2*:    *Determine how many standard errors on either side of the mean the interval needs to be.* For 90% confidence, we need 1.65 standard errors.

*Step 3*:    *Finish up.* 1.65 standard errors equals $1.65 \times 1.5$, or 2.475 k$\Omega$. So we are 90% confident that the population mean is within 2.475 k$\Omega$ of 487—that is, in the interval 484.525 to 489.475 k$\Omega$.

### Determining sample size for a given confidence level and margin for error

Now let's consider a slightly different problem involving confidence intervals. Again, we'll consider the problem through an example.

**Example 5.3**   Suppose that you wish to form a 95% confidence interval for the population mean resistance of 500-k$\Omega$ resistors but you require greater precision in the estimate than we obtained above. Specifically, you want to be 95% sure that you will get a sample mean that is within 1 k$\Omega$ of the population mean. How do you proceed?

The trick here is to realize that *any* 95% confidence interval based on a normal sampling distribution will be 2 standard errors on either side of the sample mean. So if you want 2 standard errors to be equal to 1 k$\Omega$, the standard error will have to be equal to .5 k$\Omega$.

As we know, the standard error equals the standard deviation of the original population divided by the square root of the sample size. So to make the standard error smaller, we can either make the standard deviation smaller or increase the sample size. Making the standard deviation smaller would entail changing the manufacturing process somehow, so as to make the resistors less variable in their actual resistance. Indeed, this may be an important concern of the company's quality control engineers. However, from the point of view of estimating the mean resistance as it now stands, we would concentrate on the sample size. *To make the standard error smaller, increase the sample size.*

Now the problem reduces to simple algebra. We want the standard error to be .5 k$\Omega$; thus we have $.5 = 15/\sqrt{n}$. This means that the square root of $n$ has to be equal to 30, which means that $n$ has to be 900 to give the desired standard error. So a sample size of 900 will allow us to estimate the population mean within 1 k$\Omega$ with 95% confidence.

Devotees of algebraic notation may find the following formula helpful: To estimate $\mu$ to within $k$ units with $C$% confidence requires

$$n = \left(\frac{z_C \sigma}{k}\right)^2$$

where $z_C$ is the number of standard deviations required on each side of the mean to bracket $C$% of the area in a normal distribution.

Notice how much bigger the sample size had to be to reduce the margin for error from 3 k$\Omega$ to 1 k$\Omega$. To reduce the margin for error by a factor of 3, the sample size had to be multiplied not by 3, but by $3^2$, or 9. You can see that the more demanding we are in terms of precision, the more expensive it becomes to get enough data.

## STUDY QUESTIONS—*Section 5.3*

**5.3.1** A survey samples 100 households at random and finds that the mean number of radios in the sampled households is 5.2. The standard deviation of the number of radios is known from previous work to be 3.1 radios.
   a. Calculate and interpret a 95% confidence interval for the mean number of radios in the population of households from which the sample was selected.
   b. Now calculate and interpret a 90% confidence interval.
   c. Suppose that the survey sponsors wish to be 95% confident of estimating the mean number of radios with a margin for error of only .2 radio. How big a sample of households should be selected?

**5.3.2** A certain chemical is measured in water specimens taken from a well. From past experience, the measuring process is known to give variable readings, with a standard deviation of 2.5 ppm (parts per million) among individual specimens taken from the same lake.
   a. A sample of 25 specimens yields a mean concentration of 32 ppm. Estimate the mean concentration of the population (i.e., the well) with 90% confidence.
   b. How much confidence would you have in an estimate with a margin for error of .5 ppm (i.e., the interval 31.5 to 32.5 ppm)?
   c. If you wish to have 95% confidence in an estimate with a margin for error of .5 ppm, how many specimens should be examined?

**5.3.3** [Statpal exercise—but may be performed using a calculator] An insurance company selects 12 records of motor vehicle accidents at random from its files. Experience shows that the standard deviation of dollar claims in accidents of the type from which the sample was drawn is $2800. The amounts of damage were as follows (all in dollars):

   300   1530    200   10,260   4200   1250
   2970   3360   7710     200    940   3120

   a. Obtain the sample mean and standard deviation.
   b. Does the sample standard deviation seem to be in line with the (presumed known) population standard deviation?
   c. Obtain the standard error and give a 90% confidence interval for the mean amount of damage in the population of accidents from which the sample was selected.

# 6

# Hypothesis Testing

## 6.1 The CLT can also be applied to perform hypothesis tests concerning averages.

### The idea of hypothesis testing

We have already encountered hypothesis testing in several study questions, without bothering to identify it as such. Now let's consider it in detail. Although hypothesis testing is generally presented as a separate topic from estimation, the two inferential procedures are really applications of the same basic reasoning.

Let's discuss hypothesis testing in the context of an example. When you buy a box of crackers labeled ''net weight: 2 lb,'' you expect the box to contain at least the indicated weight. But have you ever checked it on a scale? There is bound to be some variation in the weight of the crackers from batch to batch, and presumably from box to box. For our purposes, we will assume that the standard deviation of box weights is known to be equal to .12 lb (a little under 2 ounces).

Suppose that a consumer group wants to find out if a certain company is systematically short-weighting its boxes of crackers. They purchase a random sample of, say, 100 boxes of crackers, being careful to cover a wide span of both time and geographical area, and weigh the contents of each box on an accurate scale.

Chances are that none of the boxes will contain *exactly* 2 lb of crackers, at least when weighed on a very accurate scale. After all, if you are measuring to the nearest, say, hundredth of an ounce, it is quite unlikely that you would hit any prechosen weight exactly; there are thousands of possible numbers. But if the company is really trying to live up to the label, the *mean* net weight of all the boxes produced by the company should be at least 2 lb. (The boxes should not vary too much either—but that is another problem, of concern more to the company's quality control people.)

Note that the consumer group's main interest is not simply in estimating the population mean, but rather in determining whether the evidence is strong enough to conclude that the population mean is less than 2 lb. That is, the consumer group has a particular hypothesis about the population mean in mind: that the mean weight is less than 2 lb. If they cannot produce convincing evidence, they cannot conclude that the mean weight is less than 2 lb.

To put it another way, the consumer group seeks statistical ''proof'' for the hypothesis that $\mu$ is less than 2 lb. If they cannot produce such proof, they will reserve judgment on their suspicions and assume that $\mu$ is at least 2 lb. Thus the consumer group wishes to choose between two competing hypotheses: If the evidence is strong enough, they will decide that $\mu$ is less than 2 lb; otherwise, they will decide that $\mu$ is at least 2 lb.

### The terminology of hypothesis testing

The two competing hypotheses in a hypothesis test are called the *null* and the *alternative*. It is important to be able to tell which is which, because they are not simply mirror images of each other.

*The hypothesis for which you are seeking proof is the alternative.* In our example, the null hypothesis was that the average weight was at least 2 lb. The alternative hypothesis was that the average was less than 2 lb. Notice that the consumer group required proof before they could assert that the company was guilty of short-weighting. They were not seeking proof of the company's innocence, since that would be assumed until proven otherwise. The parallel with a criminal trial is direct: *We assume the null to be true unless the alternative is proven beyond a certain degree of doubt.*

### How hypothesis testing works

The way the test works is to figure out how likely results such as you got would be if the null were true. If such results would have been unlikely, you conclude that the null must not be true after all, and you have proven (in a statistical sense) the alternative.

Returning to our example, suppose the consumer group finds that the mean weight of the 100 boxes in the sample is 1.98 lb. This is indeed less than 2 lb. But, of course, the sample average came from a random sample. Is it possible that the population average is in fact what it is supposed to be, and the sample average was low just by chance? In other words, if the null were true, how likely is it that we would observe a sample mean that far below 2 lb?

If the consumer group finds that a sample mean of 1.98 lb or less is very unlikely if the population mean were really 2 lb, they will reject the idea that the population mean could be 2 lb—or, in formal terms, they will "reject the null hypothesis." On the other hand, if a sample mean of 1.98 lb or less is not so unlikely with a population mean of 2 lb, they will not reject the null hypothesis.

Make sure that you understand the logic: If what you actually observed would have been unlikely if the null were true, then the null is not very plausible and is rejected in favor of the alternative. Notice that if you *fail* to reject the null, that does not mean that you have proven it, even in a statistical sense. In a trial, if the jury fails to convict the defendant, that does not constitute positive proof of innocence; it simply means that the evidence was not strong enough to overcome the *presumption* of innocence.

## STUDY QUESTIONS—*Section 6.1*

**6.1.1**  In testing a new drug, a researcher compares the effects of the new drug to the effects of existing treatments. The researcher seeks proof that the new drug works more effectively than do existing treatments.
  a.  State the researcher's alternative hypothesis.
  b.  State the researcher's null hypothesis.
  c.  Suppose that the researcher's evidence does not allow for the rejection of the null hypothesis. Does this constitute proof of the null hypothesis? Why not?

**6.1.2**  Considering the cracker box example discussed in this section, recall that the standard deviation of the box weights is known to be .12 lb.
  a.  Calculate the standard error of the mean (using $n = 100$).
  b.  If the null hypothesis were true (i.e., if $\mu$ were equal to 2 lb), what would be the chance of getting a random sample of 100 boxes that gives a mean of 1.98 lb or lower (as the consumer group did)?
  c.  In view of your answer to part b, how plausible is it that the population mean is really 2 lb (i.e., that the null hypothesis is true)?

## 6.2   Rejecting a true null hypothesis is called a type I error.

### Type I error

When you finish your test, you will be announcing (to yourself, if to no one else) whether or not you reject the null hypothesis in favor of the alternative. Suppose you find, as in our example (see Exercise 6.1.2), that the null is implausible, because if the null were true, the chance of getting the sample results you got is small. Thus you reject the null. In so doing, you might be wrong—maybe the null is true and you were unlucky enough to get highly improbable sample results purely by chance.

If this happens—if you reject the null when it is really true—you have committed a *type I error*. Naturally, we want to keep the chance of making a type I error small, which simply means that we insist that the null hypothesis be very implausible before we reject it.

## Controlling the chance of making a type I error

How implausible? It depends on how cautious you want to be. Suppose the sampling distribution says that sample results such as you got had a 20% chance of occurring if the null were true. That is not unlikely. One-fifth of the possible samples would have given you results as discrepant from the null hypothesis as those you observed in your sample, even if the null were true. You would not want to reject the null hypothesis on that sort of evidence unless you were willing to take a 20% risk of making a type I error.

But how small a risk should you insist on? A frequently used standard is 5%—if the risk is 5% or less, reject the null hypothesis. For some purposes, we might insist on even less risk—say, 1%. On the other hand, if we are doing an "exploratory" study, we might be willing to accept a 10% or even higher risk.

## Significance level and p value

Whatever degree of risk you decide on is called the *significance level* of the test, and is often denoted by the Greek lowercase letter alpha ($\alpha$). The idea is that you set $\alpha$ in advance, before you look at the data. Then you look at the data, and see if rejecting the null would entail a degree of risk that is bigger or smaller than your preset $\alpha$. If rejecting the null based on your data would entail a level of risk that is smaller than $\alpha$, you reject the null.

For example, suppose you decide that 5% is an acceptable degree of risk of making a type I error. Thus $\alpha$ is 5% (or .05). Now you look at your sample data and determine what the degree of risk would be if you rejected the null based on your particular data. That degree of risk is called the *p value* for the data. Thus if the *p* value is less than $\alpha$, you reject the null hypothesis.

## Determining the p value for the data

How do you determine the *p* value for the data? All you have to do is look at the sampling distribution as it would appear if the null were true. The *p* value for your data is the proportion of possible sample statistics that are at least as far away from the null hypothesis as the one you actually obtained. If your sample gave you a statistic far out in a tail of the distribution—which means it is an unlikely value if the null were true—the *p* value will be small and you will reject the null hypothesis.

Do not lose sight of the basic reasoning of hypothesis testing: If your observed data give you a statistic that would be unlikely if the null were true, you reject the null. Thus *a small p value leads to rejection of the null hypothesis*. The smaller the $p$ value, the less plausible the null hypothesis.

### STUDY QUESTIONS—*Section 6.2*

**6.2.1**  A researcher is willing to allow a 5% risk of rejecting a true null hypothesis. What conclusion would the researcher draw if
   a.  The statistical evidence has a 10% chance of occurring if the null hypothesis were true?
   b.  The statistical evidence has a 1% chance of occurring if the null hypothesis were true?

**6.2.2**  A jury convicts a person of a crime. Later it is discovered that another person actually committed the crime. Which type of error did the jury make?


## 6.3  *P* values come in two flavors: one-tailed and two-tailed.

### An informal definition of p value

We noted in Section 6.2 that the $p$ value for a test is the proportion of possible sample statistics that are at least as far away from the null hypothesis as the one you observed. To put it informally in probability terms: *The p value for a test is the probability of getting what you got if the null hypothesis were true.* In this context, "getting what you got" means getting something at least as discrepant from the null hypothesis as what you have observed.

Note that the $p$ value is a kind of probability (indeed, "$p$" stands for probability). But the $p$ value is *not* the probability that the null hypothesis is true, but rather, the probability of seeing the sort of evidence we obtained if the null hypothesis were true. This may seem like a fine distinction, but it is important. In calculating the $p$ value, we *assume* that the null is true and then look at the sample evidence. If the sample evidence is very improbable under our assumption, something is wrong with our assumption and we reject it.

### One-tailed versus two-tailed p values

One important consideration at this point is: What do we mean by a statistic "at least as discrepant" from the null hypothesis as the one we observed? Do we mean as far away *in the same direction* or as far away in either direction? The answer depends on whether we care about being off in either direction, or in only one direction.

In our cracker box example (Section 6.1) we would call it a type I error only if we erroneously accused the company of short-weighting; we would not be

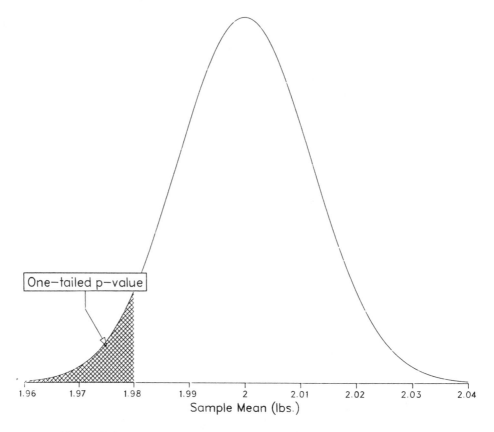

**Figure 6.1**  Distribution of all possible sample means in cracker box example if null hypothesis were true: one-tailed *p* value.

worried about overweighting. So we would only be concerned about one tail—the lower tail—of the sampling distribution. The *p* value is therefore the proportion of possible sample averages *less than* or equal to the one we observed (see Figure 6.1). This would be called a *one-tailed p* value.

However, the company would be concerned about being off either way—it would be just as concerned about overweighting as underweighting, and it would not want to commit a type I error in either direction. So the company would worry about the possibility of getting a statistic in *either* tail of the distribution, and the *p* value would be the sum of the proportions in both tails (see Figure 6.2). This is called a *two-tailed p* value. Note that the two-tailed *p* value is simply 2 times the one-tailed *p* value, because of the symmetry of the normal distribution.

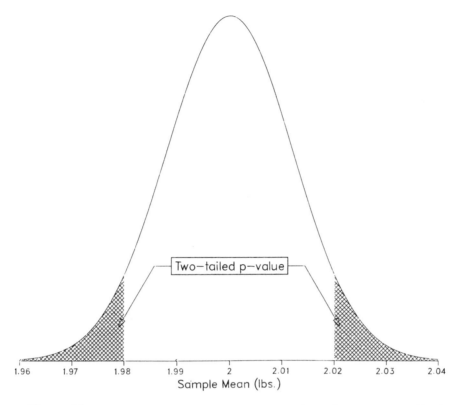

**Figure 6.2**   Distribution of all possible sample means in cracker box example if null hypothesis were true: two-tailed *p* value.

This means that it is harder to reject the null hypothesis if you are doing a two-tailed test than if you are doing a one-tailed test.

### One-tailed or two-tailed?

Of course, which one you use is (supposedly) decided in advance, based on the nature of the hypotheses. In general, if you care about deviations from the null hypothesis in either direction, use a two-tailed test. If only one direction matters, so that a deviation in the other direction might as well be included in the null hypothesis, you can use a one-tailed test.

### STUDY QUESTIONS—*Section 6.3*

**6.3.1**   In a criminal trial in the United States (and other countries that follow the English system), a defendant is presumed innocent, and the burden of proof is on the prosecution. What is the null hypothesis in a criminal trial? What is the alternative?

**6.3.2**   In the cracker box example (Section 6.1), what was the *p* value for the data?

**6.3.3**   If the company were doing the test, what would the *p* value be?

**6.3.4**   Suppose that the consumer group did not want to accept *any* risk of a type I error. Would it be reasonable to set $\alpha$ equal to zero? What would be the consequences?

**6.3.5**   In each of the following situations, determine if a one-tailed or a two-tailed test is called for.

   a.   A researcher wishes to determine if the mean concentration of lead in blood samples from a population of inner-city children is greater than the (known) mean for children in the United States as a whole.

   b.   A researcher wishes to determine if the mean change in SAT score (after minus before) for the population of students taking an SAT preparation course is greater than zero (i.e., the researcher wonders if the course improves scores on the average).

   c.   A different researcher wishes to determine if an SAT preparation course has any effect in either direction (positive or negative).

   d.   A researcher wishes to test the null hypothesis that $\mu$ equals 0 (on some variable) versus the alternative hypothesis that $\mu$ is not equal to 0.

   e.   A researcher wishes to test the null hypothesis that $\mu$ equals 0 (on some variable) versus the alternative hypothesis that $\mu$ is less than 0.

## 6.4   Failing to reject a false null hypothesis is called a type II error.

### Type II error

We do not want to reject the null hypothesis when it is true (a type I error), but neither do we want to fail to reject it when it is false (a *type II error*). In our cracker box example in Chapter 6, a type II error would mean that consumers would continue to be shortchanged. A type I error would mean that the company's good name would be unfairly besmirched. The consumer group needs to be careful about both types of error.

### Controlling the risk of a type II error

You can make the risk of type I error as small as you like, but only at the cost of increasing the risk of type II error. For example, the legal department of the consumer group might insist that the risk of type I error—falsely accusing the company—be kept extremely small, say, 1 in 100,000 (or .00001). How could we do that? Simple—all we have to do is stay with the null unless the sample average is way below 2 lb. How far below? So far that it would have occurred in only one hundred-thousandth of the possible samples if the population average were really 2 lb.

From a more extensive normal table than that provided in this book, we would find that a $z$ score of about 4.26 cuts off an area close to .00001. Thus we would stay with the null unless we get a sample mean over 4.26 standard errors below 2 lb (our null hypothesis mean). Recalling (from Exercise 6.1.2) that the standard error of the sample mean in this example is .012 lb, we determine that a sample mean of 1.9489 lb or less would lead us to reject the null hypothesis.

But what if the population average really is less than 2 lb? By insisting on such a low risk of type I error, we are increasing the risk of a type II error. Suppose, for example, that the population average were really 1.97 lb (unbeknownst to us, of course). That is less than 2 lb, and we would like our test to tell us so. But if we were insisting on a sample average of 1.9489 or less before we would reject the null, we would find the chance of our getting such a sample would be quite small—only about .04, or 4% (verify this!). So our test would be essentially incapable of detecting the company's short-weighting.

## Power

The term *power* refers to the ability to detect a false null hypothesis. So *power is the ability of a test to avoid making a type II error.* In more formal terms, the power of a test is the probability that the test will lead us to reject the null hypothesis, given that the null hypothesis is in fact false. Just as the probability of a type I error (the significance level) is referred to by the Greek lowercase letter $\alpha$, the probability of a type II error is referred to by the Greek lowercase letter beta ($\beta$). Since power is the probability of not making a type II error, we can write power as $1 - \beta$.

## Factors affecting the power of a test

As you might expect, the power of a test will depend partly on how ''false'' the null hypothesis is. If the null is just a little off from reality, a test generally will not have much power. This is because our sample results will probably seem to support the null hypothesis. On the other hand, if the null is way off—if, say, the company were putting 1 lb of crackers in boxes marked ''2 lbs''—then our sample results will tend to reflect that, and our test will be powerful.

Another factor in the power of a hypothesis test is sample size: The larger the sample, the more powerful the test, all other things equal. This is because the larger the sample, the closer the statistic will tend to be to the parameter.

Since sample size affects power in a systematic way, statisticians can determine what sample size is needed for a certain degree of power, given (1) the risk of type I error you are willing to accept, and (2) how big a discrepancy you want to be able to detect. In general, we would like to be able to get the job done with as small a sample as possible.

*Power versus p value*

Be careful not to mix up "power" with "*p* value." As we noted above, the "*p*" in "*p* value" comes from the word "probability," related to the probability of a *type I* error. Power relates to *type II* error.

## 6.5   Do not confuse statistical significance with practical significance.

*Statistical versus practical significance*

In some situations the null hypothesis implies that some type of experimental manipulation has no effect. Rejecting the null then means that there was an effect. Such a conclusion is usually stated something like this: "The difference between treatment and control groups was statistically significant at the 5% level." This means that the risk of type I error is at most 5%. Another way to say it is "*p* was less than .05."

Note that statistical significance does not necessarily mean practical significance. The distinction is especially important when you have the luxury of working with a large sample. Since a large sample makes your test powerful, you might find that even a very small effect ends up being statistically significant. A very small effect, even if you can be very confident in its reality, may be of no consequence in the real world.

### STUDY QUESTIONS—*Sections 6.4 and 6.5*

**6.5.1**   A researcher concludes that a new teaching method yields better results than do existing methods. Subsequent research in the same area fails to confirm the original results. Which type of error did the researcher make?

**6.5.2**   A state health department concludes that the available evidence concerning the dangers of a certain chemical is insufficient to warrant banning the chemical's use. Subsequently, it is determined that the chemical causes cancer. Which type of error did the health department make?

**6.5.3**   Give an example of when (a) a type I error might be considered of greater consequence than a type II error, and (b) vice versa.

**6.5.4**   A reporter, in discussing the results of a public opinion survey, stated: "The proportions of people in the sample agreeing with the president on this issue did not differ significantly between the two major parties. It is surprising that the proportions for the two parties are the same." Criticize the reporter's statement.

**6.5.5**   A psychologist studying the relationship between birth order and IQ examines the IQ scores of a random sample of 10,000 children from families with at least two children. In each family, the psychologist calculates the difference in IQ between the first-born child and the second-born child, doing the calculation in such a way that a positive number represents a higher IQ for the first born, while a negative number represents a higher IQ for the second born (i.e., first minus second). The sample yielded a mean of 1.2 points difference, with a standard deviation of 15 points.

a.   The psychologist wishes to see if there is evidence to support the hypothesis that first-born children have a higher mean IQ than second-born children. State the null and alternative hypotheses, and carry out the test.

b.   The psychologist reports the results as representing a ''significant difference'' between first-born and second-born children's mean IQ scores. Comment on this terminology.

**6.5.6**   Here is a handy way to tell a mathematician from a statistician. Suppose that you flipped a fair coin (i.e., one that is as likely to come up heads as tails) 50 times, and the coin came up heads every time. You are about to flip the coin again. What is the probability that the coin will come up heads? Answer before you read further.

Well, what did you say? If you are a mathematician at heart, you said the chance was 50%, because it is a fair coin and has no ''memory.'' (Besides, you probably heard it in high school.) If you said the coin was more likely to come up tails after all those heads, you probably believe in magic, and you are neither a mathematician nor a statistician at heart.

So what would a statistician say? A statistician would say that it is incredibly unlikely that you could have thrown 50 heads in a row with a fair coin—but since you did throw 50 heads in a row, it must not be a fair coin! Do you see that our statistician simply did a hypothesis test? Let's follow the logic.

a.   What are the statistician's null and alternative hypotheses?

b.   The 50 throws can be considered a random sample from an infinite population of throws. What sampling distribution did the statistician consider in doing the hypothesis test?

c.   The statistician might be wrong, of course. If so, she has made a type I error. What is the risk that this has happened? (The answer happens to be about .000000000000001—somewhere around one quadrillionth. How was this calculated?)

d.   Suppose that you had only flipped the coin three times instead of 50. Would the statistician have rejected the null hypothesis even if you had thrown heads every time? Why not?

Notice that this question involves power. Even if you were flipping a two-headed coin, the statistician could not have rejected the null hypothesis based on only three throws (unless you let her look at the coin!), because three heads in three throws is not all that unlikely even with a fair coin. How many throws would she need to come up heads before she would be willing to reject the null? How would you figure this out?

# 7

# Drawing Inferences About a Population Mean

## 7.1 The normal distribution can be used to test hypotheses about μ when σ is known.

*Hypothesis tests about a population mean when σ is known*

Now that we have described the basic concepts of hypothesis testing, we shall use our resistor example (Example 5.1) to illustrate how it is done.

**Example 7.1** Consider our random sample of 100 resistors made by a certain company, each resistor marked "500 kilohms." As before, we assume that the standard deviation of the true resistances is equal to 15 kΩ. Recall that our sample had a mean of 487 kΩ. Is this enough evidence to conclude that the population mean resistance is less than the nominal value of 500 kΩ?

*Step 1*: *State the null and alternative hypotheses.* Recall that the hypothesis for which we seek proof is the alternative. In this case we seek proof for the assertion that the population mean is less than 500 kΩ. Thus the null and alternative hypotheses may be stated as follows:

*Null hypothesis*: The population mean resistance is equal to 500 kΩ; symbolically, $\mu = 500$ kΩ.
*Alternative hypothesis*: The population mean resistance is less than 500 kΩ; symbolically, $\mu < 500$ kΩ.

*Step 2*: *Assume that the null hypothesis is true and draw a picture of the sampling distribution.* The idea, as we know, is to figure out how likely it would be to observe a sample mean of 487 kΩ (from a sample with $n = 100$) if in reality the population mean were 500 kΩ. We have already noted that the sampling distribution is normal (by the CLT), with mean 500 kΩ and standard error 1.5 kΩ (see Figure 7.1).

*Step 3*:   *Find the p value for the observed data.* Recall what we mean by a *p* value: the probability of observing a statistic such as the one we observed if the null hypothesis were true. In this case, what we need is the probability of observing a sample mean of 487 kΩ or less, given a population mean of 500 kΩ and a standard error of 1.5 kΩ.

This is a simple application of the methods of Chapter 4. We have already drawn our picture. To find the *z* score, we note that 487 kΩ is 8.67 standard errors below 500 kΩ, giving a *z* score of $-8.67$. The area we seek is the area to the left of $-8.67$. This area is, of course, virtually zero (such extreme *z* scores do not even appear in the table). So the *p* value is essentially zero. The chance of observing a sample mean of 487 kΩ or lower if the null hypothesis were true is nearly zero.

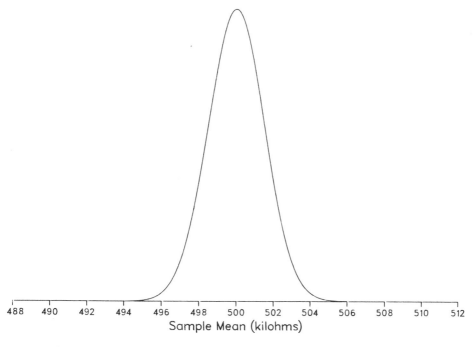

**Figure 7.1**   Sampling distribution of sample mean if null hypothesis were true (Example 7.1).

***Step 4***:  *State the conclusion.* Consider the evidence. Our *p* value tells us that it is virtually impossible to get a sample mean as low as 487 kΩ if the null hypothesis were true. But we *did* get a sample mean of 487 kΩ. Hence we reject the null hypothesis. In doing so, we can rest assured that our chance of making a type I error—the *p* value—is far lower than anything we might worry about. It is certainly much lower than the conventional threshold of .05 (5%). In light of our sample evidence, it is simply not plausible that the null hypothesis could be true.

**Example 7.2**  This time we are sampling "500-kilohm" resistors from a different company. However, we will still assume that the standard deviation of true resistances is known to be 15 kΩ. Suppose that we want to identify a discrepancy in either direction—that is, we want to know if the sample evidence is sufficient to conclude that the population mean resistance differs from 500 kΩ, be it too high or too low. We measure the resistances for a random sample of 30 resistors and find a sample mean of 503 kΩ. Can we say that the population mean is different from 500 kΩ?

***Step 1***:  *State the null and alternative hypotheses.* As the question is stated, the alternative hypothesis must be two-tailed—that is, we want to identify a discrepancy from 500 kilohms in either direction. Thus the hypotheses are as follows:

> *Null hypothesis*: The population mean resistance is equal to 500 kΩ; symbolically, $\mu = 500$ kΩ.
> *Alternative hypothesis*: The population mean resistance is different from 500 kΩ; symbolically, $\mu \neq 500$ kΩ.

***Step 2***:  *Assume that the null hypothesis is true and draw a picture of the sampling distribution.* Of course, the sampling distribution is still normal, and if the null hypothesis is true, its mean is 500 kΩ. However, this time the standard error is larger than in Example 7.1 because the sample size is smaller. Specifically, the standard error equals $15 \div \sqrt{30}$, which turns out to be 2.74 kΩ. The picture is shown in Figure 7.2.

***Step 3***:  *Find the p value for the observed data.* Since, in advance, we were concerned about discrepancies from 500 kΩ in either direction, our *p* value is defined as the chance of getting a sample mean as far away from 500 kΩ—in either direction—as the one that we actually observed. So the *p* value is the proportion of possible sample means that are at least 3 kΩ away from 500 kΩ *in either direction*. This is the area shaded in Figure 7.2.

How big is the shaded area? We again apply the techniques described in Chapter 4. Since the two pieces have the same area, all we have to do is find the area of one of them and double it. The *z* score for 503 kΩ is $3 \div 2.74$, or 1.09. From Table D.1, the area between the mean and a *z* score of 1.09 is .3621, so the

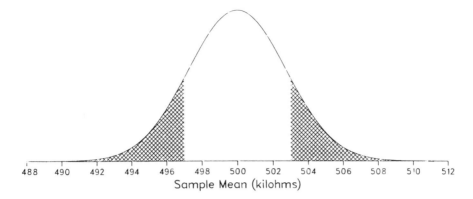

**Figure 7.2**   Sampling distribution of sample mean if null hypothesis were true (Example 7.2).

area of the right-hand part of the shaded region is .5000 − .3621, or .1379. Thus the area of the entire shaded region is .1379 × 2, or .2758.

So if the null hypothesis were true, about 28% of possible sample means would be as far away from 500 kΩ as the one we observed. That is, the *p* value for the observed data is about .28. This is, of course, much greater than the conventional α value of .05.

*Step 4*:   *State the conclusion.* The evidence does not seem to conflict with the null hypothesis at all seriously. Indeed, we would expect to see a sample mean at least 3 kΩ away from 500 kΩ over one-fourth (28%) of the time if the null hypothesis were true. Thus we do not reject the null hypothesis. We have not *proven* that the population mean is 500 kΩ—but we do not have any strong evidence to the contrary.

### Considerations of power

Note that in Example 7.2 we were not able to reject the null hypothesis. As we noted previously, when we fail to reject the null hypothesis we might be making a type II error; that is, the null might really be false, but we have failed to detect it.

Why might this occur? There are two reasons why we might make a type II error. One reason is that the null hypothesis, while not *exactly* true, might be *almost* true. For example, perhaps the mean for the resistors made by the company in Example 7.2 is not 500 kΩ but 500.01 kΩ. (Indeed, it is almost certain that the population mean is not *exactly* 500 kΩ.) As long as the true population mean is very close to the null hypothesis value, samples will tend to reflect that fact—and we will be unlikely to reject the null hypothesis.

The other reason we might make a type II error is that our standard error is large. To see why this makes a difference, think about what has to happen for us to reject the null hypothesis. For example, if we are doing a two-tailed test (as in Example 7.2) and we insist that the *p* value be .05 or less before we reject the null hypothesis, we are in effect requiring that the sample mean be at least 2 standard errors away from the null hypothesis mean. (Think about the last sentence and make sure that you understand it!)

So the larger the standard error, the farther out (in the original units) we require the sample mean to be before we will reject the null hypothesis. Now given that we know the standard deviation of the original population, what affects the standard error? The answer, of course, is sample size. With a sample size of 100, we had a standard error of 1.5 kΩ. Two standard errors was therefore equal to 3.0 kΩ, and any sample mean that was at least 3 kΩ away from 500 kΩ would have led us to reject the null hypothesis in favor of a two-sided alternative. But with a sample size of 30, the standard error was 2.74 kΩ. We would therefore have required a sample mean at least $2 \times 2.74$, or 5.48, kΩ away from 500 kΩ before we could reject the null hypothesis.

All other things equal, the larger the sample, the better the ability of a test to detect a false null hypothesis. Simply put, larger samples give tests with greater power.

## STUDY QUESTIONS—*Section 7.1*

7.1.1   In the examples involving electrical resistors, nowhere was it stated that the population of individual resistances has a normal distribution. Indeed, the methods presented *do not require* that the population from which a sample is to be drawn have a normal distribution.

a.   Why did the methods shown in the examples not require that the original population have a normal distribution?

b.   Let's consider a sample of 100 resistors. If you were to draw a histogram of the resistances of the 100 resistors in the sample, what should it look like? (*Hint*: It would *not* necessarily look normal!)

c.   What *did* have a normal distribution? (*Answer*: The sampling distribution of the sample mean. What's that?)

d.   Recall that σ, the standard deviation of the population, was known to be 15 kΩ, and hence the standard error of the sample mean was known to be 15 divided by the square root of the sample size. When *n* was 100, the standard error worked out to be 1.5 kΩ. Now consider the standard deviation of the *sample*, denoted by *s*. If we had calculated *s* for our sample of 100 resistors, what would it have been, approximately? Would it have been closer to 15 kΩ or to 1.5 kΩ?

e.  How about *s* for our sample of 30 resistors? Should it be closer to 15 kΩ or to 2.74 kΩ (the standard error)?

f.  By now it should be clear that *s* should look like $\sigma$, not like $\sigma_{\bar{x}}$. In fact, the larger the sample, the closer we would expect *s* to be to $\sigma$. Now suppose that we had not known $\sigma$ in advance. As you might expect, this is the usual situation. How would you estimate it? Given an estimate of $\sigma$, how would you estimate $\sigma_{\bar{x}}$, the standard error?

g.  If you had to estimate $\sigma_{\bar{x}}$ instead of knowing it for certain, would you be as sure of your inferences? What type of adjustment might you make in your statements to account for the relative lack of information?

h.  Specifically, suppose that you were forming a 95% confidence interval for the population mean. If you know $\sigma$, and therefore $\sigma_{\bar{x}}$, your confidence interval will be 2 standard errors on either side of the sample mean. But what if you have to estimate $\sigma$ using *s*. Are you still 95% certain that the population mean is with 2 *estimated* standard errors of the sample mean? You shouldn't be. After all, what if you underestimated $\sigma$? Then 2 estimated standard errors will be less than 2 real standard errors—and your confidence interval will be narrower than it should be. So if you still want 95% confidence in your interval, what type of adjustment will be needed? (That is, will you have to make the interval larger or smaller?)

7.1.2   In a random sample of oranges, the mean diameter of the fruit was 9.4 cm. The standard deviation of the diameters of the population of fruit from which the sample was taken is known to be 1.2 cm. The oranges were picked from trees that had been subjected to a frost, and the question of interest was whether the mean diameter of the oranges had decreased from the previous year's mean of 9.8 cm.

a.  State the null and alternative hypotheses.

b.  Suppose that the sample size had been 16. Perform the hypothesis test and state your conclusion.

c.  Now suppose that the sample size was 36. Perform the hypothesis test and state your conclusion.

d.  Explain the discrepancy between your conclusions in parts b and c.

7.1.3   In a random sample of 100 kindergarten classrooms, the mean number of children per class was found to be 20.8, with a standard deviation of 3.6 children. (With a sample of this size, it is reasonable to treat the sample standard deviation as if it were the population standard deviation.) A teacher's union official asserts that the mean number of children per class in the population of classes from which the sample was drawn exceeds the desired value of 20 children. The state education commission demands statistical proof of the assertion.

a.  State the null and alternative hypotheses.

b.  Perform the hypothesis test and state your conclusion.

7.1.4   A physicist studying the effects of ultrasound on living tissue measures the temperature change associated with a fixed dose of ultrasound on a sample of four specimens. The mean temperature change for the four specimens was $+.35°C$ (i.e., an increase of .35°C). The standard deviation of temperature change measurements in this general range of doses is assumed to be .80°C.

    a.   Calculate and interpret a 95% confidence interval for the mean temperature change at this dose level.

    b.   Is there enough evidence to conclude that this dose level causes an increase in mean temperature?

**7.1.5**  [Advanced exercise] A researcher is designing a study to test the null hypothesis that a population mean equals zero versus the alternative hypothesis that the population mean is greater than zero. She wishes to be 90% sure of rejecting the null hypothesis if the population mean is actually .5 standard deviation above zero, but also be 95% sure of *not* rejecting the null hypothesis if the population mean is zero (i.e., if the null hypothesis is true).

    a.   What value of $\alpha$ (the probability of a type I error) is the researcher willing to accept?

    b.   What value of $\beta$ (the probability of a type II error) is the researcher willing to accept, given the stated condition?

    c.   What sample size would be necessary to provide the desired $\alpha$ and $\beta$? *Hint*: Draw a picture of the null hypothesis sampling distribution; then draw a line cutting off 5% of the distribution on the right. This line represents the boundary between values of the sample mean that lead to rejecting the null (to the right) and values that do not (to the left). Now superimpose a drawing of the sampling distribution if the population mean were really .5 standard deviation above zero, scaling the drawing such that 90% of the distribution falls to the right of the cutoff line. Observe that the distance between the means of the two distributions (that is, .5 standard deviation) can be expressed as 1.65 standard errors (the left part) plus 1.28 standard errors (the right part). Plug in the formula for the standard error and solve for $n$.

    d.   Given the sample size, what would be the value of $\beta$ if the population mean were actually one-fourth (.25) of a standard deviation above zero?

## 7.2  If we must estimate the standard error from the data, we lose some certainty.

### Estimating the standard error

Perhaps you were a bit bothered in the two preceding sections by our glib assumption that we knew the population standard deviation—and hence could calculate the standard error—even though we did not know the mean. You had reason to be bothered. Although there are some circumstances in which it is reasonable to claim that we know $\sigma$, more often we have to use sample data to estimate $\sigma$.

Just as we use the sample mean to estimate the population mean, so we use the sample standard deviation (denoted $s$) to estimate the population standard deviation ($\sigma$). Since the real standard error equals $\sigma/\sqrt{n}$, *our estimate of the standard error is $s/\sqrt{n}$.*

The algebraic notation that we use for an estimated standard error is analogous to what we use for the real standard error. In both cases the notation reflects the fact that a standard error is simply the standard deviation of all possible values of a statistic. So we use the symbol for a standard deviation with a subscript to show what statistic we are talking about. Since the symbol for the real standard error is $\sigma_{\overline{X}}$, the analogous symbol for the estimated standard error is $s_{\overline{X}}$.

## Drawing inferences about μ when σ must be estimated

Now that we have a way of estimating the standard error from the data, why can't we simply proceed with our inferences as before? The reason is that we would be ignoring the fact that estimating, rather than knowing, the standard error introduces additional uncertainty.

For example, suppose that we have underestimated the standard error. If so, 2 estimated standard errors will be less than 2 real standard errors—and a 95% confidence interval will be narrower than it should be. In other words, we would be deluding ourselves, claiming greater precision than is warranted. The same problem applies to a hypothesis test. We might *think* the $p$ value for the data is less than .05, when in reality it is greater, if our estimate of the standard error is too low.

Or suppose that we have overestimated the standard error. In that case we would be understating the correct degree of precision in a confidence interval, and overstating the $p$ value in a hypothesis test. Either way, our inferences are suspect.

The problem is, of course, that we have no way of knowing whether we are underestimating or overestimating the standard error in any particular case. However, we *do* know—based on some mathematical theory beyond the scope of this book—that the general tendency is to underestimate. That is, on the average, we are more likely to underestimate the standard error than to overestimate it. (More technically, $s_{\overline{X}}$ is negatively biased as an estimator of $\sigma_{\overline{X}}$.)

So if we want to form a 95% confidence interval, and therefore need to estimate how big 2 real standard errors is, we had better use somewhat more than 2 estimated standard errors. Similarly, we had better require that a sample mean be more than 2 estimated standard errors away from a null hypothesis mean before we will reject the null hypothesis (in favor of a two-sided alternative). For other levels of confidence we need to make similar adjustments.

How much of an adjustment? Unfortunately, there is no general rule. But in one very important situation, there is a way of knowing how much of an adjustment to make. That situation is *when the original population itself has a normal distribution*. Let's look at how it is done.

## 7.3   The family of *t* distributions is used to draw inferences when σ is unknown.

### Using t scores instead of z scores

First, let's define a term. Recall that a *z* score is simply the number of standard deviations—or in the case of a statistic, standard errors—above or below the mean. So, for example, the *z* score for a sample mean of 60 from a normal sampling distribution with mean 50 and standard error 5 is +2.0. Our new term is *t score*. A *t* score is the same as a *z* score, except that we are talking about *estimated* standard errors. Thus a *t score is the number of estimated standard errors above or below the mean.*

As we discussed above, we know that estimated standard errors tend to be lower than the real standard errors they estimate. What does that imply about *t* scores? It implies that *t* scores tend to be bigger than the corresponding *z* scores. This in turn says that we are more likely to observe a *t* score greater than, say, +2.0, than we are to observe a *z* score greater than +2.0. What we are saying is that *the distribution of possible t scores is more spread out than the distribution of possible z scores.*

### The family of t distributions

Now let's consider specifically how much more spread out *t* scores tend to be than *z* scores from a normal distribution. As we have already discussed, the reason *t* scores tend to be larger than *z* scores is that the estimated standard error tends to be smaller than the real standard error. But the severity of this tendency to underestimate will depend on how big the sample is. After all, a very large sample should give us a much better estimate of the standard error than will a smaller sample.

Indeed, for large enough samples, the estimated standard error should be so close to the real standard error that *t* scores based on such samples should look very much like *z* scores. However, for small samples, the tendency to underestimate the standard error gets worse, and so will the tendency for *t* scores to look larger than *z* scores.

What this means is that there is not just one distribution of *t* scores but rather a "family" of distributions, depending on how large a sample is available to estimate the standard error. Moreover, for a large enough sample size, the *t* distribution will be almost indistinguishable from the normal distribution.

### Historical note, and an important restriction

We owe our knowledge of the family of *t* distributions to W. S. Gosset, who published under the pseudonym "Student." Thus the *t* distributions are sometimes referred to as *Student's t distributions.* Gosset worked for Guinness Brewery and often encountered situations involving small sample sizes in experiments with barley and other grains. As we noted in the preceding section, the problem of how much to adjust for underestimation of the standard error has not been solved for *all* situations. But thanks to Gosset, we do have the problem solved for situations in which the original population is itself normally distributed.

*It is important to be aware of this condition!* Although the methods based on the *t* distributions seem to remain fairly accurate even if the original population is not precisely normal, there are some circumstances in which the *t* distributions should *not* be used. In particular, methods based on *t* scores should not be used with samples of, say, 10 or fewer unless it is known that the population from which the sample is selected has a normal distribution. Also, such methods should be avoided even with larger samples if the original population is "heavy in the tails." In such cases it is best to use a method that does not require an assumption of normality (such as a nonparametric procedure, some of which we will discuss later), or to consult a statistician for advice.

### The t table

Now let's look at the *t* distributions and see how they are used. Table D.3 in Appendix D is a typical version of a *t* table. Notice that the *t* table is set up differently from the standard normal table. Instead of giving the area between the mean and a given *t* score, the *t* table gives the *t* score that cuts off a given amount of area. This is done because, as we have noted, there are many different *t* distributions, and it would take too many pages to give as much detail for each *t* distribution as for the standard normal distribution. Besides, we generally do not need more than what is given in the table. Many computer programs will provide more detailed information about whatever *t* distribution you are working with.

Each row of the *t* table represents a different *t* distribution. Instead of indexing the various *t* distributions by sample size, they are indexed by a related quantity called *degrees of freedom.* There is not much to be gained at this point from contriving an intuitive explanation of the term "degrees of freedom," so for now we will simply accept it. We abbreviate degrees of freedom as *df.*

Whenever you use the *t* table, you need to look in the row with the correct number of *df.* Different types of problems call for different *df,* and we will always mention how to figure out *df* for each type of problem. In general, however, it is always true that the larger the sample size, the larger the *df.*

Notice the last row of the table, with "infinite" *df.* Do you recognize the numbers in that row? They are simply the *z* scores from the standard normal distribution that cut off the areas given at the top of the columns. For example, the number in the column headed ".025" is 1.96. Compare that with the standard normal table; you will see that the area between the mean and a *z* score of 1.96 is .4750, leaving .0250 cut off to the right. The *z* score 1.96 is what we have been rounding off to 2 in our discussions; it is the *z* score that together with its negative, brackets the middle 95% of the scores in the normal distribution. For infinite *df*, the *t* distribution is simply the standard normal distribution.

Now let's move up the column headed ''.025.'' What is happening to the entries in the *t* table? They are getting larger. For example, with 20 *df* the *t* score that cuts off the rightmost 2.5% is 2.086. For 10 *df*, it is 2.228. By the time we get to 1 *df*, it takes a *t* score of over 12 to cut off the rightmost 2.5%.

### STUDY QUESTIONS—*Sections 7.2 and 7.3*

7.3.1   Use the *t* table (Table D.3) to find the *t* score that cuts off
    a.   5% of the *t* distribution with 12 *df*
    b.   5% of the *t* distribution with 20 *df*
    c.   5% of the *t* distribution with 56 *df* (do not bother to interpolate)
    d.   2.5% of the *t* distribution with 30 *df*
    e.   1% of the *t* distribution with 30 *df*
7.3.2   The following questions concern the *t* distribution with 20 *df*.
    a.   Find the *t* scores that bracket 90% of the distribution in a symmetric interval about zero.
    b.   Find the *t* scores that bracket 95% of the distribution in a symmetric interval about zero.
    c.   Find the *t* score that cuts off the lowest 5% of the distribution.
    d.   Find the *t* score that cuts off the lowest 2.5% of the distribution.
7.3.3   The following questions concern the *t* distribution with 25 *df*.
    a.   What proportion of *t* scores are greater than 1.3163?
    b.   What proportion of *t* scores are less than $-2.0595$?
    c.   Approximately what proportion of *t* scores are less than $-2$?
    d.   Approximately what proportion of *t* scores are greater than 2.5?

## 7.4   The *t* distribution with $n - 1$ *df* is used to draw inferences about a mean.

### *Degrees of freedom in calculating s*

As we have discussed, using the *t* table requires that we know how many degrees of freedom (*df*) we have. The problem of determining *df* is of a technical nature and we will not bother going into the mathematical details. We simply note that in the calculation of *s*, the number of *df* equals the sample size minus one (i.e., $n - 1$). So in problems in which we use *s* to estimate $\sigma$, and therefore use $s_{\bar{x}}$ to estimate $\sigma_{\bar{x}}$, we have $n - 1$ *df*. Now we are ready to summarize the procedure for estimating and doing hypothesis tests about a population mean using the *t* distributions.

## Confidence intervals for a population mean, σ unknown

Before we start, remember the condition needed for the use of the $t$ table: Technically, we require that the population from which we are sampling have a normal distribution. As a practical matter, we are safe in using the $t$ table if the population distribution is not too spread out, even if it is not strictly normal. The larger the sample size, the safer we are. For sample sizes of less than 10, do not use the $t$ table unless the population distribution is known to be close to normal. *If in doubt, talk to a statistician.*

The general procedure for forming a confidence interval for $\mu$ using a $t$ distribution is very simple: The interval is some number of estimated standard errors on either side of the sample mean. How many estimated standard errors? However many it takes to encompass our desired confidence level in the appropriate $t$ distribution. Let's go through the steps in the context of an example.

**Example 7.3**  A random sample of 28 specimens of copper wire is selected from a large spool of the wire. Each specimen is stretched until it breaks, and the amount of force required to break the specimen (in kilograms) is recorded. The distribution of breaking force may be assumed to be at least approximately normal. The sample is found to have a mean breaking force of 226 kg, with a standard deviation ($s$) equal to 3.5 kg. Give a 95% confidence interval for the mean breaking strength of the entire spool (the population).

*Step 1*:  *Calculate the estimated standard error.* This, we know, is equal to $s/\sqrt{n}$, which is $3.5/\sqrt{28}$, or .66 kg.

*Step 2*:  *Find the appropriate t score from the table.* We have 27 $df\,(28 - 1)$, so we consult the line of the $t$ table for 27 $df$. Since we want a 95% confidence interval, we want the $t$ score that cuts off 2.5% on the right of the distribution. (Be sure that you see why!) The entry in the column headed ''.025'' for 27 $df$ is 2.052 (rounded to 3 decimals).

*Step 3*:  *Finish up.* So the 95% confidence interval for the population mean is the sample mean plus and minus 2.052 estimated standard errors. This is $226 \pm 2.052 \times .66$, or $226 \pm 1.35$ kg. We are therefore 95% confident that the population mean is between 224.65 and 227.35 kg.

**Example 7.4**  Find a 99% confidence interval for the population mean, using the same data as Example 7.3.

The only difference from Example 7.3 is the $t$ score. For a 99% confidence interval, we need the entry in the column headed ''.005''—that is, the $t$ score that cuts off one-half of 1% of the area. The entry for 27 $df$ is 2.771. Thus a 99% confidence interval for the mean is $226 \pm 2.771 \times 0.66$, or $226 \pm 1.83$ kg. We are 99% confident that the population mean is between 224.17 and 227.83 kg.

## Hypothesis tests concerning a population mean, σ unknown

As you might expect, the procedure for doing hypothesis tests using $t$ scores is analogous to the procedure using $z$ scores. The steps are outlined in the examples below. Since the procedure uses $t$ scores based on a single sample, the procedure is often called the *one-sample t test*.

**Example 7.5**   Continuing with the copper wire example, suppose that the wire is intended to have a nominal mean breaking strength of 220 kg. The manufacturer would like to have reasonable proof that the actual mean breaking strength is greater than the nominal value. Is the sample evidence (sample mean 226 kg, sample standard deviation 3.5 kg, $n = 28$) sufficient for the manufacturer to conclude that the population mean is greater than 220 kg?

*Step 1*:   *State the null and alternative hypotheses.* We seek proof for the hypothesis that the population mean is greater than 220 kg. Thus:

*Null hypothesis*: The population mean is equal to 220 kg.
*Alternative hypothesis*: The population mean is greater than 220 kg.

*Step 2*:   *Find the estimated standard error.* As we showed in Example 7.3, the estimated standard error is .66 kg.

*Step 3*:   *Assume that the null hypothesis is true and find the t score for the sample mean.* Recall that a $t$ score is simply the number of estimated standard errors away from the mean of a distribution. In this case we are looking for the number of estimated standard errors our sample mean (226 kg) is away from the null hypothesis mean of 220 kg. With a standard error of .66 kg, 226 kg is 6/.66, or 9.09 estimated standard errors above the null hypothesis mean, for a $t$ score of +9.09.

*Step 4*:   *Find the p value for the observed data.* With a sample size of 28, we have 27 *df*. In this example we have a one-tailed alternative hypothesis. So the $p$ value is the proportion of $t$ scores in the $t$ distribution with 27 *df* that are +9.09 or greater. A look at the $t$ distribution and a moment's thought convinces us that $t$ scores with 27 *df* are virtually never +9.09 or greater. The $p$ value is, for practical purposes, zero.

*Step 5*:   *State the conclusion.* If the null hypothesis were true, our sample mean of 226 kg would be over 9 estimated standard errors above the population mean. That is so unlikely that we must reject the null hypothesis. The sample data strongly indicate that the population mean is greater than 220 kg, and the manufacturer can rest assured.

**Example 7.6**   Another sample of wire specimens from a different manufacturing plant is selected at random. This time, 19 specimens are available and yield a mean breaking force of 222.5 kg, with a standard deviation of 6.4 kg. Is there enough evidence to indicate that the population mean is greater than the nominal mean of 220 kg?

*Step 1*:    The null and alternative hypotheses are as in Example 7.5.

*Step 2*:    *Find the estimated standard error*. This time the estimated standard error is $6.4/\sqrt{19}$, or 1.47 kg.

*Step 3*:    *Assume the null hypothesis is true and find the t score for the sample mean*. The sample mean, 222.5 kg, is 2.5 kg above 220. This is $2.5 \div 1.47$, or 1.70 estimated standard errors above the null hypothesis mean. Thus the $t$ score for the sample mean is $+1.70$.

*Step 4*:    *Find the p value for the observed data*. With a sample size of 19, we have 18 *df*. Unless we have a more detailed table, we cannot find the exact $p$ value. However, we can observe that our $t$ score of $+1.70$ is between two of the $t$ scores in Table D.3 for 18 *df*: 1.3304 (in the ".10" column) and 1.7341 (in the ".05" column). Thus the proportion of $t$ scores that are 1.70 or greater—that is, the $p$ value for our data—must be between .10 and .05.

*Step 5*:    *State the conclusion*. The $p$ value is greater than .05. So if we are insisting on the conventional cutoff of .05 or less, we cannot reject the null hypothesis. The evidence is not sufficient, at the .05 level, to conclude that the mean breaking force is greater than 220 kg.

## STUDY QUESTIONS—*Section 7.4*

**7.4.1**    For each of the following examples, calculate a 95% confidence interval for the population mean. You may assume in each case that the individuals in the sample were drawn randomly from a population with a normal distribution. *Note*: If the number of *df* for a problem is not given in the $t$ table, just use the line in the table that comes as close to the correct number of *df* as possible.

a.    Sample mean $= 26$, sample standard deviation $= 12$, $n = 14$
b.    Sample mean $= 2$, sample standard deviation $= 6$, $n = 10$
c.    Sample mean $= 10$, sample standard deviation $= 5$, $n = 10,000$
d.    Sample mean $= 10$, sample standard deviation $= 5$, $n = 100$
e.    Sample mean $= 0$, sample standard deviation $= 10$, $n = 100$

**7.4.2**    For each of the examples in Exercise 7.4.1, determine if the sample evidence is sufficient to conclude that the population mean is different from zero. (*Note*: This calls for a two-tailed test.) In drawing your conclusion, use $\alpha = .05$; that is, reject the null hypothesis if the $p$ value is .05 or less. Then look back at the 95% confidence intervals you calculated for Exercise 7.4.1. When you rejected the null hypothesis, what did the confidence interval look like?

**7.4.3**    Let's consider what happens if we *cannot* assume that the population from which we are sampling has a normal distribution. Suppose that you select a random sample of six freshman psychology students and measure their pulse rates before and after watching a certain movie. The measure of interest is the change in pulse rate from before to after the movie. Unfortunately, you have no way of knowing what the distribution of such changes might look like. So if you wanted to test the hypothesis that pulse rates for freshman psychology students tend to, say, increase upon viewing the movie, you could not safely use a *t* test.

But suppose that all six students showed some amount of increase in pulse rate. Even though you should not use a *t* test, you can still perform a legitimate hypothesis test based on the information you do have. How? By asking the same kind of question we always ask in performing a hypothesis test: *If the null hypothesis were true, how likely would our sample results be?*

Of course, the null hypothesis in this case is that pulse rates do not show a tendency to increase. If that were true, we would expect to see some students with an increase and some with a decrease. But in our sample, six out of six showed an increase. How likely is that result if the null hypothesis were true? (*Hint:* The variable of interest is the number of increases in six independent trials, where the probability of an increase in one trial if the null hypothesis is true is .5.)

# 8

# Further Topics in Inference About Single Populations

## 8.1 Proportions can be treated as a special kind of mean.

### Inferences about proportions

Suppose that we wish to estimate the proportion of people in the United States who identify themselves as affiliated with a political party. In this situation our interest is not in an average of some variable, but simply in the fraction of the population showing a particular characteristic. We can select a random sample and determine the proportion showing the characteristic in the sample. But how do we then draw inferences about the population proportion?

By using a clever trick, we can apply the techniques we have learned for drawing inferences about a mean to the problem of drawing inferences about a proportion. The trick is to recognize that a proportion is itself a mean—the mean of a special kind of variable. Once we recognize that, techniques for drawing inferences about proportions can be viewed simply as special cases of techniques for drawing inferences about means.

### Dummy variables

We have already encountered this trick in Chapter 4: the trick is to assign every individual in the sample a score on a *dummy variable*. (Other terms often used for dummy variables include *yes-no variables* and *indicator variables*.) Recall that a dummy variable is a variable that has only two possible scores: 0 or 1. If the individual has the characteristic of interest, that individual gets a score of 1; otherwise, the individual gets a score of 0.

How does this turn a proportion into a mean? The easiest way to convince yourself is to try it. Consider a sample of 60 people, of whom 20 say they are affiliated with a political party. Now assign those 20 people a score of 1, and the other (unaffiliated) people a score of 0. What is the mean of the 60 scores? 20 ÷ 60, or .33—which is none other than the proportion of the sample that identify themselves as affiliated. This, of course, is no accident. The mean of a dummy variable is simply the proportion of 1's. So to draw inferences about a population proportion based on a sample, we can assign a score on a dummy variable to every member of the sample and proceed with our inference as with any other variable.

### The catch

There is only one thing to watch out for in using techniques for means to draw inferences about proportions. Since a dummy variable can have only two possible scores, it cannot possibly have a normal distribution. For large samples, this makes no practical difference, because of the CLT: the mean of *any* variable has a normal sampling distribution (at least approximately) if the sample size is large enough. But for small samples, the non-normality of a dummy variable can cause difficulties. This is particularly true if the dummy variable has a very skewed distribution—that is, if the proportion being estimated is close to either zero or 1.

So we should use techniques such as the *t* test on proportions only if the sample is large enough. What constitutes a large enough sample? This amounts to the same question that we asked in Chapter 4 in considering how large $n$ must be to permit using the normal approximation for a binomial probability. As we noted, a handy rule of thumb is to require that both "$n \times p$" and "$n \times (1 - p)$" be at least 5, where "$p$" refers to the proportion being estimated. By this rule, if the population proportion were equal to .5, a sample size of 10 is sufficient. However, if the population proportion were equal to .9, we would require a sample size of 50.

What do we do if we do not have a large enough sample? We would then have to make use of the binomial distribution directly—a simple extension of the techniques we discussed in Chapter 4. We shall present a guided example of this in a study question to show how it is done. For now, we concentrate on situations for which the sample size is sufficient [i.e., both $np$ and $n(1 - p)$ at least 5].

### Testing a hypothesis about a population proportion

Let's look at an example.

**Example 8.1**  A candidate for public office is to make a policy speech and wants some assurance that the group before which he will be speaking is friendly to his views. He is willing to consider the group friendly if more than 75% of the group members agree with him. A random sample of 50 members of the group reveals that 42 agree with the candidate's positions on most issues. Is this sufficient evidence for the candidate to conclude that the group as a whole is "friendly"?

The sample proportion of people who agree with the candidate's positions equals 42 ÷ 50, or .84. If we assign dummy variables, we can calculate $s$ to be .370. (Or you could use the shortcut formula given below.) Thus the estimated standard error equals .370 ÷ $\sqrt{50}$, or .052. The sample proportion is therefore .09 ÷ .052, or 1.73, standard errors above the null hypothesis proportion.

Now, do we look in the $t$ table or the standard normal table? The answer is that it makes little practical difference. If we have a sample size large enough to justify using this procedure, the $t$ table is virtually the same as the standard normal table anyway. Usually, we would use the standard normal table if we are doing it by hand.

Whichever technique we use in this case, the $p$ value (one-tailed) is less than .05. Thus the candidate is safe in concluding that the group is "friendly."

## A shortcut formula for the standard deviation of a dummy variable

The standard deviation of a dummy variable, like that of any variable, can be computed as the root-mean-squared deviation. But there is a simple shortcut that applies in the case of dummy variables. It turns out that the standard deviation of a dummy variable is simply the square root of the quantity $p \times (1 - p)$. (To be picky, this is actually the formula for the *population* standard deviation—that is, dividing by $n$ instead of $n - 1$—of a dummy variable, but with a large sample it makes little difference.)

This means that you can easily calculate the standard deviation of a dummy variable simply by knowing its mean (i.e., the proportion of 1's). For example, if 40% of a sample have a certain characteristic, so that the mean of the associated dummy variable is .4, the standard deviation would be the square root of (.4 × .6), or .49. Once you have the standard deviation, calculating the standard error is easy—it is the standard deviation divided by $\sqrt{n}$.

## A conservative 95% confidence interval for a population proportion

The biggest the standard deviation will ever be is when $p$ equals .5, in which case the standard deviation is also .5. (Try it!) Thus the biggest the standard error will be is also when $p$ equals .5. This handy fact allows us to form a "conservative" confidence interval for a population proportion (conservative in the sense that it will be at least as wide as it has to be): We calculate the standard error as .5 ÷ $\sqrt{n}$, regardless of what the sample standard deviation is. In this way we can be sure that we are not underestimating the standard error, since the standard deviation can never be greater than .5.

*A conservative 95% confidence interval for a population proportion (with a large enough sample, as discussed above) is therefore the range of proportions within 2 standard errors of the sample proportion, that is, the sample proportion* $\pm 1/\sqrt{n}$. For example, if 37% of a sample of size 64 shows a certain characteristic, a 95% confidence interval for the population proportion is .37 $\pm$ 1/8, or the interval from .245 to .495. We can be at least 95% certain that the population proportion showing the characteristic is between 24.5% and 49.5%.

### STUDY QUESTIONS—*Section 8.1*

**8.1.1**  A sample of 144 eggs contains 24 broken eggs. Calculate and interpret a conservative 95% confidence interval for the proportion of broken eggs in the population from which the sample was taken.

**8.1.2**  Of 25 swimming pools randomly selected from all pools in a certain state, seven had bacterial counts exceeding the recommended maximum. Is there enough evidence to conclude that more than 10% of the pools in the state have bacterial counts exceeding the recommended maximum?

**8.1.3**  [Continuation of Exercise 8.1.2] Calculate and interpret a conservative 95% confidence interval for the proportion of swimming pools in the state with bacterial counts exceeding the recommended maximum, using the data in Exercise 8.1.2.

**8.1.4**  [Statpal exercise—Coronary Care data set (see Appendix A)]
   a.  Use Statpal to find the mean and standard deviation of the variable HADMI, which is a dummy variable indicating whether or not the patient was diagnosed as having a myocardial infarction (abbreviated MI—i.e., a heart attack). A value on HADMI of 0 indicates that the patient did not have an MI, and a value of 1 indicates that the patient did have an MI.
   b.  Now use the HIstogram procedure to find how many 0's and 1's there were among the patients on the variable HADMI.
   c.  Note the relationship between the proportion of 1's as shown in the HIstogram output and the mean.
   d.  What about the standard deviation? Use the formula given in the text [i.e., $\sqrt{p(1 - p)}$, where $p$ is the proportion of 1's] to calculate the standard deviation, and observe how close the result is to the standard deviation from Statpal's DEscriptive statistics procedure. Why is it not exactly the same?

**8.1.5**  To find out how many households with televisions had sets tuned to a particular program at a particular time, a random sample of 300 households was selected. Each household was contacted by telephone at the time of interest, and the person answering the phone was asked whether a television set in the household was tuned to the program. Of the 300 respondents, 130 (or 43.3%) said there was a television set in the household tuned to the program at the time of the call.
   a.  Calculate a conservative 95% confidence interval for the proportion of all households in the population with a television set tuned to the program at the

time the calls were made. (*Note*: The square root of 300 is 17.32, and 1 ÷ 17.32 is about .058, or 5.8%.)

    b.  Suppose that the sponsor of the program wants an estimate of the population proportion that has a margin for error of only 3%, with 95% confidence. How big a sample size would be necessary to get such an estimate? (*Hint*: The sponsor's requirement means that two standard errors must be equal to 3%, or .03. Recall that for a conservative 95% confidence interval for a proportion, 2 standard errors equals $1/\sqrt{n}$. Now solve for *n*.)

**8.1.6**  In a certain county, 40% of all adult residents are black. A group concerned with voting rights selects a random sample of 200 people from the list of registered voters and determines that 56 (28%) are black.

    a.  Is there sufficient evidence to conclude that blacks are underrepresented on the list of registered voters?

    b.  Calculate a conservative 95% confidence interval for the proportion of registered voters who are black.

**8.1.7**  [Continuation of Exercise 8.1.6] Suppose that the group selects only 12 people for their random sample of registered voters and finds that three (25%) are black. The question remains the same as in part a: Is there sufficient evidence to conclude that blacks are underrepresented on the list? Note that the sample size is not big enough to meet our rule of thumb for using the *t* distribution, because $np = .40 \times 12 = 4.8$, which is less than 5. So we will use the binomial distribution to address the question.

    a.  Suppose that the null hypothesis is true (that blacks are not underrepresented, and therefore that the checklist does contain 40% blacks). What would be the probability of observing 25% or fewer blacks out of 12 randomly selected voters? Note that this is the same as asking for the probability of 3 or fewer yeses out of 12 independent trials, with $p = .40$.

    b.  Based on your answer to part a, can the null hypothesis be rejected?

    c.  Now perform the test using the *t* distribution method and compare your results.

## 8.2  The sign test requires few assumptions.

### *Parametric versus nonparametric tests*

In our discussion of techniques that make use of the *t* table, we mentioned that the techniques assume that the population from which individuals are being sampled has a normal distribution. When we knew the population standard deviation and used *z* scores instead of *t* scores, it was not necessary to make that assumption, but we still had to require that the sample size be large enough to guarantee that the sampling distribution of the sample mean was approximately normal. The less normal the original population, the greater the sample size would have to be to ensure an approximately normal sampling distribution.

Tests that rely on assumptions about the distribution of the population or its parameters are known as *parametric tests*. Both the $z$ test and the $t$ test are parametric tests. If we have a small sample and are not sure that the original population distribution is roughly normal, such tests can be misleading.

The tests we discuss in this section and the next are called *nonparametric tests*, because they do not require much in the way of assumptions about the population distribution. Another term sometimes used for tests that do not require assumptions about distributions is *distribution-free*. The terms "nonparametric" and "distribution-free" tend to be used synonomously, although some technical distinctions have been proposed.

We shall use the term "nonparametric" descriptively rather than technically, to refer to tests that require few (if any) assumptions about the population or populations from which we are sampling. In practice, nonparametric tests have another attractive characteristic: they are usually easy to understand.

Both of the nonparametric tests we discuss in this chapter are used to test hypotheses about central tendency, but they concentrate on the median instead of the mean. Specifically, they test the null hypothesis that the population median is equal to some specified value, versus either a one-tailed or a two-tailed alternative.

### The sign test

This is one to keep handy in your toolkit, because it is very easy to apply. As usual, we present the details in the context of an example.

**Example 8.2**  Several companies market "5-year" light bulbs, claiming on the label that the bulbs work for an average of 3000 hours before burning out. To test whether such bulbs really do tend to last at least 3000 hours, a random sample of seven bulbs is obtained. Each bulb is left on until it burns out and the number of hours the bulb worked is recorded. The results are as follows:

| Bulb | Hours until failure |
|------|---------------------|
| 1    | 3215                |
| 2    | 3630                |
| 3    | 2880                |
| 4    | 6345                |
| 5    | 3135                |
| 6    | 4270                |
| 7    | 3010                |

Is there sufficient evidence to conclude that most bulbs last at least 3000 hours?

*Step 1*: *State the null and alternative hypotheses.* Note that the question is asking whether "most" bulbs last at least 3000 hours. In other words, we are seeking evidence that the median lifetime of the bulbs is greater than 3000 hours. Thus we have the following hypotheses:

*Null hypothesis*: The median lifetime of the bulbs is 3000 hours.

*Alternative hypothesis*: The median lifetime is greater than 3000 hours.

*Step 2*: *Count how many scores are above the null hypothesis value and how many are below.* For our sample of seven bulbs, we have six above 3000 hours and one below 3000 hours. In terms of the sign of the difference between each score and the null hypothesis median, we have six pluses and one minus.

*Step 3*: *Assume that the null hypothesis is true and find the p value.* First, let's consider what we would expect to see, on the average, if the null hypothesis were true. If the median lifetime were really 3000 hours, we would expect about half of our sample scores to be above 3000. But our sample yielded six above. The *p* value, then, is simply the probability that at least six out of seven bulbs will last longer than 3000 hours, given that the chance of any one bulb lasting that long is 50%.

This is none other than a binomial probability, with $n = 7$ and $p = .5$. We can obtain the probability of six or more yeses from the binomial table by adding up the probabilities for six and seven yeses. The answer (as you should verify) works out to be .063, or a little over 6%.

Since this is a one-tailed test (in that we are seeking proof that the median is *greater than* 3000, not *different from* 3000), our *p* value is just what the table told us: .063. (For a two-tailed test, we would double the tabled *p* value.)

*Step 4*: *State the conclusion.* Since the *p* value is greater than 5%, we would generally not reject the null hypothesis. We cannot claim that the median lifetime of the bulbs is greater than 3000 hours.

So six pluses out of seven is not sufficient to reject the null hypothesis at the 5% level of significance using the sign test. Would seven out of seven have been sufficient? A look at the binomial table should convince you that it would have been, with a (one-tailed) *p* value of .008.

Nowhere in the reasoning of the sign test do we require any assumptions about the form of the distribution of the population from which we are sampling. The only condition we need is that all the differences be either positive or negative— in other words, that none of the scores have *exactly* the null hypothesis value.

If we are working with data for a continuous (or nearly continuous) variable, this presents no problem. But if we are working with discrete data, it is quite possible that some of the scores will have exactly the null hypothesis value and therefore have a difference of zero. What do we do if we have such scores? In the sign test we simply ignore them altogether—that is, we leave them out of the analysis. We use only those scores that are above or below the null hypothesis median.

### The sign test with large samples

The binomial table includes sample sizes up through 20. What if you are using a larger sample size? Since the $p$ value for the sign test comes from the binomial distribution, we can make use of the normal approximation discussed in Chapter 4.

As we noted in Chapter 4, to use the normal approximation, all we need to know is the mean and standard error of the statistic we are using (in this case, the number of pluses). Under the null hypothesis, we have $p = .5$. Thus applying the formulas for the mean and standard deviation of a binomial variable from Chapter 4, we have

$$\mu = np = .5n$$

and

$$\sigma = \sqrt{np(1 - p)} = \sqrt{n(.5)(.5)} = .5\sqrt{n}$$

Thus when the null hypothesis is true, the mean of the sampling distribution of the number of pluses (our statistic for the sign test) is equal to half of $n$, with a standard error of half of $\sqrt{n}$. Let's consider an example to show how the sign test may be used with a large sample.

**Example 8.3**   This time, 36 light bulbs were tested. Of these, 25 lasted longer than 3000 hours. Is there sufficient evidence to conclude that the median lifetime of the population of light bulbs is greater than 3000 hours?

*Step 1*:   *State the null and alternative hypotheses.* The hypotheses are the same as in Example 8.2.

*Step 2*:   *Count how many scores are above the null hypothesis value and how many are below.* We have already been provided with this information in the question. (In fact, that is all the information we have about the sample.)

*Step 3*:   *Assume that the null hypothesis is true and find the p value.* Here is where we must make use of the normal approximation. If the null hypothesis were true, the mean number of pluses out of 36 tries would be $36 \div 2$, or 18. The standard error of the number of pluses would be half of $\sqrt{36}$, or 3. So if the null hypothesis were true, we would be picking a random number of pluses from a normal sampling distribution with mean 18 and standard error 3. The $p$ value is therefore the proportion of scores in the distribution to the right of our observed number of pluses, which was 25.

Applying the techniques we learned previously, we draw a picture, find the $z$ score for 25 (which works out to be 2.33), and determine from the table of the standard normal distribution that the $p$ value for our data is $.5000 - .4901$, or $.0099$. Thus the $p$ value is a bit less than 1%.

***Step 4***:   *State the conclusion*. With a *p* value less than 1%, we would reject the null hypothesis and conclude that the median lifetime of the bulbs is greater than 3000 hours.

Notice that the sign test considers only the *sign* of the difference between each score and the null hypothesis value. It ignores the *magnitude* of the differences. In particular, in Example 8.2 it ignored the fact that the one bulb that burned out before 3000 hours still got pretty close to 3000, while several of the bulbs that lasted longer than 3000 hours lasted *much* longer than 3000. So the sign test did not take advantage of some relevant information in the sample. In the next section we consider another test that works much like the sign test but listens to more of the information in the sample data.

### STUDY QUESTIONS—*Section 8.2*

**8.2.1**   A random sample of 15 machined parts selected from a large lot contains 10 parts that exceed the nominal size. Is there enough evidence to conclude that the median size of the parts in the lot is greater than the nominal size?

**8.2.2**   Twelve pairs of identical twins are selected to participate in a study of a new teaching method. One member of each pair is assigned to the new method, while the other member is assigned to the existing method. At the end of the study period, each participant's learning was assessed using a special test. In nine of the pairs, the score for the new-method twin exceeded that for the existing-method twin. Is there enough evidence to conclude that the new method is associated with higher test scores on the average?

**8.2.3**   Of 64 men over age 60, 39 are taller than 5 ft 8 in. Assuming that the men were randomly selected from the population of all men over 60, is there evidence to conclude that the median height of that population is greater than 5 ft 8 in.?

**8.2.4**   One hundred randomly selected people were asked to rate the performance of a public official on a scale of 1 to 5, with 3 labeled ''acceptable.'' Fifty-seven of the people gave the answer 1 or 2, indicating a rating of less than acceptable. A political opponent claims that this represents a clear statement that ''most people'' feel the official's performance is less than acceptable. Test the validity of the opponent's claim and write a statement summarizing your findings.

## 8.3   The Wilcoxon signed rank test works with ranks instead of the original units.

### *Transforming data into ranks*

In the sign test, we threw away everything but the sign of the difference between each score and the null hypothesis median. In the test we discuss in this section, called the *Wilcoxon signed rank test*, we keep not only the sign of the differences, but also the relative magnitudes of the differences. We do this by transforming the difference scores into ranks, with 1 being the rank of the smallest difference score and *n* being the rank of the largest.

Many nonparametric procedures use the technique of rank transformation. While working with ranks instead of the original scores does entail a loss of information, it also allows us to work with statistics that have known sampling distributions, regardless of the form of the original population distribution. In many cases, the amount of information lost by going to ranks is surprisingly little. Let's look again at the light bulb example, this time using the Wilcoxon signed rank test instead of the sign test.

**Example 8.4**   Let's repeat the data on light bulb life from Example 8.2.

| Bulb | Hours until failure |
|------|---------------------|
| 1    | 3215                |
| 2    | 3630                |
| 3    | 2880                |
| 4    | 6345                |
| 5    | 3135                |
| 6    | 4270                |
| 7    | 3010                |

The question remains as before: Is there sufficient evidence to conclude that the median lifetime of the population of light bulbs is greater than 3000 hours?

*Step 1*:   *State the null and alternative hypotheses.* As we have seen, the null hypothesis is that the median lifetime is 3000 hours. The alternative hypothesis is that the median lifetime is greater than 3000 hours.

*Step 2*:   *Find the difference between each score and the null hypothesis value, keeping the sign.* The difference scores are therefore as follows:

| Bulb | Difference from 3000 hours | Rank |
|------|----------------------------|------|
| 1    | $3215 - 3000 = +\ 215$     | 4    |
| 2    | $3630 - 3000 = +\ 630$     | 5    |
| 3    | $2880 - 3000 = -\ 120$     | 2    |
| 4    | $6345 - 3000 = +3345$      | 7    |
| 5    | $3135 - 3000 = +\ 135$     | 3    |
| 6    | $4270 - 3000 = +1270$      | 6    |
| 7    | $3010 - 3000 = +\ \ 10$    | 1    |

**Step 3**: *Rank the differences, regardless of sign, from smallest to largest.*
The ranks are shown next to the difference scores. Note that the sign of the score
is ignored at this step; thus bulb 3, with a difference of $-120$, gets rank 2,
because 120 is the second largest absolute difference. If there had been ties in the
data, we would have assigned the average rank to the tied scores. For example, if
two tied scores should occupy ranks 4 and 5, we would assign them both a rank
of 4.5. As in the sign test, we would simply ignore differences of zero (i.e.,
bulbs that burned out at exactly 3000 hours would be excluded from the sample
entirely).

**Step 4**: *Add up the ranks for the plus differences and minus differences,
respectively.* Let's do the minus difference first, because there is only one: The
sum of the minus ranks is 2. To get the sum of the six plus ranks, we could add
up the numbers and get 19. Or we could observe that the sum of the ranks of
seven numbers is always 21 and simply subtract the sum we already got for the
minus ranks.

What would we expect to see if the null hypothesis were true? If the median
lifetime were really 3000 hours, we would expect about half of the difference
scores to be plus and about half to be minus. (That, of course, is the reasoning
behind the sign test.) Moreover, we would expect the positive differences and the
negative differences to be roughly equal in magnitude, so that the two sums of
ranks would be roughly equal.

Thus, if the null hypothesis were true, we would expect the positive and
negative rank sums each to be close to 10.5. In our example data, we observed 2
for the negatives and 19 for the positives, indicating a preponderance of positive
differences.

A conventional notation for the sum of plus ranks is $W+$, and for the minus
ranks $W-$. Of course, $W+$ and $W-$ supply exactly the same information; if you
know one, you know the other. We use the notation $W$ for the sum that should be
*smaller* if the alternative hypothesis is true. In our example, $W$ is $W-$, or 2.

**Step 5**: *Find the p value.* In other words, what is the chance of observing a
rank sum for the negatives of 2 or less (or equivalently, a rank sum for the
positives of 19 or more) if the null hypothesis were true? We again have a table
available to make it easy: the table for the Wilcoxon signed rank test, Table D.4.

The table gives the values of $W$ that give $p$ values of (at most) 5% and 1%. In
our example we have a sample size of 7, so we looked in the row labeled "$n =
7$." Under the column headed ".05" we find the number 3. This means that a $W$
of 3 or less would occur by chance less than 5% of the time if the null hypothesis
were true. Since our value of $W$ was in fact 2, our $p$ value must be less than 5%.

Notice that the column headed ".01" has no entry for $n < 7$. That is because
even a $W$ of zero could occur by chance more than 1% of the time if the null
hypothesis were true. If we are insisting on a $p$ value of 1% or less, there is no
point doing a signed rank test with a sample size of less than 7, because even the

most extreme evidence would not let us reject the null hypothesis. In other words, with $n < 7$ and $\alpha$ set at .01, the signed rank test has zero power.

*Step 6*:   *State the conclusion.* Since our $p$ value is less than 5%, we can reject the null hypothesis and conclude that the median lifetime of the bulbs is greater than 3000 hours.

Thus the Wilcoxon signed rank test led us to reject the null hypothesis, whereas the sign test did not. The signed rank test was more sensitive because it used more of the information in the sample data—specifically, it considered not only the signs of the difference scores, but also their relative magnitudes.

### A normal approximation for the Wilcoxon signed rank test

Table D.4 for the Wilcoxon signed rank test includes sample sizes up through 50. For larger sample sizes, a normal approximation may be used: $W$ has a mean equal to $(W+ \text{ plus } W-)/2$, and a standard error equal to

standard error of $W$

$$= \sqrt{\frac{1}{24}\left[n(n+1)(2n+1) - \frac{1}{2}\sum_{i=1}^{g} t_i(t_i-1)(t_i+1)\right]}$$

where
$\quad n$ = number of nonzero difference scores
$\quad g$ = number of groups of tied scores
$\quad t_i$ = number of scores in the $i$th group of ties
If there are no tied scores, this formula simplifies to

$$\sqrt{\frac{n(n+1)(2n+1)}{24}}$$

In practice, if the sample size is greater than 50, you can almost always use a $t$ test safely.

### STUDY QUESTIONS—*Section 8.3*

**8.3.1**   For the 12 machined parts mentioned in Exercise 8.2.1, the nominal size was 150 mm. The actual sizes in centimeters of the 12 parts were as follows:

142   151   153   152   140   149   154   153   155   152   151   152

Using the Wilcoxon signed rank test, is there enough evidence to conclude that the machined parts in the lot from which the sample was taken have a median size greater than the nominal value of 150 cm?

**8.3.2**   [Statpal exercise] Use Statpal to perform a sign test and a Wilcoxon signed rank test for the data in Exercise 8.3.1. To do this, create a Statpal system file with two variables: SIZE (using the numbers given) and NOMINAL, with the value 150 for

each case. Then go into the NOnparametric tests procedure and perform the two
tests.

**8.3.3**   The test scores of the twins in the study described in Exercise 8.2.2 were as
follows:

| Pair | New-method twin's score | Old-method twin's score |
|------|-------------------------|-------------------------|
| 1    | 83                      | 79                      |
| 2    | 63                      | 65                      |
| 3    | 87                      | 77                      |
| 4    | 90                      | 88                      |
| 5    | 54                      | 66                      |
| 6    | 32                      | 31                      |
| 7    | 71                      | 70                      |
| 8    | 86                      | 80                      |
| 9    | 57                      | 53                      |
| 10   | 69                      | 66                      |
| 11   | 93                      | 80                      |
| 12   | 74                      | 67                      |

Using the Wilcoxon signed rank test, is there enough evidence to claim that the
new method is associated with higher scores on the average? (*Hint*: The first thing
to do is change the scores for each pair into a difference score—i.e., the new-
method score minus the old-method score. Then proceed as in Example 8.4.)

**8.3.4**   [Statpal exercise] Use Statpal to perform a sign test and a Wilcoxon signed rank
test for the data in Exercise 8.3.3. To do this, create a Statpal system file with two
variables: NEWMETH and OLDMETH. (Note that a "case" in this situation
represents a pair of twins.) Then go into the NOnparametric tests procedure and
perform the tests.

**8.3.5**   [Statpal exercise—Coronary Care data set (see Appendix A)] Perform a Wilcoxon
signed rank test to determine if the median age of the population of patients from
which the sample was selected is greater than 60 years. To do this:

1.  Use the TRansformation utility to create a new variable called CONST60
    with the constant value 60. A TRansformation statement to do this is

    CONST60 = 60

2.  Now go back to the Main Menu, ask for the NOnparametric tests procedure,
    and perform the test, specifying the variables AGE and CONST60.

3.  Interpret the output and state your conclusion.

**8.3.6** A new biological assay procedure has been developed for measuring the concentration of phosphorus in samples of lake water. To test the accuracy of the assay procedure, quantities of water with a *known* concentration of phosphorus are assayed using the new procedure. Any assay procedure will be subject to some degree of variability; the question of immediate concern is whether the new procedure tends to report concentrations that are systematically too high or too low. Nine assays were performed on water known to have a concentration of 20 parts per billion (ppb) of phosphorus. The assay results were as follows:

| Trial | Assayed concentration of phosphorus (ppb) |
|-------|-------------------------------------------|
| 1     | 24                                        |
| 2     | 19                                        |
| 3     | 27                                        |
| 4     | 213                                       |
| 5     | 18                                        |
| 6     | 21                                        |
| 7     | 44                                        |
| 8     | 26                                        |
| 9     | 30                                        |

a. Is there enough evidence to indicate that the new assay gives a median result different from the known concentration of 20 ppb? Use the sign test to find out. (Note that this question specifies a two-tailed alternative hypothesis; to find the *p* value, follow the procedures shown in the text to obtain a one-tailed *p* value, and double it.)

b. Now use the Wilcoxon signed rank test to answer the same question. Do you come to the same conclusion as with the sign test? What if you were willing to risk a 10% chance of a type I error, instead of a 5% chance—would you still come to the same conclusion using the two tests? (*Note*: For a two-tailed test, *W* is defined as the smaller of $W+$ and $W-$.)

**8.3.7** [Continuation of Exercise 8.3.6] If you had some assurance that the distribution of assay results is approximately normal, you could do a *t* test for this problem. (If the normality assumption is valid, the mean is the same as the median, so you would still be testing the same hypothesis.)

a. Cite sample evidence that should dissuade you from doing a *t* test.

b. Now throw caution to the winds and do the *t* test anyway. For your convenience, here are some handy calculations already worked out for you:

Sample mean:                   46.889 ppb
Sample standard deviation:     62.774 ppb
Estimated standard error:      20.925 ppb

Can you reject the null hypothesis, even allowing a 10% risk of a type I error?

c.  What causes the *t* test to be less sensitive in this situation?

**8.3.8**  [Advanced exercise—Continuation of Exercise 8.3.7] We have covered hypothesis testing using the sign test and Wilcoxon signed rank test. How about estimation? Suppose that we wish to calculate a 95% confidence interval for the median phosphorus concentration. Let's consider a nonparametric approach to the problem. The key is to understand the relationship between a confidence interval and a null hypothesis. What is a confidence interval? It is a range of values which, we believe with a certain degree of confidence, contains the parameter we are trying to estimate. But we can also define a confidence interval in a different but logically equivalent way: *A confidence interval is the range of null hypothesis values we would not reject, based on our data.* Let's use the Wilcoxon signed rank test to form a 95% confidence interval for the median.

a.  Using Table D.4, determine what values of *W* would lead to not rejecting the null hypothesis with alpha = .05 (two-tailed). (*Answer*: Any *W* greater than or equal to 6. That would mean that neither *W*+ nor *W*− could be less than 6.)

b.  Arrange the nine scores in order from smallest to largest. Now suppose that the null hypothesis value were midway between 18 and 19. Calculate *W*. (*Answer*: *W* equals 1.5.)

c.  Now move the cut point down to the midpoint of 19 and 21—that is, 20—and recalculate *W*. (*Answer*: *W* equals 3.5.)

d.  Keep moving the cut point and recalculating *W* until *W* is 6 or greater. (*Answer*: When the cut point is at 22.5, *W* equals 11.) This is the lower limit of the 95% confidence interval.

e.  Keep moving the cut point and recalculating *W* until *W* is again less than 6. (*Answer*: When the cut point is at 128.5, *W* equals 1.5.) This is the upper limit of the 95% confidence interval. Thus a nonparametric 95% confidence interval for the median concentration is the interval 22.5 through 128.5 ppb.

# 9

# Drawing Inferences About Group Differences

## 9.1 Many scientific hypotheses can be stated in terms of group differences.

### Experiments versus observational studies

In broad terms, an *experiment* is a scientific activity in which procedures and conditions are controlled as carefully as possible so as to isolate the effect of some kind of deliberate manipulation. The key word is "controlled"—to perform an experiment, the scientist must have control over the conditions that might affect the results.

For example, if you wish to determine whether rats given saccharine face an increased likelihood of developing bladder cancer, you might design an experiment in which two groups of rats are used. One group of rats, called the *experimental* or *treatment group*, is given saccharine in their diets. The other group, called the *control group*, is treated in exactly the same way as the experimental group, except that the saccharine in their diet is replaced by some substance believed to be biologically inert. After a certain length of time, the rates of bladder cancer in the two groups are compared. Of course, there would probably be some difference in the rates just by chance, even if saccharine has no cancer-causing effect. But if the difference is greater than what might reasonably be expected to occur by chance, we infer that the difference must be due to the effect of saccharine, since the two groups were identical—supposedly—in every other respect.

What if we wished to find out if saccharine use is associated with an increased rate of bladder cancer in people rather than rats? Now a controlled experiment is out of the question. For obvious ethical, to say nothing of practical reasons, it would be impossible to assign people to "control" and "treatment" groups to study the effect of a potential carcinogen.

But even though we cannot *assign* people to such groups, we can *observe* the fact that some people have used saccharine and some people have not. If we can obtain information about samples of both kinds of people, we can then compare the sample bladder cancer rates. Again, even if the two populations (saccharine users and nonusers) were identical in their cancer rates, we would expect to see some difference in the *sample* rates simply by chance. However, if the difference in the sample rates is greater than we could reasonably attribute to chance, we conclude that the rates in the two populations differ.

It is at this point that a controlled experiment differs most from an observational study. Whereas in the case of the experiment, we inferred that saccharine must have caused the difference, we cannot make the same inference from the observational study. Why? Because in the observational study, *there may have been other systematic differences between the two groups besides use or nonuse of saccharine.* Perhaps people who choose to use saccharine would be more prone to bladder cancer, regardless of whether they actually use saccharine or not. (Lest this sound farfetched, consider the fact that people who choose to use saccharine sometimes do so because of other conditions, such as diabetes, which may in turn entail their own risks.) A well-designed observational study can control for this kind of problem to some extent, but the possibility always remains that the groups differ in some important way besides the "treatment" of interest.

Whether a study involves a true experiment, an observational study, or something in between, the statistical methods for comparing groups are essentially the same. It is the *interpretation* that differs.

## Differences in central tendency versus differences in variability

Most studies are concerned with detecting differences in *central tendency* among groups, and the methods we discuss in this chapter will largely be concerned with that problem. That is, we consider methods for testing hypotheses about whether certain populations differ in terms of some variable *on the average*.

But as we have discussed, central tendency does not tell the whole story. Some studies are more concerned with detecting differences in *variability* among groups. For example, suppose that we are comparing the sizes of parts produced by two manufacturing processes, one of which is substantially less costly. Even if the cheaper process produces parts that have the same *average* size, it is quite possible that the size could be more *variable* than that of the parts produced by the more expensive process. This would be of concern to the people responsible for quality control.

Another reason we may be concerned with comparing variability among groups is somewhat less direct. It turns out that some of the methods for comparing group *means* require the assumption that the populations being compared have equal *standard deviations*. Thus to use those methods we need to have some way of testing whether the assumption of equal group variability is tenable.

## 9.2 For two paired samples, the problem reduces to the one-sample case.

*Paired versus independent samples*

Suppose that you wish to study whether college students gain weight, on the average, during their first year on campus. (This phenomenon is known to believers as ''the freshman ten''—pounds, that is.) There are two ways in which you might gather sample data to address this question.

One way would be to obtain a random sample of college-bound high school seniors and another random sample of college students just completing their first year, and use one of the methods described later in this chapter to compare their average weights. The two samples would be *independent*, in the sense that the selection of people for one sample was in no way affected by the selection of people for the second sample.

The other approach would be to obtain a random sample of college-bound high school seniors, weigh them, wait until after their first year in college, and weigh them again. You still have two sets of weights, just as in the case of independent samples—but this time, the two sets are *paired* rather than independent, because each weight in the first set is associated with a weight in the second set, and vice versa.

How should we proceed with our paired data to compare the ''before'' and ''after'' average weights? We have already covered the necessary methods—all we have to do is apply a little trick. The trick is to *turn the two paired samples into a single sample of difference scores*. That is, instead of considering the ''before'' and ''after'' weight for each student, we simply consider the ''after minus before'' difference for each student.

Now we can use the methods of Chapter 7 to test our hypothesis. Our original null hypothesis that ''average weight does not change during the first year of college'' now becomes ''the average difference score equals zero.'' *Once we have changed the data into difference scores, we simply have one sample.*

## Advantages of paired designs

Which of the two designs would be better for determining whether average weight actually does increase during the freshman year? The independent-sample design has the advantage of allowing us to take a "snapshot"—that is, we could get all our data at one time, without having to wait a year as we would for the paired design. However, the paired design would almost certainly be more sensitive, all other things equal.

Why should the paired design be more sensitive? Because when we change the original "before" and "after" scores to difference scores, we are focusing on the data most pertinent to our question. In particular, with a paired design we can ignore the variability among individuals at each time point and concentrate on the variability of the difference scores. To the extent that people who start heavy tend to stay heavy (relative to other people), we will be reducing the variability in the statistic we are looking at, thereby increasing the power of the procedure.

## Other examples of paired designs

Besides "before" and "after" studies using the same subjects, there are other common situations in which paired designs are used. For example, many studies make use of identical twins, with one of each set of twins being placed in a control group, the other in a treatment group. Once the data are gathered, a difference score is calculated for each set of twins by subtracting, say, the control twin's score from the treatment twin's score. The difference scores can then be analyzed using the methods of Chapter 7.

Of particular scientific value are identical twins who were separated at birth. Since the members of each pair are genetically identical, any differences between them must be due to nongenetic factors—"nurture" rather than "nature." An international registry exists of such individuals, and it may be presumed that they are hounded by eager researchers as soon as their status becomes known. Studies involving them, of course, use a paired design.

Other examples include studies of husbands and wives, older and younger siblings, and animals born of the same litter. In each case, the key point is that there is some one-to-one correspondence between the members of the two groups. In a design with independent samples, there is no such correspondence.

## STUDY QUESTIONS—Sections 9.1 and 9.2

9.2.1   In an effort to control computer use, a computer center implemented a new set of rates charged to users. The director of the center wondered whether the new rate structure was having any impact on the average amount of central processing unit

(CPU) time used per user in a month. A random sample of 10 users was selected, and the amount of CPU time (in hours) used by each user was recorded for the month immediately before and immediately after the new rates went into effect. The results were as follows:

| User | CPU hours used during the month: | |
| | Before new rates | After new rates |
| --- | --- | --- |
| 1 | .37 | .20 |
| 2 | .90 | .71 |
| 3 | 15.83 | 15.60 |
| 4 | 1.26 | 2.90 |
| 5 | .44 | .10 |
| 6 | 3.19 | 2.62 |
| 7 | .23 | .18 |
| 8 | 106.12 | 103.55 |
| 9 | .85 | .61 |
| 10 | 2.60 | 1.94 |
| Mean | 13.18 | 12.84 |
| Std. dev. | 32.99 | 32.21 |

Mean difference score (before minus after): .34
Std. deviation of difference scores: 1.01
Std. error of mean difference score: $1.01/\sqrt{10} = .32$

a.  Are the two sets of monthly CPU times paired or independent?
b.  The director wishes to test the null hypothesis that the population average CPU time per user remained the same against the alternative that the average CPU time decreased. Let's restate these hypotheses in terms of difference scores. If the null hypothesis were true, what would the population average *difference* score be? What would it be if the alternative hypothesis were true?
c.  How many standard errors away from the null hypothesis average is the sample average difference score? In other words, what is the $t$ score for the sample average if the null hypothesis is true?
d.  Does this $t$ score lead you to reject the null hypothesis?

**9.2.2** [Continuation of Exercise 9.2.1]
a.  Now use the Wilcoxon signed rank test to test the same hypotheses as in Exercise 9.2.1 (in terms of the median instead of the mean). Recall the procedure: First find the difference scores, then rank them regardless of sign, then find the sum of the ranks for the positive and negative differences, and

finally use the table to determine if the $p$ value is less than, say, 5%. (Consult Section 8.3 to review the procedure, if necessary.) What conclusion do you reach?

    b.  Why does the $t$ test fail to reject the null hypothesis, whereas the Wilcoxon signed rank test rejects the null hypothesis in this case?

## 9.3   The two-sample $t$ test is used for problems involving independent samples.

### Statistics calculated from other statistics

Before we go into the details of the two-sample $t$ test, let's get our bearings. The statistic we will be looking at in the two-sample $t$ test is itself calculated from some other statistics. The same holds true of the standard error of the statistic. This fact tends to make the procedure for doing the test look complicated.

But it is really just the same kind of procedure we have already seen in the one-sample case: We figure out a $t$ score for a statistic. What is a $t$ score? It is the number of (estimated) standard errors a statistic is away from a null hypothesis value. All we have to do is figure out the statistic and its standard error, and we can do the test.

In the case of the two-sample $t$ test, the statistic is easy: It is the difference between the two sample means. Calculating the standard error of that difference is more involved, but once we have it, the test is simple. The same situation obtains in many different statistical procedures. There may be complicated calculations involved in getting the statistic or its standard error, but once that is done, the rest is easy. In these days of easy access to computers, even complicated calculations need not be intimidating.

### Assumptions required by the two-sample t test

Recall that for the one-sample $t$ test, we required that the original population from which the sample was drawn have (at least approximately) a normal distribution. If this assumption is not valid, we cannot be sure that the $t$ distribution would give the correct $p$ value, and other methods may be needed.

Similarly, the two-sample $t$ test requires some assumptions. One is directly analogous to the normality assumption for the one-sample $t$ test; the other two are meaningful only in the context of a problem involving more than one sample. The three assumptions are (1) independence, (2) normality, and (3) homogeneity. Let's consider each in turn.

The *independence* assumption relates to the design of the study: The score obtained by each individual must be unrelated to the score obtained by every other individual. This is necessary both between samples and within samples. The independence assumption can be a problem if there are relationships among individual scores.

The *normality* assumption is the same as that for the one-sample *t* test, extended to two populations: Both populations must have (at least approximately) normal distributions.

The *homogeneity* assumption refers to the degree of variability of the populations: The two populations must have the same standard deviation. As a practical matter, this means that the two sample standard deviations (which, after all, serve as estimates of the respective population standard deviations) should be pretty close to each other.

### Checking the assumptions

Draw pictures! Draw pictures! Draw pictures! What pictures? Histograms of the two samples! Good computer programs make it easy. Both histograms should be to the same scale, so you can compare them.

By looking at the histograms, you can get some idea of how well the normality and homogeneity assumptions are met. Of course, the assumptions are concerned with the populations, not the samples; but the samples are, we hope, sufficiently representative of the population to give us an idea of what the population distributions look like. If the samples are at least mound-shaped, the normality assumption is met for all practical purposes. If the samples seem to have a similar degree of spread, the homogeneity assumption is probably reasonable. If you are still in doubt, you can then use fancier techniques (some of which are described briefly in Appendix C) to check for normality and homogeneity. But the first thing to do is look at pictures.

When both samples are big enough, the normality assumption becomes less critical. (This is simply another manifestation of the fact that given enough data, the *t* test behaves as if it were a *z* test.) But the homogeneity assumption can be important even with large samples. Appendix C gives a brief description of some ways of dealing with the problem of lack of homogeneity.

How about the independence assumption? For independence, we must rely on the design and execution of the study. For example, if the two samples are paired, we clearly do not have independence, and we should analyze the data accordingly. Even if there is not explicit pairing, but still some degree of relationship among some of the individuals in the samples, we could have problems with independence. Randomization is a good way to guarantee independence of individuals, provided that it is performed correctly. Special methods exist for handling situations that fall between perfect independence and pairing. If you encounter such a situation, consult a statistician—preferably *before* you gather the data!

### Overview of the two-sample t test

Let's look at an example.

**Example 9.1** Do mothers of low-birth-weight infants show higher levels of stress than mothers of normal-birth-weight infants? Measures of stress during the year prior to delivery were obtained for 25 mothers whose infants had low birth weights and 30 mothers whose infants had normal birth weights. The two groups of mothers were each selected at random from among the mothers who delivered their babies at a certain hospital in a given year. The stress measure was a scale that assigned weighted points to certain "life events." Results for the two samples were as follows.

Sample 1   Stress scores for mothers of low-birth-weight infants:

| 103 | 133 | 112 | 148 | 114 | 143 | 103 | 102 | 101 | 96 |
|-----|-----|-----|-----|-----|-----|-----|-----|-----|-----|
| 126 | 132 | 116 | 90  | 106 | 122 | 143 | 129 | 112 | 102 |
| 81  | 107 | 105 | 96  | 102 |     |     |     |     |     |

Mean = 112.96     Standard deviation = 17.43     $n = 25$

Sample 2   Stress scores for mothers of normal-birth-weight infants:

| 103 | 107 | 81  | 104 | 58  | 127 | 108 | 71 | 127 | 75 |
|-----|-----|-----|-----|-----|-----|-----|----|-----|-----|
| 112 | 85  | 120 | 106 | 66  | 120 | 123 | 90 | 76  | 94 |
| 76  | 85  | 37  | 64  | 84  | 105 | 104 | 76 | 98  | 99 |

Mean = 92.70     Standard deviation = 22.08     $n = 30$

Histograms of the two samples are shown in Figure 9.1. Is there sufficient evidence to conclude that the population of mothers of low-birth-weight infants has a higher mean stress score than the population of mothers of normal-birth-weight infants?

*Step 1*:   *State the null and alternative hypotheses.* The null hypothesis is that there is no difference between the means of the two populations (or equivalently, that the mean of population 1 minus the mean of population 2 equals zero). The alternative is that the mean for population 1 is higher than the mean for population 2 (or equivalently, that the mean of population 1 minus the mean of population 2 is greater than zero). Symbolically:

*Null hypothesis*: $\mu_1 = \mu_2$, or $\mu_1 - \mu_2 = 0$
*Alternative hypothesis*: $\mu_1 > \mu_2$, or $\mu_1 - \mu_2 > 0$

*Step 2*:   *Check the assumptions.* Provided that we do not have more than one mother from the same family (or even two deliveries of the same mother) in our data to any great extent, and provided that the samples were selected randomly, we can be confident that the independence assumption is met. The two histograms both look mound-shaped and the standard deviations are close together, so we can be comfortable about the normality and homogeneity assumptions as well.

*Step 3*:   *Calculate the difference between sample means and the (estimated)*

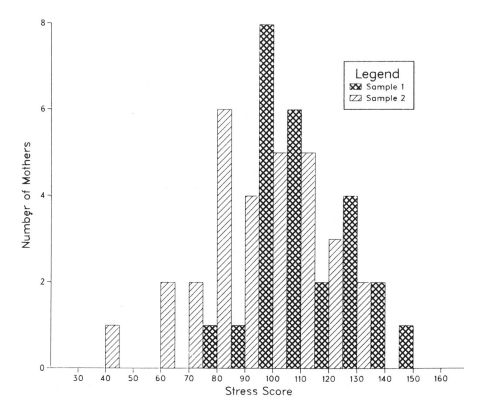

**Figure 9.1**   Stress scores for two samples of mothers of low-birth-weight infants.

*standard error of the difference.* For convenience, we calculate the difference as sample mean 1 minus sample mean 2, rather than the other way around, because it is easier to work with positive numbers than negative numbers: Sample mean 1 minus sample mean 2 = 112.96 − 92.70 = 20.26. Thus the mean stress score for the mothers in sample 1 exceeded that for mothers in sample 2 by 20.26 points.

To get the standard error of the difference, we would almost always use a computer these days. We discuss the details later in this section, because it is important to understand where the calculations come from. For now, we simply accept the result: The (estimated) standard error of the difference between the sample means works out to be 5.45 points.

Remember what the standard error indicates: If we could consider *all possible* differences between sample means and could calculate their standard deviation, we would have the standard error. We estimate that quantity to be 5.45. Loosely, we can say that we would expect a sample difference to be off (from the population difference) by 5.45 points, on the average.

**Step 4**: *Assume that the null hypothesis is true and find the t score for the difference between sample means.* Since the null hypothesis says that the difference between population means is zero, our *t* score will be the number of standard errors our sample difference is away from zero. How many standard errors is 20.26 away from zero? With a standard error of 5.45, 20.26 is 20.26 ÷ 5.45, or 3.72, standard errors away from zero. Thus, the *t* score is 3.72.

**Step 5**: *Find the p value.* We need the number of degrees of freedom (*df*) for the *t* score. *The way to figure df in this case is to add up the df contributed by each of the two samples.* Each sample contributes $n - 1$ *df*, where *n* is the sample size; thus we have 24 *df* from sample 1 and 29 *df* from sample 2. The total *df* for our *t* score is therefore 24 + 29, or 53.

Looking in the *t* table with 50 *df* (the closest entry), we find that the biggest entry is 2.6778, which cuts off .005 of the *t* distribution. Our sample *t* score was 3.72, which is even farther out than 2.6778. Thus the *p* value for our data is less than .005. That is, the probability of observing a difference between sample means as great as 20.26 points by chance, if in reality the population means were equal, is less than one-half of 1%.

**Step 6**: *State the conclusion.* Since the *p* value is very small, we reject the null hypothesis. We conclude that mothers of low-birth-weight infants do have a higher mean stress score than mothers of normal-birth-weight infants.

### Where the standard error comes from

Before we simply respond "from the computer!", let's consider where the standard error really does come from. We will discuss the ideas intuitively rather than mathematically; the formula will follow our discussion. First, let's reiterate what the standard error of any statistic means. The standard error of a statistic gives us an idea of how far off we would expect to be if we used the statistic to estimate the parameter of interest. A large standard error means that if we repeated our study many times, we would find that our statistic varies quite a bit.

What affects how much a statistic would vary in repeated studies? In general, two things: (1) the variability of the populations from which samples are drawn, and (2) the sample size. The more variable the populations, the more variable the statistic. The bigger the sample size, the less variable the statistic. We have already seen how both of these factors operate in the case of the standard error of the mean of a single sample.

In the case of the two-sample *t* test, the parameter of interest is the difference between two population means, and we are estimating that parameter by the difference between two sample means—one statistic derived from two. What determines how variable that one statistic should be? As you might expect, the variability of the one statistic will depend on the variability of each of the two statistics from which it was calculated. In fact, it turns out that the variance of the difference between two independent sample means is simply the *sum* of the variances of the two sample means. Each of the two sample means, in other words, contributes its own amount of uncertainty into the composite. (Recall that the ''variance'' is the square of the standard deviation of a distribution—in the case of a sampling distribution, the square of the standard error.)

To end up with the formula shown below, we take account of one other fact: We have assumed that the standard deviations of the two populations are equal. Since we have two samples from populations assumed to have equal standard deviations, the two sample standard deviations are two estimates of the same quantity, and it makes sense to combine them into one ''pooled'' estimate before we proceed. How? By taking a weighted average (actually of the variance)— where the weight for each sample is its *df*. The idea is that we want to give the bigger sample more weight in our estimate.

The net effect of this is the algebraic derivation shown below. If you are not algebraically skilled, it suffices to understand that the standard error of the difference comes from a process of summing the variabilities of the two sample means, taking into account the fact that the two population standard deviations are equal.

### Formula for the standard error of the difference between sample means

***Step 1*:**   We note first that the variance of the difference between sample means 1 and 2 (assuming independent samples) is equal to the sum of the variances of the two sample means. Since sample mean $\bar{X}_j$ has variance $\sigma_j^2/n_j$, we have

$$\sigma_{\bar{X}_1 - \bar{X}_2}^2 = \frac{\sigma_1^2}{n_1} + \frac{\sigma_2^2}{n_2}$$

***Step 2*:**   Next, we incorporate the assumption that the standard deviations of the two populations are equal; that is,

$$\sigma_1 = \sigma_2$$

We shall call this common standard deviation $\sigma$ with no subscript. We therefore have

$$\sigma_{\bar{X}_1 - \bar{X}_2}^2 = \frac{\sigma^2}{n_1} + \frac{\sigma^2}{n_2}$$

$$= \sigma^2 \left( \frac{1}{n_1} + \frac{1}{n_2} \right)$$

***Step 3***:   Since $\sigma^2$, the common population variance, is unknown, we need to estimate it from the data. We have two independent estimates of this quantity available: the two sample variances. To combine our two estimates into one, we take an average, weighting each estimate by its degrees of freedom. Thus we define our pooled estimate of the population variance as follows:

$$s^2_{\text{pooled}} \equiv \frac{(n_1 - 1)s_1^2 + (n_2 - 1)s_2^2}{(n_1 - 1) + (n_2 - 1)}$$

***Step 4***:   Having defined an estimate of $\sigma^2$, we now substitute it into the formula derived at step 2 to arrive at the formula for the estimated variance of the difference between sample means:

$$s^2_{\bar{X}_1 - \bar{X}_2} = s^2_{\text{pooled}}\left(\frac{1}{n_1} + \frac{1}{n_2}\right)$$

The square root of the estimated variance is the estimated standard error:

$$s_{\bar{X}_1 - \bar{X}_2} = s_{\text{pooled}}\sqrt{\frac{1}{n_1} + \frac{1}{n_2}}$$

## STUDY QUESTIONS—*Section 9.3*

**9.3.1**   The computer output (from Statpal) shown below comes from a study comparing fish caught in two different lakes. The variable WEIGHT indicates the weight of the fish.

   a.   Is there sufficient evidence to conclude that the mean weights of the fish that might have been caught differ between the lakes? (You may assume that the assumptions are valid.) Note that the output contains several statistical results that we have not yet discussed. Based on what we have discussed, however, you should be able to find the relevant numbers and report the results.

---

Statpal output for Exercise 9.3.1:

Dependent variable: WEIGHT      Weight of the fish
Grouping variable:  LAKE        Lake in which the fish was caught (of 2)

|  | Mean | Std. dev. | Minimum | Maximum | N |
|---|---|---|---|---|---|
| LAKE equals 1 | 210.6823 | 28.2373 | 158.7590 | 269.5990 | 35 |
| LAKE does not equal 1 | 231.1863 | 37.5687 | 172.2830 | 310.5380 | 40 |

*F* test for equality of variances:
   *F* score: 1.7701 with 39 and 34 *df*      Two-tailed significance = .093

Mean of group 1 minus mean of group 2: $-20.5039$

Pooled variance estimate $t$ test:
  Standard error of difference $= 7.7646$
  $t$ score $= -2.6407$ with 73 $df$      Two-tailed significance $= .010$

Separate variance estimate $t$ test:
  Standard error of difference $= 7.6201$
  $t$ score $= -2.6908$ with 71 $df$      Two-tailed significance $= .009$

Valid cases: 75      Missing cases: 0

---

b.  Using the numbers on the output for the means of the two groups and for the standard error of the difference between the means using the pooled variance estimate, confirm the $t$ score reported by Statpal. Then use the $t$ table to confirm the $p$ value reported by Statpal.

c.  Using the numbers on the output for the standard deviations of the two groups and the formula given in the text, confirm the standard error of the difference between the means reported by Statpal.

9.3.2   [Statpal exercise] Use Statpal to perform the $t$ test shown in Example 9.1. The steps are as follows:

a.  Create a Statpal system file with two variables: STRESS (the stress score) and GROUP (1 for the low-birth-weight group, 2 for the normal group). Note that data for both groups (i.e., all 55 mothers) go into the same system file; the variable GROUP will be used to distinguish the two samples.

b.  Use the SElect facility to select cases for which GROUP is 1; then use the HIstogram procedure to get a histogram of the sample. Then do the same for cases in the other group. Observe the general shape of the two histograms.

c.  Go into the TTest procedure, specifying a ''grouped'' (not a paired) $t$ test. The ''dependent'' variable is STRESS and the ''grouping'' variable is GROUP. The code for GROUP that defines group 1 is 1 (all others automatically being put into group 2).

d.  Interpret the results, confirming what was reported in the text. Note that Statpal reports a two-tailed significance level (i.e., $p$ value), while the problem requires a one-tailed significance level; simply divide the two-tailed $p$ value in half to get a one-tailed $p$ value.

9.3.3   [Statpal exercise—Coronary Care data set (see Appendix A)] Which patients are older on the average, men or women? Use the variables AGE and SEX (for which $1 =$ male, $2 =$ female) to perform a $t$ test of the null hypothesis that the mean age of patients is the same for the two sexes, versus a two-tailed alternative. Be sure to check the assumptions, at least informally.

9.3.4   [Statpal exercise—Electricity Consumer Questionnaire data set (see Appendix B)] Which group has a longer average membership in the cooperative, proponents of nuclear power or opponents? Use the variables YRSMEMBR (number of years of membership) and LIKENUKE (1 for proponents, 2 for opponents) to perform a $t$ test. Use a two-tailed alternative hypothesis—that is, test the null hypothesis that the groups have the same mean versus the alternative hypothesis that the means differ. Be sure to check the assumptions, at least informally.

## 9.4   The Wilcoxon rank sum test is another method for comparing two groups.

*A nonparametric approach to comparing two independent samples*

As the name implies, the *Wilcoxon rank sum test* uses ranks rather than the original data as the data on which the comparison is based. The test may be used when the two samples are *independent*, not paired. Since the data are converted to ranks, the normality assumption is not necessary for the Wilcoxon rank sum test; the only assumption necessary is that the two population distributions have (at least approximately) similar shapes. (If they do not, it may not be meaningful to test differences in central tendency anyway.)

The idea of the test is very simple. We start by arranging the scores from smallest to largest, forming one large list that includes all the scores from both samples. Then we assign ranks to the scores, with rank 1 for the smallest score. Next, we add up the ranks for the scores in each of the two samples. If the two samples came from populations with the same median (the null hypothesis), we would expect that the two samples would have rank sums roughly proportional to their sample sizes. That is, there would not be any tendency for one of the samples to have the large ranks and the other to have the small ranks—the scores from the two samples would be mingled on the ordered list. If the rank sums are very different from what we would expect under the null hypothesis, we reject the null hypothesis and conclude that the populations must have different medians. Let's go over the details by example.

**Example 9.2**   In an effort to improve traffic flow through a busy intersection, two different timing schemes are proposed for the traffic lights along the streets leading to the intersection. To compare the two schemes, researchers randomly assigned one or the other scheme to each of 11 days—5 days with scheme A and 6 days with scheme B—and recorded the amount of time required to traverse a standard route through the intersection at 5:00 P.M. each day. (Why 11 days rather than 10 or 12 with equal sample sizes? Probably because something went wrong on one of the days—perhaps the light malfunctioned, or the researcher's car broke down, or there was an accident along the route.)

The results were as follows:

Times required to traverse the route under scheme A (minutes):

   15   19   20   17   18

Times required to traverse the route under scheme B (minutes):

   23   20   21   24   19   22

Is there sufficient evidence to conclude that the median times for the two schemes differ?

*Step 1*:   *State the null and alternative hypothesis.* The null hypothesis is that the two population medians are equal. The alternative is that they differ. (Thus we have a two-tailed alternative in this example.)

*Step 2*:   *Arrange all the scores in order and assign ranks from smallest to largest.* The 11 scores are ranked as follows:

| Score: | 15 | 17 | 18 | 19 | 19 | 20 | 20 | 21 | 22 | 23 | 24 |
|--------|----|----|----|-----|-----|-----|-----|----|----|----|----|
| Sample: | A | A | A | (A and B) | (A and B) | B | B | B | B |
| Rank: | 1 | 2 | 3 | 4.5 | 4.5 | 6.5 | 6.5 | 8 | 9 | 10 | 11 |

Note that the tied scores are assigned the average of the ranks they occupy.

*Step 3*:   *Calculate W, the sum of the ranks for the smaller sample.* The smaller sample in this case is sample A (with $n = 5$ versus 6 for sample B), so we calculate the rank sum for sample A: $1 + 2 + 3 + 4.5 + 6.5 = 17$. So, for our example, $W$ equals 17.

If the samples had been of equal size, either rank sum could have been used. The smaller sample is used when the sample sizes differ because the table used for this test (discussed below) was prepared for the rank sum for the smaller sample. Of course, the rank sum for one sample gives the same information as that for the other sample, since the two rank sums necessarily add up to the total sum of the ranks from 1 to 11, which is 55.

*Step 4*:   *Find the p value.* If the null hypothesis were true, what would we expect for the rank sum of sample A? We would expect sample A to partake of about 5/11 of the total rank sum of 55, or 25. (Do you see why?) But our rank sum for sample A was actually 17. To find the *p* value, we need to find how likely it would be to observe a sample result as far away from 25 as 17 is if the null hypothesis were true.

To calculate this probability requires some fancy combinatorics. But fortunately, we have a table we can use (Table D.5). The table is arranged according to the sample sizes being used. In this case we look at the part of the table for sample sizes of 5 and 6.

Note that there are two columns of $W$ values given. The column headed ''L'' gives the values of $W$ that cut off a specified proportion of the lower tail of the sampling distribution; the column headed ''U'' gives values of $W$ that cut off the same proportion of the upper tail. For our sample sizes of 5 and 6, the numbers for a two-tailed *p* value of .05 are 18 and 42. Thus if the null hypothesis were true, there would be less than a 2.5% chance of observing a $W$ of 18 or less and less than a 2.5% chance of observing a $W$ of 42 or more. A $W$ of 18 or less, or 42 or more, translates into a *p* value (two-tailed) less than or equal to 5%. Of course, a one-tailed *p* value would simply be half the two-tailed *p* value. Since our value of $W$ was 17, our $p$ value (two-tailed) is indeed less than 5%. In fact, we see from the next line of the table that the $p$ value is approximately 2%.

*Step 5*:  *State the conclusion.* Since the $p$ value is less than 5%, we reject the null hypothesis and conclude that the two population medians do differ, apparently in favor of less time required under scheme A.

### One-tailed tests

Our example was for a two-tailed test—that is, we were looking for a difference in either direction. For a one-tailed test, we would look only for a difference in a particular direction. For example, if our alternative hypothesis had been that scheme A times have a smaller median than scheme B times, we would only have considered the lower tail of the sampling distribution, and our $p$ value would have been about .01 instead of about .02.

### The Wilcoxon rank sum test for large samples

Table D.5 of the Wilcoxon rank sum test covers sample sizes only up to 20 and 20. For larger samples, we again have a normal approximation available. The formulas for the mean and standard deviation of $W$ under the null hypothesis are as follows (Ott, 1977):

$$\text{mean of } W = \frac{n_1(n_1 + n_2 + 1)}{2}$$

standard error of $W$ (large sample)

$$= \sqrt{\frac{n_1 n_2}{12}\left[(n_1 + n_2 + 1) - \frac{\sum_{i=1}^{g} t_i(t_i^2 - 1)}{(n_1 + n_2)(n_1 + n_2 - 1)}\right]}$$

where

$n_j$ = number of scores in the $j$th sample, with the samples defined such that $n_1$ is less than or equal to $n_2$

$g$ = number of groups of tied scores

$t_i$ = number of scores in the $i$th group of ties

If there are no ties, the standard error formula simplifies to

$$\sqrt{\frac{1}{12}n_1 n_2(n_1 + n_2 + 1)}$$

### The Mann-Whitney U test

Some books and computer programs use a procedure called the Mann-Whitney U test in situations involving two independent samples. The Wilcoxon rank sum test and the Mann-Whitney U test are completely equivalent—you will always arrive at the same conclusion with one test as with the other.

STUDY QUESTIONS—*Section 9.4*

**9.4.1**  In a recent study of microcomputer use among elementary school children, researchers recorded the amount of time spent using a school microcomputer in a 2-day period for a class of nine boys and six girls. The results (in hours) were as follows:

Boys:

4.2   3.6   3.8   4.8   4.2   3.2   2.5   2.2   3.4
Mean: 3.54 hours        Standard deviation: .83 hour

Girls:

0.1   0.8   3.2   2.1   1.5   2.0
Mean: 1.62 hours        Standard deviation: 1.08 hours

Difference (mean for boys minus mean for girls):          1.92 hours
Standard error of the difference:                          .49 hour

The researchers wanted to know if there is sufficient evidence to conclude that girls use microcomputers less than boys, on the average.

a.   If the researchers simply wanted to know about the children in the particular class being studied, there would be no need for a statistical test. But the researchers want to make a broader statement. Is it reasonable to generalize from this one class? What issues must the researchers consider before they draw inferences about children in general based on the data from this one class?

b.   Now assume that it is reasonable to generalize, and state the null and alternative hypotheses.

c.   What assumptions are necessary for a two-sample *t* test? Do the assumptions seem reasonable? (*Note*: It will be difficult to check for normality with such small samples, but draw pictures anyway. Can you think of any independence problems that might occur in this kind of situation? For example, what if there is only one microcomputer available to the class?)

d.   Perform a two-sample *t* test, and state your conclusion.

e.   Perform a Wilcoxon rank sum test for the same data.

**9.4.2**  [Continuation of Exercise 9.4.1] Now suppose that we wanted to estimate the difference between the population mean for boys and the population mean for girls rather than simply testing whether the difference is zero. As long as the assumptions are reasonable, we can calculate a confidence interval for the difference between means, just as we did in Chapter 7 for a single mean.

Recall the reasoning: We know in advance that we have a 95% chance of getting a sample statistic within a certain number of standard errors of the parameter. Thus we are 95% confident that the parameter is within that number of standard errors of the sample statistic we actually got. How do we get the number of standard errors? From the *t* distribution with the appropriate *df*.

So a 95% confidence interval for the parameter is the interval *t* standard errors around the statistic, where "$t$" comes from the $t$ table. Let's apply the same reasoning to find a confidence interval for the difference between population means, using as our statistic the difference between sample means.

  a.  Using the $t$ table, find the $t$ score needed to encompass 95% of the area of the distribution (i.e., the score that cuts off 2.5% of the area from each end).

  b.  The difference between the sample means and the (estimated) standard error have already been calculated for you. Calculate a 95% confidence interval for the difference between population means.

  c.  Interpret the result. (That is, finish the sentence, "We are 95% confident that. . . .")

**9.4.3**  Statpal output for a Wilcoxon rank sum test using the same data as in Exercise 9.3.1 is shown below. Using the output, determine if the evidence leads to the conclusion that the median weights of the fish in the two lakes are different.

---

Statpal output for Exercise 9.4.3:

Dependent variable: WEIGHT        Weight of the fish
Grouping variable:  LAKE          Lake in which the fish was caught (of 2)

Group 1: LAKE equals 1
   Rank sum = 1113      $n = 35$

Group 2: LAKE does not equal 1
   Rank sum = 1737      $n = 40$

Test statistic $W = 1113$      Sig. (two-tailed, using normal approx.) = .0212
Valid cases: 75     Missing cases: 0

---

**9.4.4**  Five girls and four boys are ranked on a sociometric scale. Rank 1 was assigned to the least popular child, up to rank 9 for the most popular child. The results were as follows:

| Child: | A | B | C | D | E | F | G | H | I |
|--------|---|---|---|---|---|---|---|---|---|
| Sex:   | M | F | F | M | F | M | M | F | F |
| Rank:  | 3 | 5 | 2 | 1 | 9 | 4 | 6 | 8 | 7 |

Assuming that the children represent random samples of girls and boys, does the sample evidence point to differences in popularity between girls and boys?

**9.4.5**  [Statpal exercise—continuation of Exercise 9.4.4] Use Statpal to perform the test using the data in Exercise 9.4.4. To do this, create a Statpal system file with the two variables SEX (coded, say, 1 for boys and 2 for girls) and RANK. Then select the NOnparametric tests procedure, and choose RS (for the Wilcoxon rank sum test) from the nonparametric tests menu. Specify RANK as the "dependent" variable and SEX as the grouping variable, with code 1 defining group 1. Interpret the results.

**9.4.6**  [Statpal exercise] Use Statpal to perform the test shown in Example 9.2. To do this, create a Statpal system file with the two variables TIME and SCHEME (coded 1 for scheme A and 2 for scheme B). Then select the NOnparametric tests procedure, and choose RS (for the Wilcoxon rank sum test) from the nonparametric tests menu. Specify TIME as the "dependent" variable and SCHEME as the grouping variable, with code 1 defining group 1. Confirm that the results are as in the text description.

**9.4.7**  [Statpal exercise—Electricity Consumer Questionnaire data set (see Appendix B)] Use the variables YRSMEMBR (number of years of membership) and LIKE-NUKE (1 for proponents of nuclear power, 2 for opponents) to perform a Wilcoxon rank sum test of the null hypothesis that proponents and opponents have the same median length of membership against a two-tailed alternative.

# 10

# One-Way Analysis of Variance

## 10.1  Analysis of variance is a general method used in analyzing group differences.

### Analyzing variance to detect differences among means

Despite the name, *analysis of variance* is used to draw inferences about *means*, not variances. It is particularly useful as a way of determining if there are differences among the means of more than two populations: The null hypothesis is that the means of all populations are equal, and the alternative hypothesis is that at least one population mean differs. The term "analysis of variance" is commonly abbreviated "ANOVA."

### Partitioning variability

The basic idea of analysis of variance is to partition (i.e., divide up) the variability observed in the data into two parts: variability that can be accounted for by group membership, and variability that cannot. What do we mean by partitioning variability? There is both a mathematical and an intuitive way of looking at it. As usual, we concentrate on the intuitive way.

Let's consider a simple example. A certain room contains 400 schoolchildren. Figure 10.1 shows a histogram of the children's heights. Note that their heights range from about 95 cm (just over 3 ft) to 145 cm (about 4 ft 9 in.). The mean absolute deviation (MAD) of the heights is 7.45 cm—that is, the heights deviate from their mean, on the average, by 7.45 cm.

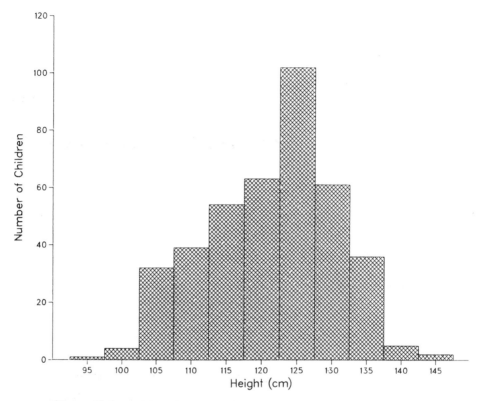

**Figure 10.1**   Heights of 400 children.

Why do the children's heights vary? In the absence of any other information, we cannot "explain" the variability. We simply chalk it up to individual variation, which occurs randomly.

But suppose that we are now told the age of each child, and it turns out that there are 100 children each of age 5, 6, 7, and 8. Figure 10.2 shows histograms for each of the four age groups. Notice that each of the four distributions is less spread out than the distribution of all the children together. For comparison, the MAD is 4.25 cm for the 5-year-olds, 4.47 cm for the 6-year-olds, 3.14 for the 7-year-olds, and 4.32 for the 8-year-olds, all considerably less than the MAD of 7.45 cm for the group as a whole.

Now we still cannot explain why the children in each age group vary in height. We would need to know a great deal more about the children to do that, and no matter how much we knew, given the state of scientific understanding of what regulates growth, there would still be some unexplained variability. Nevertheless, the amount of unexplained variability is less than it was before we knew

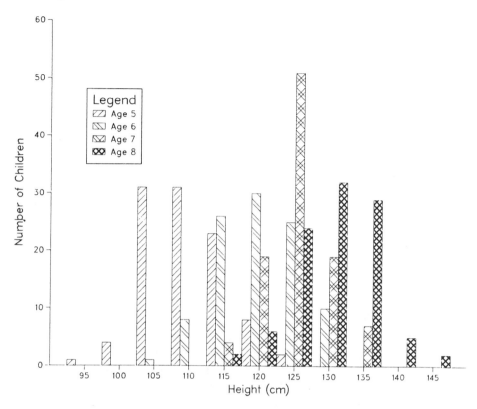

**Figure 10.2** Heights of 100 children each of ages 5 through 8.

about the children's ages. Why? Because part of the original variability in the children's heights is attributable to differences among the average heights of the age groups.

Of course, if the average heights of all four age groups had been about the same—as it might, for example, if we had been talking about fully grown adults rather than children—then knowing which age group a person was in would not have reduced the unexplained variability very much. Each of the four group distributions would have shown about the same amount of variability as the whole group did.

## Using ANOVA to test for differences among group means

Now let's consider the problem of statistical inference concerning the means of several populations. Suppose that we have a sample of scores from each of several populations, with the samples all being independent of each other. We wish to test the null hypothesis that the four population means are equal against the alternative that at least one population mean differs.

The logic of partitioning variance provides us with an approach to this problem. The idea is to look at the total amount of variability in the scores, without regard to group membership, and then see how much of it we can explain using group membership. Of course, in sample data there would be some differences among the sample means just by chance even if the null hypothesis were true. It would therefore look like we could explain some of the variability just by chance. But if the amount of variability explained by group membership is greater than we would expect to see by chance, we conclude that the null hypothesis is not plausible, and we reject it. It boils down to this: If the variability *within groups* is substantially less than the variability *overall*, we conclude that the group means are different.

## 10.2   One-way ANOVA can be used to test for differences among more than two groups.

### Overview of one-way ANOVA

One-way ANOVA is used to test for differences among group means. The "one-way" part of the name means that we are considering only one type of grouping for each subject—for example, age group. A two-way ANOVA would consider two different groupings, such as age and sex. We consider two- and higher-way ANOVA methods in Chapter 14.

The null hypothesis tested in a one-way ANOVA is that the means of several populations are equal. The alternative hypothesis is that at least one mean differs from the others. Note that the one-way ANOVA is inherently a two-tailed test, in the sense that we are looking for any type of difference, in any direction, among the population means.

If we have only two populations of interest, the one-way ANOVA is equivalent to the two-sample *t* test. In fact, when we have only two samples the *F* statistic used in the one-way ANOVA (discussed below) is simply the square of the *t* score from the two-sample *t* test. This suggests that the assumptions for the one-way ANOVA had better be the same as those for the two-sample *t* test—and indeed they are.

### Assumptions required for one-way ANOVA

Just as for the two-sample *t* test, three assumptions are required for one-way ANOVA. First, we require that all individual scores be *independent* of one another, both within and among samples. Second, we require that all populations have *normal* distributions. Third, we require that all populations have the same standard deviation (*homogeneity*).

Assumptions should be checked by looking at pictures of the sample histograms, by inspecting the sample standard deviations, and when in doubt, by applying one of the fancier techniques described briefly in Appendix C. Violations of the normality or homogeneity assumptions can sometimes be made less severe by transforming the data, using one of the methods described in Appendix C.

## Sums of squares

Before we get into it, note that this discussion is directed toward understanding rather than computational detail. Nowadays, we almost always use a packaged computer program to carry out the calculations, right down to the test statistic.

As we discussed in Section 10.1, the basic idea of ANOVA is to look at how much of the variability in the scores we can attribute to differences among the group means. If we can attribute a larger amount than would occur by chance if the null hypothesis were true, we conclude that the null hypothesis must be false. The first thing we need, then, is a way of measuring how much of the variability in the scores can be accounted for by group membership.

How do we measure the amount of variability that we can account for by group membership? We start with a simple index of variability called a *sum of squares*, which is short for *sum of squared deviations*. We have already seen one type of sum of squares: In calculating a standard deviation, we first found the sum of squared deviations from the mean, then divided by $n - 1$ to find the average squared deviation (also called the variance), then took the square root of the variance to give the standard deviation. The more variable the scores, the bigger the sum of squares (and hence the bigger the standard deviation).

If we consider the sum of squared deviations from the overall mean of all the scores, regardless of group membership, we have an index of the *total* variability in the scores. This index is called SST, for *sum of squares, total*. So SST measures variability about the overall mean. In mathematical notation,

$$\text{SST} \equiv \Sigma (X - \overline{X})^2$$

Now suppose that we sum the squared deviations from the group means. That is, we find each score's deviation from the mean of its own group, square it, and sum up all such squared deviations. By using the group means rather than the overall mean, we have taken group membership into consideration. The sum of squared deviations about these group means therefore reflects variability we still cannot account for using group membership. Since this sum of squares reflects variability within groups rather than overall variability, we call it SSW, for *sum of squares, within*. In notation,

$$\text{SSW} \equiv \Sigma (X - \overline{X}_j)^2$$

where $\overline{X}_j$ is the mean of group $j$.

Now we have SST, which measures total variability, and SSW, which measures variability that we cannot account for using group membership. So how much *can* we account for using group membership? The answer is the difference between SST and SSW—that is, SST minus SSW. Since this difference measures the amount of variability that can be explained by differences *between* groups, we call it SSB, for "*sum of squares, between*". (Of course, with more than two groups it should be "among," not "between"—but nobody says it that way!)

Let's look at what happens to SSB as group means become more and more different. First, suppose that all the group means are exactly equal—that is, the sample data shows no difference at all among the group means. Then all the group means would simply be the same as the overall mean, and SSW would therefore be precisely the same as SST. SSB, of course, would then be zero. We would have accounted for *none* of the variability using group membership.

Now suppose that the group means are different from each other, and moreover, that every score is exactly equal to the mean of its group. In this case, group membership would explain *all* the variability. SSW would be zero, and SSB would therefore equal SST.

Thus an SSB of zero indicates no differences among group means, whereas SSB equal to SST indicates that group membership explains all the variability. In real situations, of course, SSB falls somewhere between zero and SST. The ratio of SSB to SST—that is, SSB divided by SST—serves as a handy index of how much group membership explains. This index, called *eta-squared* or *R-squared* (depending on the context) is particularly important in a related set of techniques called "regression." We discuss it in more depth in Chapter 12.

### The test statistic used in one-way ANOVA

We now have a measure of how much variability we can account for using group membership: SSB. Now let's get back to the problem of testing the null hypothesis of equal population means against the alternative of at least one inequality.

Even if the null hypothesis were true, we would still expect to see some differences among the sample means just by chance. So we would expect to get an SSB greater than zero just by chance, even if the null hypothesis were true. However, if SSB is greater than we would expect to see by chance, we conclude that the population means must really be different. What we need, then, is a way of judging how likely a given SSB would be if the null hypothesis were true.

Instead of using SSB directly as the test statistic, we use a related statistic called $F$. We use $F$ rather than SSB because the sampling distribution of $F$ under the null hypothesis is known (provided that the assumptions mentioned above are valid). $F$ is related to SSB in monotonic fashion—that is, the bigger SSB is (for a given SST), the bigger $F$ is.

What is $F$? It is the ratio of two statistics, called *mean squares*. A mean square is a sum of squares divided by its degrees of freedom. (So, for example, a sample variance is a type of mean square.) For one-way ANOVA, $F$ is calculated as MSB (for *mean square, between*) divided by MSW (for *mean square, within*). MSB is SSB divided by its *df*, and MSW is SSW divided by its *df*.

The *df* for SSB is the number of groups minus 1. The *df* for SSW is a bit more complicated: It is the number of scores ($n$), minus the *df* for SSB, minus 1. A quick calculation shows that the *df* for SSW can be expressed as the number of scores minus the number of groups. We will not bother trying to explain the *df*, but notice that the *df* for SSB and the *df* for SSW add up to $n - 1$. What is $n - 1$? It is the *df* for SST. So just as SSB plus SSW equals SST, the *df* for SSB plus the *df* for SSW equals the *df* for SST. To summarize, if there are $k$ groups, then

$$MSB = \frac{SSB}{k - 1}$$

$$MSW = \frac{SSW}{n - k}$$

### The rationale behind the F statistic

So much for alphabet soup. Now let's consider the rationale of the $F$ statistic. As we mentioned above, $F$ equals MSB/MSW. Let's look at MSW first. MSW is the mean squared deviation of the scores from their group means. Since we are assuming that all populations have the same variance, MSW will simply be an estimate of that common population variance.

Now consider MSB. If the null hypothesis is true, variability among the sample means is occurring only by chance. The greater the population variance, the more variability we would expect to see in the sample means. It turns out that MSB is calculated so as to be another estimate of the common population variance if the null hypothesis is true.

So if the null hypothesis is true, both MSB and MSW are estimates of the same quantity (the common population variance). We would expect $F$ to be somewhere around 1 in this case. If, however, the null hypothesis is false, and there are in fact differences among the population means, we would expect MSB to be *larger* than MSW, because MSB will reflect the additional variability among group means. So values of $F$ greater than 1 will tend to support the alternative hypothesis.

How big must $F$ be before we will reject the null hypothesis? The answer, as usual, is that a table is available to tell us. In general, the smaller the sample sizes, the bigger an $F$ value we insist upon before we reject the null hypothesis. We discuss the use of the $F$ table in the context of an example.

**Example 10.1** Three groups of schoolchildren, all from the same school, are asked to report everything they have eaten in the past 24 hours. Group A consists of 32 children who live in single-family homes with vegetable gardens. Group B consists of 47 children who live in single-family homes with no vegetable garden. Group C consists of 59 children who live in multifamily buildings with no vegetable garden. The reports of the 138 children are analyzed to determine the quantity of various nutrients, together with the total number of calories, consumed during the 24 hours.

For total number of calories, the results were as follows:

| Group | $n$ | Mean | Std. dev. |
|-------|-----|---------|-----------|
| A | 32 | 1928.72 | 175.99 |
| B | 47 | 2146.55 | 149.09 |
| C | 59 | 1951.32 | 169.86 |

Figure 10.3 shows histograms for the three samples.

If we consider the three groups of children to be random samples from their respective populations, can we conclude that the three populations differ in their mean calorie intake?

*Step 1*: *State the null and alternative hypotheses.* The null hypothesis is that the three populations have the same mean calorie intake. The alternative hypothesis is that at least one of the populations differs from the others.

*Step 2*: *Check the assumptions.* The histograms give us no reason for concern about either the normality or the homogeneity assumptions; all three samples have normal-like distributions with about the same degree of spread. We need to give some thought to the independence assumption in this case, however. Children from the same family would be expected to have similar diets. If there are many sets of children from the same family represented in the data, we might be overstating our effective sample size. This could lead us to have greater confidence in our results than we should. One way around this problem would be to include only one score from each family in the sample, by averaging the scores for all children in any one family. We shall assume that this is not a problem for purposes of this example.

*Step 3*: *Compute and display the ANOVA table.* As we noted above, nowadays we would use a computer to do this. Here are the results:

| Source | $df$ | SS | MS | $F$ |
|--------|-----|-----------|-----------|-------|
| Between groups | 2 | 1,290,041 | 645,020.5 | 23.82 |
| Within groups | 135 | 3,655,973 | 27,081.3 | |
| Total | 137 | 4,946,014 | | |

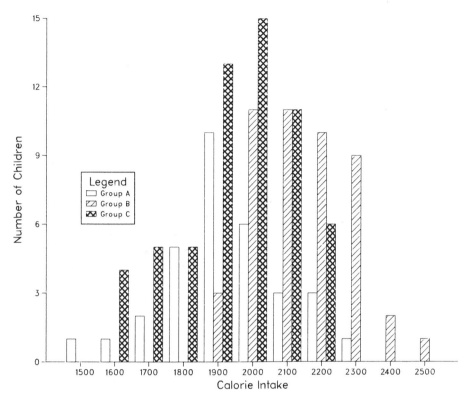

**Figure 10.3**   Calorie intake of three groups of children.

This tabular format, or something similar, is a standard way of reporting ANOVA results. Notice how it works: SSB is 1,290,041, SSW is 3,655,973, and so on. Actually, all you really need from the computer is any two of SSB, SSW and SST, and you could figure out the rest of the table using nothing more complicated than division.

***Step 4***:   *Find the p value.* At this point, we need to use the $F$ table (Table D.6). To use the $F$ table, we need not one but two numbers for degrees of freedom: a "numerator" $df$ and a "denominator" $df$. In this case the numerator $df$ equals 2 and the denominator $df$ equals 135. (Do you see why?)

Consulting the $F$ table with 2 and 120 $df$ (the closest denominator $df$ to 135), we find 4.77 in the row labeled ".01". This is the $F$ value needed for a $p$ value of 1%. Since our $F$ of 23.82 is much greater than 4.77, the $p$ value must be less than 1%.

***Step 5*:**   *State the conclusion.* Since the *p* value is less than 1%, we reject the null hypothesis. It appears that the caloric intakes of the three populations are not the same.

## Determining which populations are different from which

Note that the *F* test tells us only whether or not we reject the null hypothesis that all population means are equal. If we do reject the null hypothesis (as in the example), we still have not determined which groups are different from which.

In our example, the answer seems clear simply by looking at the histograms and sample means: groups A and C are pretty close together, while group B is decidedly different. But in many cases, the answer is not as clear. In such cases we can use special techniques for comparing individual groups with one another. If we have determined in advance which groups we will want to compare (rather than basing our decision on the data), we may use *a priori* comparison techniques. If, however, we have looked at the data and then decided to compare the groups that seem to be different, we must use *a posteriori* techniques. Several methods for comparing individual groups are briefly described in Appendix C.

## STUDY QUESTIONS—*Sections 10.1 and 10.2*

**10.2.1**   Use the *F* table to answer the following questions.
   a.   What value of *F* is needed for a *p* value of .05 with 3 and 20 *df*?
   b.   What value of *F* is needed for a *p* value of .01 with 1 and 60 *df*?
   c.   What is the probability of obtaining an *F* of 2.49 with 6 and 25 *df*?
   d.   What is the approximate probability of obtaining an *F* of 3.5 with 4 and 40 *df*?
   e.   What is the approximate probability of obtaining an *F* of 7.5 with 4 and 20 *df*?

**10.2.2**   [Statpal exercise] A wildlife biologist studying bird populations counted the number of birds of a certain species in a total of 19 plots each of 100 m². Each plot was classified as one of three types: clear-cut, uncut, and partially cut. The data were as follows:

| Clear-cut plots: | | 5 | 6 | 3 | 10 | 6 | 7 | |
|---|---|---|---|---|---|---|---|---|
| Uncut plots: | | 10 | 15 | 8 | 12 | 11 | 9 | |
| Partially cut plots: | 16 | 11 | 13 | 9 | 15 | 18 | 17 | |

Do the three types of plots differ significantly in the mean number of birds? Use Statpal to perform an analysis of variance, as follows:
   a.   Create a Statpal system file with two variables: NBIRDS (number of birds) and PLOTTYPE (coded 1, 2, or 3 for the three types of plots). (*Note*: Save this system file for a later exercise.)
   b.   Use the BReakdown procedure to get means and standard deviations of the bird counts for each of the three plot types.
   c.   Go into the ANalysis of variance procedure, specifying NBIRDS as the

dependent variable and PLOTTYPE as the grouping variable. The codes for the grouping variable are, of course, 1, 2, and 3.

d.   Interpret the results.

**10.2.3**   [Statpal exercise—Coronary Care data set (see Appendix A)] The variable RISK indicates which of three groups the patient is in: (1) low-risk, as evidenced by normal or unchanged ECG reading upon admission; (2) high-risk, as evidenced by abnormal or changed ECG; or (3) indeterminate. Do the three risk groups differ significantly in their mean age? Use BReakdown to find the mean ages and ANalysis of variance to perform the test. Interpret your results.

**10.2.4**   [Statpal exercise—Coronary Care data set (see Appendix A)]

a.   Use the ANalysis of variance procedure to determine if there is a significant difference in mean age between the two sexes. To do this, specify AGE as the dependent variable and SEX as the grouping variable with codes 1 and 2.

b.   Now use the TTest procedure to test the same hypothesis.

c.   Square the $t$ score (from the pooled variance estimate) you got in part b. What is the resulting number the same as?

**10.2.5**   Complete the following ANOVA table, derived from an analysis involving five groups with a total of 50 individuals.

| Source | SS | df | MS | F | p value |
|---|---|---|---|---|---|
| Between groups | _____ | ____ | _____ | _____ | _____ |
| Within groups | 18,975.20 | ____ | _____ | | |
| Total | 36,140.85 | ____ | | | |

**10.2.6**   A researcher with an interest in astrology wishes to determine if there are differences in personality among people born under different "sun signs." Of particular interest to the researcher is the idea that people born under some signs might show a greater tolerance for pain than people born under other signs.

A random sample of 10 people born under each of the 12 signs of the zodiac is selected (for a total of 120 people). For each subject, measures of pain tolerance were obtained on a standard measure, with higher scores representing a higher tolerance. The means and standard deviations of the 12 samples were as follows:

| Sign | Mean | Std. dev. | Sign | Mean | Std. dev. |
|---|---|---|---|---|---|
| Aries | 43.72 | 10.09 | Libra | 38.52 | 5.48 |
| Taurus | 45.64 | 12.06 | Scorpio | 37.16 | 10.06 |
| Gemini | 42.16 | 8.47 | Sagittarius | 41.04 | 7.63 |
| Cancer | 48.84 | 10.26 | Capricorn | 40.03 | 10.24 |
| Leo | 39.68 | 6.38 | Aquarius | 43.11 | 9.31 |
| Virgo | 43.66 | 9.73 | Pisces | 40.75 | 5.76 |

An ANOVA was performed on the data, with the following results:

| Source | df | SS | MS | F |
|---|---|---|---|---|
| Between groups | 11 | 1143.5 | 104.0 | 1.28 |
| Within groups | 108 | 8674.0 | 81.1 | |
| Total | 119 | 9907.5 | | |

a. Is there enough evidence to reject the null hypothesis that the mean pain tolerances of people in the 12 populations are equal? (You may assume that the required assumptions are satisfied.)

b. Having failed to reject the null hypothesis, the researcher nevertheless noticed that some of the group means seemed to be quite different. In particular, the means for Cancer and Scorpio seemed far apart. The researcher decided to do a two-sample $t$ test to test for a significant difference between Cancer and Scorpio. The standard error of the difference between sample means works out to be 4.55. Perform the $t$ test and state the conclusion.

c. Which conclusion do you believe? Why was the $t$ test incorrect in this situation?

d. Might it have made a difference if the researcher had hypothesized a difference between Cancer and Scorpio *before* looking at the data? Why?

## 10.3   The Kruskal-Wallis test is a nonparametric analog of one-way ANOVA.

### *Overview of the Kruskal-Wallis test*

Just as the Wilcoxon rank sum test is a nonparametric analog of the two-sample $t$ test, so the Kruskal-Wallis test is a nonparametric analog of one-way ANOVA. Like the Wilcoxon rank sum test, the Kruskal-Wallis test works with ranks instead of the original scores.

The Kruskal-Wallis test is used when we have *independent* samples from three or more populations and we wish to test the null hypothesis that the samples came from populations with identical distributions. The alternative hypothesis is that the samples come from populations that have different centers. As with the Wilcoxon rank sum test, we assume that the shape of the population distribution is the same for all populations.

The procedure starts by ranking all scores, regardless of group, from smallest to largest. Ties, as usual, are assigned the average rank for the places occupied. Next, the rank sum for each group is calculated. If each sample partakes of the total rank sum roughly in proportion to its sample size, we have support for the null hypothesis. If the rank sums deviate substantially from what we would expect to see under the null hypothesis, we reject the null hypothesis.

### The Kruskal-Wallis H statistic

As we see in the example below, the test statistic for the Kruskal-Wallis test, called H, is a function of the degree to which the rank sums deviate from what we would expect them to be under the null hypothesis. The greater the deviation, the bigger H is. So large values of H will lead us to reject the null hypothesis. All we need to know is how large a value of H is required for a given $p$ value. The formula for H is as follows (Siegel, 1956):

$$H = \frac{12}{n(n+1)} \sum_{j=1}^{k} \frac{R_j^2}{n_j}$$

where

$k$ = number of groups
$n_j$ = number of scores in the $j$th sample
$n$ = number of scores in all samples combined
$R_j$ = sum of the ranks in the $j$th sample

Although tables of H do exist for small sample sizes, we usually make use of a different table, which comes close enough to the distribution of H for practical purposes. This table, called the *chi-square table* (Table D.7), is of great importance in other contexts as well (some of which we discuss in Chapter 15). We can treat H as if it were a chi-square value—that is, we can use the chi-square table to find the $p$ value for a given value of H.

As with many other tables, to use the chi-square table we need to know the number of degrees of freedom that applies to the problem. For the Kruskal-Wallis test, the number of $df$ is the number of groups minus 1. Let's demonstrate the Kruskal-Wallis test through an example.

**Example 10.2** In a study of the effect of splenectomy (removal of the spleen) on resistance to bacterial infection, 20 mice were randomly assigned to one of four groups (five mice per group). Mice in group A had their entire spleen removed. Mice in group B had half their spleen removed. Mice in group C had a "sham" operation—that is, their spleen was exposed but then they were sewn up again without removing the spleen. Group D had no operation at all.

After allowing for the surgical wounds to heal, all mice were exposed to an aerosol suspension of *Streptococcus pneumoniae*, a bacterium that causes pneumonia. The survival times (to the nearest quarter-day) of the mice after exposure were recorded, with the following results:

Group A:  0.25   0.25   0.5    1.0    3.0
Group B:  2.0    2.5    4.0    4.5    6.0
Group C:  2.5    3.0    3.25   5.25   7.0
Group D:  1.75   2.25   4.5    5.0    5.75

Figure 10.4 shows a plot of the data. Is there enough evidence to conclude that the treatment populations vary in distributions of survival time?

***Step 1*:** *State the null and alternative hypotheses.* The null hypothesis is that the distributions of survival time for the four populations are the same. The alternative is that they are not the same.

***Step 2*:** *Convert the data to ranks and calculate H.* The ranks for the data are as follows:

Group A:  1.5    1.5    3      4      10.5     Rank sum = 20.5
Group B:  6      8.5    13     14.5   19       Rank sum = 61.0

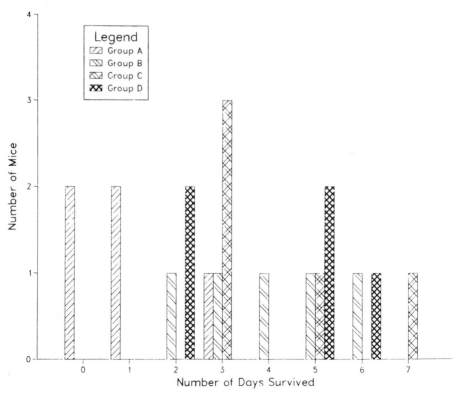

**Figure 10.4**   Survival times of four groups of mice.

| Group C: | 8.5 | 10.5 | 12 | 17 | 20 | Rank sum = 68.0 |
| Group D: | 5 | 7 | 14.5 | 16 | 18 | Rank sum = 60.5 |

Note that if the null hypothesis were true, we would expect the total rank sum (equal to 210, which is the sum of the integers from 1 through 20) to be equally divided among the four groups; thus we would expect each group's rank sum to be approximately $210 \div 4 = 52.5$. As you can verify, plugging these numbers into the formula for H yields 8.0029.

We noted above that a correction for ties is sometimes desirable if there are tied scores in the data (as above). Using Statpal, which reports both a corrected and an uncorrected value, the corrected H turns out to be equal to 8.0270, which is very close to the uncorrected value.

*Step 3*:   *Find the p value.* The chi-square table is organized in a way similar to the *t* table: the numbers for given *df* represent the chi-square values needed for certain *p* values. For this problem, we have 4 minus 1 = 3 *df*. As noted above, we treat H as if it were a chi-square value. With 3 *df*, a chi-square of 8.0029 is greater than the tabled value for ".05" (7.815) but less than that for ".01" (11.345). Therefore, the *p* value is between 5% and 1%.

*Step 4*:   *State the conclusion.* With a *p* value less than 5%, we reject the null hypothesis and conclude that the population survival time distributions differ. A look at the data shows us that splenectomy lowers survival time compared to half-splenectomy, sham operation, or no operation.

## STUDY QUESTIONS—*Section 10.3*

**10.3.1**   [Statpal exercise—Coronary Care data set (see Appendix A)] Perform a Kruskal-Wallis test to determine if the three groups defined by the variable RISK differ significantly in median AGE. Note that the specifications for the Kruskal-Wallis procedure (found within the NOnparametric tests procedure) are very similar to those for ANalysis of variance.

**10.3.2**   [Statpal exercise] Perform a Kruskal-Wallis test for the data in Exercise 10.2.2 to determine if the three plot types differ significantly in median bird count. Compare your results to those from the analysis of variance you performed in Exercise 10.2.2.

**10.3.3**   [Statpal exercise] Use Statpal to perform the Kruskal-Wallis test for the data in Example 10.2. To do this, create a Statpal system file with two variables: SURVTIME (survival time) and GROUP (coded 1, 2, 3, or 4). Then go into the NOnparametrics procedure, select the Kruskal-Wallis test, and obtain the results.

**10.3.4**   Each of the data sets below consists of three samples of data, which may be assumed to be independent of each other. For each data set, part of the ANOVA table and the Kruskal-Wallis H statistic are given. For each data set:

   a.   Sketch histograms of the samples, and consider the validity of the assumptions needed for ANOVA.

b. Perform a one-way ANOVA (which entails calculating the rest of the ANOVA table).
c. Perform the Kruskal-Wallis test.
d. Compare the results of the two tests.

Data set 1:

| Sample | A | B | C |
|--------|-------|-------|-------|
|        | 8.79  | 10.69 | 11.33 |
|        | 10.80 | 10.52 | 13.35 |
|        | 10.05 | 10.01 | 12.65 |
|        | 9.81  | 10.81 | 11.07 |
|        | 9.96  | 11.40 | 12.33 |
|        | 11.17 | 10.10 | 12.97 |
| Mean: | 10.10 | 10.59 | 12.28 |
| Std. dev. | .83 | .51 | .91 |

ANOVA table:

| | df | SS | MS | F |
|--------|----|--------|----|---|
| Between | | 15.793 | | |
| Within | | 8.880 | | |
| Total | 17 | | | |

Kruskal-Wallis H = 10.40

Data set 2:

| Sample | A | B | C |
|--------|-------|-------|--------|
|        | 8.79  | 10.69 | 11.33  |
|        | 10.80 | 10.52 | 195.00 |
|        | 10.05 | 10.01 | 12.65  |
|        | 9.81  | 10.81 | 11.07  |
|        | 9.96  | 11.40 | 12.33  |
|        | 11.17 | 10.10 | 12.97  |
| Mean: | 10.10 | 10.59 | 42.56 |
| Std. dev. | .83 | .51 | 74.69 |

ANOVA table:

| | df | SS | MS | F |
|--------|----|------------|----|---|
| Between | | 4,152.165 | | |
| Within | | 27,893.669 | | |
| Total | | | | |

Kruskal-Wallis H = 10.40

Data set 3:

| Sample | A | B | C |
|--------|-------|-------|-------|
|        | 10.27 | 10.71 | 11.44 |
|        | 9.80  | 9.66  | 12.61 |
|        | 11.70 | 11.20 | 10.54 |
|        | 8.59  | 10.86 | 11.73 |
|        | 11.85 | 10.42 | 11.63 |
|        | 7.53  | 10.87 | 12.11 |
| Mean: | 9.96 | 10.62 | 11.68 |
| Std. dev. | 1.70 | .53 | .69 |

ANOVA table:

| | df | SS | MS | F |
|--------|----|-------|----|---|
| Between | | 9.03 | | |
| Within | | | | |
| Total | | 27.37 | | |

Kruskal-Wallis H = 5.56

# 11

# Describing Relationships Between Two Variables

## 11.1 A scatterplot shows the shape of a relationship between two variables.

### Relationships between variables

Before we discuss ways of describing relationships between variables, we need to consider why we should bother. The answer is the same as it was when we considered group differences on a single variable: Many scientific hypotheses can be stated in terms of the relationship between two variables. In fact, the issue of group differences can itself be considered in terms of a relationship between two variables and some of the same techniques apply, as we will see.

What do we mean by a "relationship" between variables? *To say that two variables are related means that knowledge of an individual's score on one variable changes our best guess about the individual's score on the other variable.*

For example, yield of corn per hectare and amount of rainfall during the growing season are presumably related to each other. If we knew nothing else about a given farm (besides the fact that corn was planted there), our best guess about the yield we might expect from that farm would simply be the average corn yield. But if we were given information about the rainfall at the farm, our guess would probably be affected. If we were told that there is almost no rain at the farm, then, in the absence of irrigation, we would expect a very low yield. If we were told that there is too much rain, we would similarly expect a low yield. If we were told that the amount of rain is just right for corn, we would expect a high yield.

Figure 11.1 shows a graphical representation of this type of relationship. By drawing such a picture, we can characterize the "shape" of a relationship. In this case it is curvilinear: High corn yields are associated with moderate rainfall, with lower yields for very high or very low rainfall.

### Scatterplots

Of course, Figure 11.1 is unrealistic. Even though rainfall undoubtedly does affect corn yield, it is not the only factor. Many other things matter also. Knowing the amount of rainfall would change our best guess about corn yield, but we still would not be able to predict corn yield *perfectly*.

Figure 11.2 is a more realistic representation of the relationship. In Figure 11.2, we are representing some (hypothetical) data for a sample of farms. *Each*

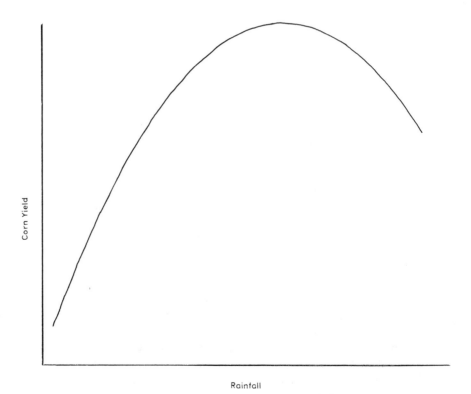

**Figure 11.1**   Curvilinear relationship.

*dot on the figure represents a farm.* We can still see the curvilinear relationship between the two variables, but we see also that it is not perfect. For any given amount of rainfall, there is variability in corn yield (and vice versa).

A drawing such as Figure 11.2, with one variable on each axis and *each individual represented by a dot*, is called a *scatterplot* (also known as a *scatter diagram* or *scattergram*). Whenever you look at a scatterplot, you should make sure you understand what the dots represent.

## Measuring degree of relationship

How strong is the relationship shown in Figure 11.2? We can see that the relationship is not perfect, but it is clear that knowledge of rainfall does modify our best guess of corn yield to some extent. How might we measure the extent?

Intuitively, we can judge the strength of the relationship in terms of the degree to which the points in the scatterplot adhere to some sort of line—in this case a curved line. If the points show very little deviation from the line, we would feel

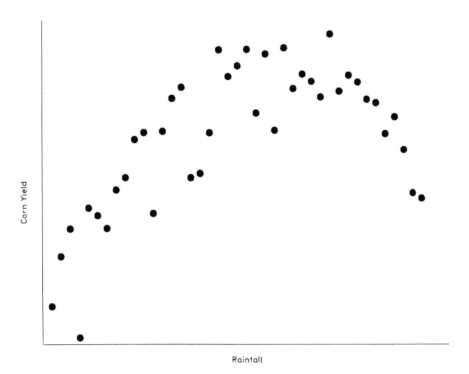

**Figure 11.2**  Scatterplot showing a curvilinear relationship.

that the relationship is strong. If the points scatter widely from the line, we would feel that the relationship is weak.

The line represents our view of the underlying relationship between the variables—a "model" of how the variables relate. Of course, how closely the points adhere to the line will depend very much on what type of line we draw. A straight line, for example, would not fit the data in Figure 11.2 very well at all, and we would judge the variables to have a weak relationship in terms of a straight-line model. This illustrates an important (and often forgotten) fact: The strength of the relationship between two variables must be judged in the context of a particular model, which we must specify. If we specify the wrong model, measures of relationship are not useful.

### Linear models

Many studies involve variables that we would expect to have, at least approximately, a straight-line relationship. In fact, even variables that have a curvilinear relationship can often be transformed into variables that have a straight-line relationship. Straight-line models are one type of a general category of model called *linear models*.

Linear models are important because of their simplicity. After all, a straight line is very easy to interpret: As one variable changes, the other tends to change in proportion. A linear relationship is simple to specify mathematically and simple to represent graphically. Even when there is a curvilinear relationship, a straight line is often a good approximation. In this chapter we concentrate on methods for describing and drawing inferences about a linear relationship between two variables.

## 11.2   Easy-*r* is a simple measure of the strength of a monotonic relationship.

### Measuring the strength of a linear relationship

Is the number of hours of television a child watches related to aggressive behavior? Suppose that we obtain a random sample of 100 children. For each child we find out how much television the child watches in an average week and obtain an aggressiveness rating from the child's teacher (on a 1 to 10 scale). Figure 11.3 is a scatterplot of the resulting data. (Make sure that you understand what each point represents!)

If we postulate a linear model relating aggressiveness and time spent watching television, to what extent do the data support the model? What we need is a way of evaluating the strength of the linear relationship between the two variables.

First, let's consider two extreme cases. Suppose that our model were perfectly accurate in the sense that aggressiveness and television time are exactly propor-

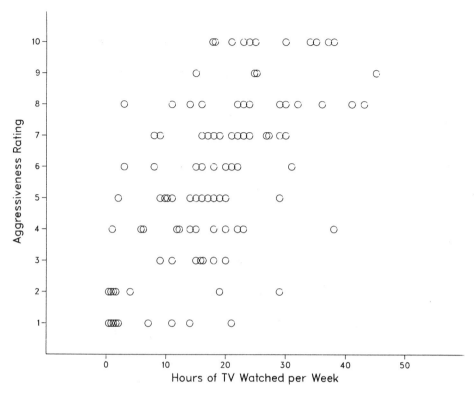

**Figure 11.3**   Hours of TV watched versus aggressiveness rating in a sample of 100 children.

tional to each other. Figure 11.4 shows what the data would look like if this were the case: the points on the scatterplot form a straight line, with no point deviating from the line at all. Children with any given amount of television watching have one specific aggressiveness score, and vice versa.

Now look at Figure 11.5. In this scatterplot there is no apparent relationship at all between the two variables. The average aggressiveness score seems to be relatively constant no matter what the amount of television watched, and vice versa. The points on the plot are scattered about in no particular pattern.

Clearly, Figure 11.3 shows a relationship somewhere between the two extremes. Although the points do not form a perfectly straight line, there is a definite upward drift in the points as we move from left to right. So although the points do not adhere as closely to a line as in Figure 11.4, neither do they deviate as widely from linearity as in Figure 11.5. How can we measure the strength of the relationship?

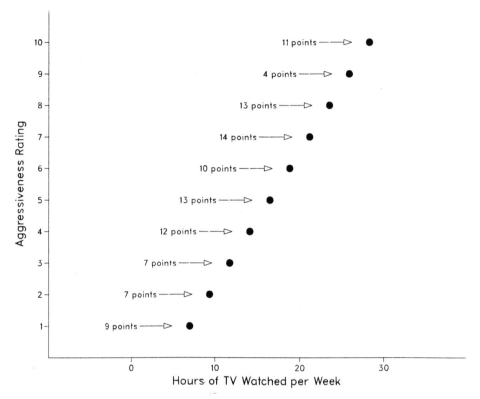

**Figure 11.4**   Perfect linear relationship.

## A simple index of relationship

The index most frequently used to measure the strength of a linear relationship is the correlation coefficient, usually denoted $r$. But before we discuss the correlation coefficient, let's consider a simpler measure that has a similar rationale. We will call it *easy-r*.

Suppose we divide the scatterplot of our TV/aggressiveness data into four quadrants, as shown in Figure 11.6. The four quadrants are formed by dividing the plot at the *median* of both variables. Thus points in quadrant I represent individuals above the median on both variables. Points in quadrant III represent individuals below the median on both variables. Points in quadrants II and IV represent individuals who are above the median on one variable but below it on the other.

Now let's consider how the points would be distributed among the four quadrants for various degrees of linear relationship. First, let's suppose that the

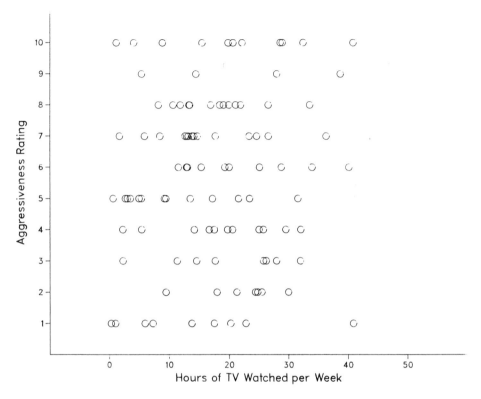

**Figure 11.5**  No linear relationship.

relationship were perfectly linear—the points form a straight line. If the relationship were positive (i.e., high scores on one variable are associated with high scores on the other variable), as in Figure 11.4, all the points would be in quadrants I and III. If the relationship were negative, all the points would be in quadrants II and IV. Either way, all the points will be in diagonally opposite quadrants. If, however, there were little or no linear relationship between the variables, as in Figure 11.5, the points would be distributed in all four quadrants with roughly equal frequency.

Suppose that we assign a score to every point in the plot, depending on which quadrant the point is in. We will assign a score of $+1$ to points in quadrants I and III, indicating positive relationship, and a score of $-1$ to points in quadrants II and IV, indicating negative relationship. Points that fall right on a dividing line (indicating a score right on the median for one or the other variable) we will simply ignore.

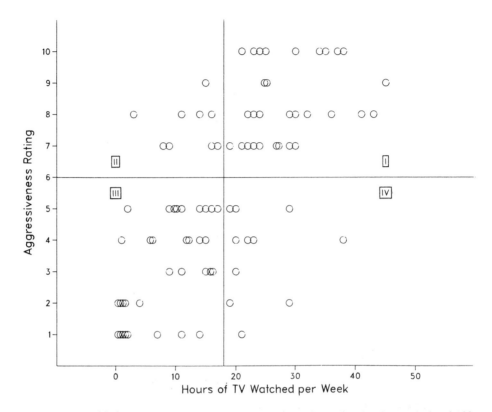

**Figure 11.6** Hours of TV watched versus aggressiveness rating in a sample of 100 children (omitting points on quadrant boundaries).

Now we add up the scores for all the points. As you can verify, there are 30 points in quadrant I, 9 in quadrant II, 34 in quadrant III, and 11 in quadrant IV. (We are ignoring a total of 16 points that fall on a dividing line.) The sum of the scores in the positive quadrants is $30 + 34 = 64$, and in the negative quadrants $(-9) + (-11) = -20$. So the sum of all the scores is $64 + (-20) = 44$.

Of course, the number 44 is hard to interpret in isolation. But if we express it as a proportion of the total number of points included in the sum, we will have an interpretable quantity. Since we included 84 points in the sum, our proportion is $44 \div 84$, or .524.

This proportion is easy-*r*. Thus *easy-r* is defined as the sum of all the "quadrant scores" (either $+1$ or $-1$ for each point) divided by the total number of points in all the quadrants. (Points on a boundary are simply ignored.) So what is easy-*r*? It is the average quadrant score.

Another way to think of it is in terms of the proportions of scores in positive versus negative quadrants. For our example, the proportion of scores in the positive quadrants exceeded the proportion in the negative quadrants by .524, or 52.4%. For comparison, let's find the corresponding easy-*r* values for the data in Figures 11.4 and 11.5. The same plots with the quadrants superimposed are shown in Figures 11.7 and 11.8, respectively.

For the perfect straight-line relationship (Figure 11.7), *all* the points are in the positive quadrants (except for 10 points that fall exactly on the median of both variables). So the sum of the scores is $90 - 0 = 90$, and easy-*r* is $90 \div 90 = 1.00$. Of course, if the line had sloped downward instead of upward, all the points would have been in the negative quadrants, so easy-*r* would have been $-1.00$.

For the very weak relationship (Figure 11.8), the points are distributed among the quadrants with roughly equal frequency: 20, 22, 24 and 24 in quadrants I

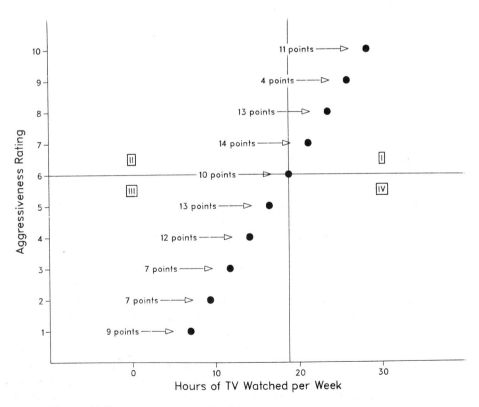

**Figure 11.7**   Perfect linear relationship.

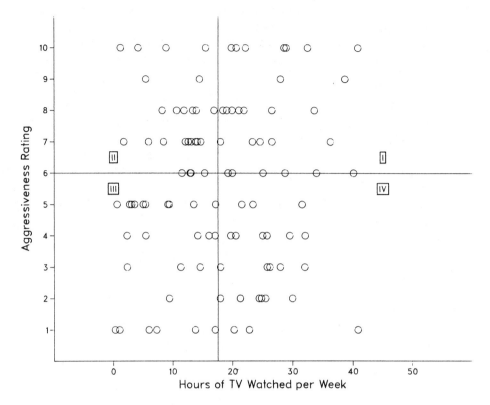

**Figure 11.8**   No linear relationship.

through IV, respectively, with 10 points on the dividing line. The sum works out to be $44 + (-46) = -2$, giving an easy-$r$ of $-2 \div 90 = -.022$. This is close to zero—there is almost no tendency for points to be in one type of quadrant more than the other.

So we have derived a simple index of relationship that ranges from $-1.00$ to $+1.00$. A perfect relationship gives an easy-$r$ of $+1.00$ (for a positive relationship) or $-1.00$ (for a negative relationship). A relationship with no particular pattern gives an easy-$r$ close to zero. Our TV/aggressiveness data gave an easy-$r$ of $+.524$—about halfway to perfect.

We should note that a easy-$r$ of $+1.00$ or $-1.00$ does not necessarily indicate that the points fall on a straight line. Any arrangement of points that puts all the points into two diagonally opposite quadrants will give a easy-$r$ of plus or minus 1.00. This will be true of any *monotonic* relationship (one in which the direction of the relationship stays the same throughout the range of scores), of which linearity is simply one example. (It will also be true of many nonmonotonic relationships, as long as the points stay in diagonally opposite quadrants.) Nevertheless, its simplicity of computation and interpretation make easy-$r$ a useful descriptive measure.

Moreover, easy-$r$, as its name implies, is in fact a simplified version of $r$, the correlation coefficient. In fact, easy-$r$ *is* the correlation coefficient if you transform the original variables into dichotomous variables—that is, variables that can have only two possible scores, like $-1$ and 1—using a median split. This implies that the techniques we will be presenting to draw inferences using $r$ are also valid using easy-$r$. It implies also that you can use any standard computer package that calculates $r$ to calculate easy-$r$. So if you understand how easy-$r$ works, you are most of the way to understanding how $r$ works.

## STUDY QUESTIONS—*Sections 11.1 and 11.2*

**11.2.1**   Fifteen subjects are questioned about their attitude toward a particular product. The response (on a scale of 1 to 20, with 20 being the most favorable) for each subject and the subject's age were as follows:

| Subject | Age | Response | | Subject | Age | Response |
|---------|-----|----------|--|---------|-----|----------|
| 1 | 52 | 24 | | 9 | 22 | 9 |
| 2 | 49 | 17 | | 10 | 44 | 18 |
| 3 | 58 | 37 | | 11 | 38 | 19 |
| 4 | 26 | 19 | | 12 | 27 | 11 |
| 5 | 14 | 8 | | 13 | 31 | 13 |
| 6 | 32 | 14 | | 14 | 41 | 27 |
| 7 | 47 | 18 | | 15 | 34 | 11 |
| 8 | 33 | 18 | | | | |

a.   Draw a scatterplot for this set of data.
b.   Find the median age and median response, and draw lines on the scatterplot dividing the plot into four quadrants at the median.
c.   Calculate easy-$r$ and describe the relationship between age and response in this set of data.

**11.2.2**   [Statpal exercise]
a.   Use Statpal to prepare the scatterplot for the data of Exercise 11.2.1. To do this, create a Statpal system file containing two variables: AGE and RESPONSE. Then use the SCatterplot procedure. (*Note:* Save this Statpal system file for use in later exercises.)
b.   Use the HIstogram procedure to help you obtain the median on each of the two variables. Then draw lines on the scatterplot to divide it into quadrants based on the medians.

c. Go into the TRansformations utility and create three new variables as follows.

    1. Create a variable called AGELOHI such that cases with scores below the median age are assigned a score of $-1$, cases with scores equal to the median AGE are assigned a score of 0, and cases with scores above the median age are assigned a score of $+1$. Statements to do this are

        AGELOHI $= -1$
        IF (AGE EQ median) AGELOHI $= 0$
        IF (AGE GT median) AGELOHI $= 1$

      where "median" is the median AGE you found above.

    2. Do the same kind of operation to create a new variable called RESPLOHI based on the variable RESPONSE.

    3. Then form a new variable called QUADSCOR by multiplying the variables AGELOHI and RESPLOHI, using the statement

        QUADSCOR $=$ AGELOHI $*$ RESPLOHI

d. Before proceeding, what does the QUADSCOR indicate in terms of the scatterplot? (Think about what happens when you multiply $-1$ times $-1$, $-1$ times 0, and so on.)

e. Use the SElect facility to select cases with QUADSCOR not equal to zero. Then use the HIstogram procedure to find out how many scores of $-1$ and $+1$ there are on the variable QUADSCOR. Based on this result, calculate easy-$r$.

f. Use the DEscriptive statistics procedure to obtain the mean of the variable QUADSCOR, and confirm that this is equivalent to your answer for easy-$r$ obtained in part e. (Leave case selection in effect.)

g. Now use the TAble procedure, specifying the variables AGELOHI and RESPLOHI. Confirm that you could have calculated easy-$r$ from these results without the necessity of creating the variable QUADSCOR.

**11.2.3** [Statpal exercise—Coronary Care data set (see Appendix A)]

a. Use the SCatterplot procedure to obtain a scatterplot of NDAYS (the number of days hospitalized) versus AGE. (Save this scatterplot for later exercises.)

b. How would you characterize the relationship between the two variables in this set of data?

c. Calculate easy-$r$ for this set of data. The HIstogram facility will help you obtain the medians of the two variables. To calculate easy-$r$, you may count up points by hand on the scatterplot, or let Statpal calculate quadrant scores using the tricks described in Exercise 11.2.2.

d. What does easy-$r$ say about the relationship between age and number of days hospitalized in this set of data?

**11.2.4** [Statpal exercise—Electricity Consumer Questionnaire data set (see Appendix B)]

a. Use the SCatterplot procedure to obtain a scatterplot of YRSMEMBR (number of years of membership) versus LIKENUKE (1 for proponents of nuclear power, 2 for opponents). Put YRSMEMBR on the vertical axis and LIKENUKE on the horizontal axis.

    b.   Calculate easy-*r* for this set of data, with one variation in procedure: For LIKENUKE, which has only two values, simply draw the line between the two values (even though one of them is the median).

    c.   Now find the median years of membership for each of the two groups as defined by LIKENUKE. You can do this any of at least three ways: (1) directly from the plot, by counting; (2) using the SElect facility followed by the HIstogram procedure, to give you the distribution of YRSMEMBR for each of the two groups defined by LIKENUKE separately; or (3) using the TAble procedure to give you the two distributions in the same table.

    d.   Put a mark on the scatterplot in the middle of the group of proponents of nuclear power, indicating their median years of membership. Do the same for the opponents of nuclear power.

    e.   Draw a line connecting the two medians. Calculate the slope of this line as the change in YRSMEMBR divided by the change in LIKENUKE (which is 1 unit). What does this slope represent?

## 11.3   The correlation coefficient tells how well a straight line describes the plot.

### The correlation coefficient

The *correlation coefficient*, also known as the *Pearson product-moment correlation coefficient* and usually denoted *r*, measures the degree to which two variables are related *linearly*. Thus whereas easy-*r* attains its limits of $+1$ or $-1$ when the relationship is generally monotonic, whether curved or not, *r* attains its limits (also $+1$ or $-1$) only when the relationship is precisely a straight line.

How does *r* manage to be sensitive to linearity rather than simply to monotonicity? The trick is that *r* considers not merely *which* quadrant a point is in but also *where* the point is in the quadrant. This is accomplished by assigning each point a score that reflects how far the point is from the means of the two variables. The average of these scores is *r*, just as the average of the ''quadrant scores'' of $+1$ or $-1$ was easy-*r*.

The score assigned to each point is the product of the distances to the means of the two variables, measured in units of standard deviation. That is, for each point, we change the scores on the two variables to *z* scores and multiply the two resulting *z* scores together. The average of these products, over all the points, is *r*. So *r* is defined as the average *cross-product* of *z* scores on the two variables. We look at some numerical examples in a study question, to show how the computations work. For now, let's consider how *r* changes for various degrees of linearity.

### What r does and does not measure

When the relationship is perfectly linear, *r* will be either $+1$ or $-1$, depending on the direction of the relationship. So *r* for the data plotted in Figure 11.4 is

+1.00, since the line slopes upward. If the line sloped downward, $r$ would be −1.00. Either way, we would characterize the relationship as perfectly linear.

When there is no particular pattern to the points, $r$ will be near zero. For the data plotted in Figure 11.5, $r$ works out to be .085, which is indeed close to zero. Our TV/aggressiveness data, plotted in Figure 11.3, yield an $r$ of .605, between no linear relationship and a perfect linear relationship.

One thing $r$ does *not* indicate is the steepness of the linear relationship. For example, the two scatterplots shown in Figure 11.9 both give an $r$ of −1.00, since in both cases the relationship is perfectly linear with the line sloping downward. The fact that one line is steeper than the other makes no difference in the calculation of $r$.

In fact, the units in which the variables are measured makes no difference in the calculation of $r$ either. If we were measuring TV time in minutes instead of hours, the value of $r$ would still be the same—as indeed it should if $r$ is to be a useful index of relationship. As long as any change in units involves adding or

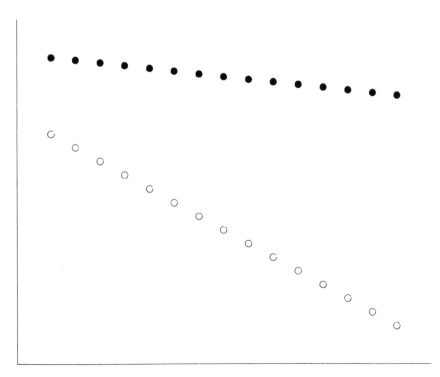

**Figure 11.9**  Two perfect negative relationships ($r = -1.00$).

subtracting a constant from all scores and/or multiplying all scores by a constant, the value of $r$ will not be affected. (However, transformations that change the scores in "nonlinear" fashion—such as taking the square root or the logarithm of the scores—*do* change $r$.)

## Interpreting the correlation coefficient

What can we say about the relationship between TV time and aggressiveness score? We noted above that $r$ equals .605. How strong a relationship does this represent?

With easy-$r$ we could interpret the value in terms of the proportions of individuals in the positive versus negative quadrants. The value of .524 for easy-$r$ meant that the proportion of individuals in the positive quadrants exceeded that in the negative quadrants by .524, or 52.4% of all the points.

With $r$, the interpretation is not quite as simple. Our value of .605 means that the average cross-product of $z$ scores is .605—but that really does not tell us much. Suppose, for example, that another variable—say, video-game playing time—had a correlation with aggressiveness of .300. Can we legitimately say that an $r$ of .605 represents a degree of linearity about twice as strong as an $r$ of .300?

The answer is that we cannot. The definition of $r$ does not support such an interpretation. However, there is a related quantity that does allow us to compare degree of correlation in ratio terms. That quantity is the *square* of $r$. The quantity $r$-squared may be interpreted as the proportion of variability in one variable that can be accounted for by a straight-line relationship with the other variable. We discuss this idea more fully in Chapter 12.

What this means is that we can compare two relationships in terms of the ratio of their $r$-squared values. So for our example, $r$-squared for the relationship between TV time and aggressiveness was .605 squared, or .366, while $r$-squared for the relationship between video-game time and aggressiveness was .300 squared, or .090. The linear relationship between TV time and aggressiveness is not merely twice as strong as that between video-game time and aggressiveness, but over four times as strong.

What about judging the strength of a linear relationship in absolute rather than relative terms? Is a $r$ of .300 (or an $r$-squared of .090) indicative of a strong relationship or a weak one? The answer depends largely on the context of the study. In the social sciences, data tend to be "fuzzy," and being able to explain .090 (i.e., 9%) of the variability in one variable using the other might be a triumph of methodology. On the other hand, if Sir Isaac Newton had said he could explain 9% of the variability in the speed of dropped objects using the square of the time in flight, it is unlikely that he would have been knighted. Different disciplines call for different judgments of the strength of relationship represented by a given $r$.

STUDY QUESTIONS—*Section 11.3*

**11.3.1**  Some psychologists have observed that birth order (whether a child is first-born, or second-born, etc.) seems to be related to IQ scores. A teacher is interested in examining the relationship between birth order and IQ among the 10 children in her fifth-grade class. The children's birth order and IQ scores were as follows:

| Child:      | 1   | 2   | 3   | 4  | 5  | 6   | 7   | 8   | 9   | 10  |
|-------------|-----|-----|-----|----|----|-----|-----|-----|-----|-----|
| Birth order:| 1   | 2   | 2   | 3  | 3  | 6   | 2   | 4   | 3   | 1   |
| IQ:         | 110 | 105 | 120 | 97 | 90 | 103 | 118 | 102 | 114 | 108 |

a. Make a scatterplot of the data. Put IQ on the vertical axis and birth order on the horizontal axis.

b. Calculate easy-*r* for the data. Recall the steps: first divide up the scatterplot by drawing lines at the medians (which are 2.5 for birth order and 106.5 for IQ, as you should verify). Then assign the points in the positive quadrants a score of $+1$, and the points in the negative quadrants a score of $-1$. Finally, average the scores. You should get the answer $-.60$.

c. Now let's calculate *r* for the data. To do this, we first convert the scores to *z* scores, then multiply together the two *z* scores for each point to form cross-products, and finally average the products. (Note that this is *not* the most efficient way to go about it, but usually we let a computer do the calculations.)

   To get the *z* scores, we need the mean and standard deviation of each variable. The mean birth order was 2.7, with a standard deviation of 1.49. The mean IQ was 106.7, with a standard deviation of 9.32. The following table shows the necessary calculations. Make sure that you see where each number comes from. In particular, make sure that you see how the *z* scores were calculated.

|       |             |     | z scores for: |       |            |
|-------|-------------|-----|---------------|-------|------------|
| Child | Birth order | IQ  | Birth order   | IQ    | Product of z scores |
| 1     | 1           | 110 | $-1.14$       | .35   | $-.3990$   |
| 2     | 2           | 105 | $-.47$        | $-.18$| .0846      |
| 3     | 2           | 120 | $-.47$        | 1.43  | $-.6721$   |
| 4     | 3           | 97  | .20           | $-1.04$| $-.2080$  |
| 5     | 3           | 90  | .20           | $-1.79$| $-.3580$  |
| 6     | 6           | 103 | 2.21          | $-.40$| $-.8840$   |
| 7     | 2           | 118 | $-.47$        | 1.21  | $-.5687$   |
| 8     | 4           | 102 | .87           | $-.50$| $-.4350$   |
| 9     | 3           | 114 | .20           | .78   | .1560      |
| 10    | 1           | 108 | $-1.14$       | .14   | $-.1596$   |

Sum of products: $-3.4438$

So the sum of the cross-products of $z$ scores is $-3.44$. Now to average, we could divide by 10, giving a correlation coefficient of $-.344$. However, just as in the case of the standard deviation, when we are working with sample data we divide not by $n$ but by $n - 1$. So if we want to consider our class of 10 children to be a sample from some larger population, we would calculate $r$ as $-3.44 \div 9$, or $-.382$.

    d.  Characterize the apparent relationship between birth order and IQ.

    e.  Why was there such a discrepancy between easy-$r$ and $r$ for these data?

    f.  Suppose that child 6 had had an IQ of 80 instead of 103. How would easy-$r$ have been affected? How do you think $r$ would have been affected? (*Note*: If we make the change, $r$ becomes $-.731$. Why does it go up so much?)

**11.3.2**  [Continuation of Exercise 11.3.1] We have discussed the calculation and interpretation of $r$, but not the problem of making *inferences* concerning a population correlation coefficient using sample data. Suppose that we consider our class of 10 children to be a random sample from the population of all fifth-grade children. (The fact that this is patently ridiculous does not stop many researchers from making similar assumptions.)

    We wish to test the null hypothesis that the population correlation coefficient equals zero versus the alternative that it is not equal to zero. Of course, even if the null hypothesis were true, $r$ calculated from a *sample* will differ from zero just by chance. (This reasoning should have a familiar ring by now.) But if our sample $r$ is farther from zero than could reasonably be attributed to chance, we will reject the null hypothesis. What we need to know, then, is the sampling distribution of $r$ under the null hypothesis.

    a.  What would you expect the sampling distribution of $r$ to look like, for large enough samples? Why? (*Hint*: $r$ is a type of average.)

    b.  What should the standard error of $r$ depend on? (Curiously enough, it does not depend directly on the variability of the population on either variable.) In other words, what should determine how variable $r$ would be if we repeatedly chose samples of children, found their birth orders and IQs, and calculated $r$?

    c.  What type of test, with what assumption, do you suppose we will use to draw inferences using $r$ with a small sample?

**11.3.3**  [Statpal exercise] Use the COrrelation procedure to calculate $r$ for the data of Exercise 11.2.1. Interpret its value and compare it to the value of easy-$r$ you calculated previously.

**11.3.4**  [Statpal exercise—Electricity Consumer Questionnaire data set (see Appendix B)]

    a.  Use the COrrelation procedure to calculate $r$ between the variables YRS-MEMBR and LIKENUKE. (Refer to the scatterplot you made in Exercise 11.2.4.)

    b.  Is this a meaningful number? (Surprisingly enough, it is!) What would a value of $+1$ indicate? A value of $-1$? A value of zero?

    c.  Now use the SCatterplot procedure to plot CONSTYPE (consumer type, where $1 = $ residential, $2 = $ farm, $3 = $ industrial) versus YRSMEMBR. Then calculate $r$ for these two variables (using the COrrelation procedure). Is $r$

meaningful here as an index of relationship? If you are not sure whether this is a meaningful number, consider what would happen if you had used a different coding scheme for CONSTYPE—say, if you had switched the codes for farm and industrial. Would this change *r*?

d.  Under what conditions can *r* be meaningfully used with "nominal" variables—that is, variables for which possible values do not have an inherent ordering?

e.  What might be an appropriate technique for determining the degree to which consumer type is related to years of membership?

**11.3.5**  [Statpal exercise—Coronary Care data set (see Appendix A)]

a.  Use the COrrelation procedure to calculate *r* between the variables NDAYS and AGE. Compare this result to the value for easy-*r* you calculated in Exercise 11.2.3.

b.  Use SCatterplot and COrrelation to obtain a scatterplot and calculate *r* between NDAYS and NCCU (the number of days in the coronary care unit).

c.  Now use the COrrelation procedure, specifying all three of the variables NDAYS, AGE, and NCCU. Notice how the output contains all three pairwise correlations among the three variables.

## 11.4  A form of *t* test can be used to draw inferences about correlation coefficients.

### *Drawing inferences about the population correlation coefficient*

Returning to our TV/aggressiveness example, we noted that the correlation between TV time and aggressiveness score, as measured by *r*, was .605. This represents a fairly strong relationship for this type of data. In the sample of 100 children, there is a tendency for children who watch more TV to have higher aggressiveness scores.

But suppose that we wish to draw an inference about the population from which the children were selected. What can we say about the *population* correlation coefficient, based on the sample value of *r*? In particular, can we infer from the sample correlation that there is indeed some degree of linear relationship between TV time and aggressiveness in the population as a whole?

The question calls for a hypothesis test. The null hypothesis is that the population correlation coefficient equals zero. The alternative hypothesis could be either one-tailed or two-tailed (a decision that needs to be made, of course, before we look at the data). If we are interested in detecting a relationship in either direction, we would specify a two-tailed alternative: the population correlation coefficient is different from zero (in either direction). If we are only interested in determining whether TV tends to increase (but not decrease) aggressiveness, we would specify a one-tailed alternative: The population correlation coefficient is greater than zero. The population correlation coefficient is usually denoted by the Greek lowercase letter rho ($\rho$), in accord with the usual convention of using Greek letters for parameters and Latin letters for statistics. So our null hypothesis is that $\rho$ equals zero.

### The t test for a correlation coefficient

Even if $\rho$ is zero, a *sample* correlation coefficient, $r$, will differ from zero just by chance. If, however, $r$ is farther from zero than we could reasonably attribute to chance, we conclude that $\rho$ must not be zero, and we reject the null hypothesis. What we need, then, is a way to determine how likely a given value of $r$ would be if the null hypothesis were true—that is, the null hypothesis sampling distribution of $r$.

Since $r$ is a kind of average, it should come as no surprise that for large enough samples, the sampling distribution of $r$ when the null hypothesis is true is approximately normal—another manifestation of the central limit theorem. Since our calculation of $r$ requires that we estimate the standard deviations of the two variables (in order to calculate $z$ scores), we are faced with the same type of problem we encountered when we were trying to draw inferences about a population mean: There is no general test procedure to use when the population standard deviations are unknown.

However, just as in the case of the mean, W. S. Gosset (''Student'') has provided us with a way to handle the problem—provided that we can make one assumption. The assumption is that at least one of the variables has a normal distribution. As long as this is a reasonable assumption, we can apply a form of a $t$ test to correlation coefficients.

The formula for the $t$ score used in this situation is as follows:

$$t = \frac{r\sqrt{n-2}}{\sqrt{1-r^2}}$$

The formula requires only two numbers: $r$ (the sample correlation coefficient) and $n$ (the sample size). As a look at the formula should convince you, the bigger the sample size, the bigger the $t$, for a given $r$. This means that the bigger the sample size, the less likely it would be to observe a large value of $r$ if $\rho$ were really zero. To put it another way, $r$ from a bigger sample should be a better estimate of $\rho$ than $r$ from a smaller sample.

To test the null hypothesis that $\rho$ equals zero, we could use the formula for $t$ and look up the value in the $t$ table. The number of degrees of freedom in this case is $n - 2$. If our value of $t$ is bigger (in absolute value) than the $t$ value from the table for a given $p$ value, we would reject the null hypothesis. However, we do not need to bother calculating the $t$ score, since a special table is available for determining the $p$ value directly from $r$. Let's use the table to determine if our $r$ of .605 for the TV/aggressiveness data should lead us to reject the null hypothesis.

## Using the r table

The *r* table (Table D.8) gives the values of *r* needed for *p* values of .10, .05, and .01 (two-tailed), or .05, .025, and .005 (one-tailed). To use the table, we first find the line of the table corresponding to the sample size (in the column headed "*n*").

For our example, we look in the line for *n* = 100 and find that .603 is larger—indeed, much larger—than the *r* for a *p* value (two-tailed) of .01, which is .256. In other words, the chance of observing an *r* as far away from zero as .256, if ρ is really zero, is about 1%. We have observed an *r* much farther out than that—so the chance is even smaller. Thus the *p* value of less than 1% (two-tailed).

We would therefore reject the null hypothesis and conclude that ρ is not zero. On the basis of our (hypothetical) data, we would conclude that TV time is correlated with aggressiveness score, with more TV time being associated with higher aggressiveness scores.

## A handy approximation

As we have noted, a *p* value of .05 is a very frequently used cutoff for determining if the null hypothesis should be rejected. There is a very simple way of approximating the value of *r* needed for a *p* value of .05: *An r of 2 divided by the square root of n is roughly what we need for a two-tailed p value of .05.* You can easily check this approximation using the *r* table. Here are some examples:

|        | r needed for $p = .05$ (two-tailed) | |
| :---: | :---: | :---: |
| *n* | Actual value (from table) | Approximate value (using $2/\sqrt{n}$) |
| 9   | .666 | .666 |
| 16  | .497 | .500 |
| 25  | .396 | .400 |
| 36  | .329 | .333 |
| 100 | .197 | .200 |

Note that the approximation is exact to two decimal places in all these examples.

Why should this approximation work? Essentially, because the standard error of *r*, if ρ is really zero, is about $1/\sqrt{n}$. That is where the square root of *n* comes from. Two standard errors about zero encompass about 95% of a normal distribution. That is where the 2 comes from. It works!

## Testing null hypotheses that ρ is something other than zero

The method we have just described, using the *r* table, can only be used for testing the null hypothesis that ρ equals zero. This is usually what we want to test, since a $\rho$ of zero means no linear relationship between the variables. If we reject the null hypothesis that $\rho$ is zero, we are saying that the sample evidence convinces us that there *is* some linear relationship between the variables.

We have not, however, addressed the issue of how strong that relationship might be. *If we wish to test the null hypothesis that $\rho$ is equal to some constant other than zero, we cannot use the r table.* Instead, we can use a special transformation (often called *Fisher's z transformation*) to change *r* into a z score. Appendix C contains the details.

### STUDY QUESTIONS—*Section 11.4*

**11.4.1**  Use the *r* table to find the value of *r* needed for
   a.  Two-tailed *p* value = .05, *n* = 12
   b.  Two-tailed *p* value = .05, *n* = 30
   c.  Two-tailed *p* value = .02, *n* = 12
   d.  Two-tailed *p* value = .02, *n* = 30
   e.  One-tailed *p* value = .025, *n* = 12
   f.  One-tailed *p* value = .01, *n* = 30

**11.4.2**  A sample of 20 cases yields a value for *r* of .40.
   a.  Find the approximate *p* value for a two-tailed test.
   b.  Does the sample evidence lead to rejecting the null hypothesis that the population correlation coefficient equals zero?

**11.4.3**  Compare the approximation $2/\sqrt{n}$ to the tabled *r* value for a two-tailed *p* value of .05 for
   a.  *n* = 4
   b.  *n* = 20
   c.  *n* = 50
   d.  *n* = 75

**11.4.4**  [Statpal exercise] In Exercise 11.3.3 you used the COrrelation procedure to calculate *r* between the two variables AGE and RESPONSE for the set of data presented in Exercise 11.2.1.
   a.  Use the *r* table to confirm the significance level given in the output from Statpal's COrrelation procedure.
   b.  What would the *p* value be for a two-tailed test rather than a one-tailed test as performed by Statpal?

**11.4.5**  [Statpal exercise—Coronary Care data set (see Appendix A)] Determine if there is a significant correlation between AGE and NDAYS. Write a sentence interpreting the results.

**11.4.6**  [Statpal exercise—Coronary Care data set (see Appendix A)]
   a.  Use Statpal's COrrelation procedure to determine if there is a significant relationship between the variables HADMI (1 if the patient had had a myocardial infarction, 2 if not) and NDAYS (number of days hospitalized). Interpret the results.
   b.  Use Statpal's TTest procedure to perform a grouped *t* test of the null hypothesis that the mean number of days hospitalized did not differ between

patients who had had an MI and patients who had not had an MI. Write a sentence summarizing the results.

c. Compare the $p$ values reported for the correlation coefficient in part a and the (pooled variance estimate) $t$ score in part b. How do they differ?

**11.4.7**  [Statpal exercise—Electricity Consumer Questionnaire data set (see Appendix B)] Is there a significant relationship between years of membership in the organization (as indicated by the variable YRSMEMBR) and rating of overall electric service (as measured by the variable RATESRVC)? In answering this question, consider the shape of the relationship as well as the correlation coefficient.

**11.4.8**  In Exercise 11.3.1 we considered the relationship between birth order and IQ, based on evidence from a sample of 10 children. Recall that $r$ for the sample was $-.382$. Easy-$r$ was $-.600$.

a. Let's consider our sample of 10 children to be a random sample from some population. Suppose that we wish to test the null hypothesis that $\rho$ equals zero. What assumption do we require? Based on the data, does the assumption seem to be reasonable?

b. Using the $r$ table, determine if the sample evidence is strong enough to conclude that $\rho$ differs from zero. (Do a two-tailed test. Note that we ignore the sign of $r$ in comparing it to the tabled value.) State the $p$ value.

**11.4.9**  [Continuation of Exercise 11.4.8] How about easy-$r$? We noted that easy-$r$ is itself a correlation coefficient—it is the correlation between the transformed scores (transformed, that is, from the original scores into $-1$ or $+1$ depending on whether the score is below or above the median).

a. When we were using $r$, the null hypothesis was that the population value of $r$ equals zero, which meant that there is no linear relationship between the variables. What would the null hypothesis be using easy-$r$, and what would it mean about the relationship between the variables?

b. Use the $r$ table to test whether our easy-$r$ of $-.600$ is sufficiently far from zero to allow us to reject the null hypothesis.

c. Another way to test the same hypothesis is to do a sign test. How would this work? Perform the test and state your conclusion. (*Hint*: If the null hypothesis were true, we would expect about half of the quadrant scores to be plus and half to be minus. We observed 2 minuses and 8 pluses.)

d. Which conclusion do you believe? Consider in particular the validity of any assumptions you needed to make in doing the $t$ test (using the $r$ table) versus using the sign test.

## 11.5  Spearman's rho and Kendall's tau are nonparametric measures of correlation.

### Nonparametric correlation coefficients

We have already considered one form of nonparametric correlation: easy-$r$. In this section we consider two more powerful nonparametric measures of correlation. In all three measures, information in the original scores is sacrificed in exchange for a test that is usable even when the assumption of normality is not met. Spearman's rho and Kendall's tau use more of the information than does easy-$r$ and are therefore more powerful. On the other hand, they are more difficult to compute, although with a computer program available this is no problem.

Note, incidentally, that Spearman's rho is a statistic, not a parameter, in spite of its Greek-letter name. It is not at all the same as $\rho$ used to mean the population correlation coefficient. The notational confusion is unfortunate, but both uses are conventional.

## Spearman's rho

We can describe Spearman's rho very simply: *Spearman's rho is r applied to ranks*. That is, to calculate Spearman's rho, we first change the scores on each variable to ranks, from lowest (l) to highest (n), and then calculate $r$ using the ranks.

Since Spearman's rho is itself a version of $r$, it ranges from $-1.00$ to $+1.00$. A perfect $+1$ or $-1$ indicates that a scatterplot of the *ranks* would form a straight line. This will be true whenever the original scores have a perfect monotonic relationship of any kind (linear or curved). Thus Spearman's rho, like easy-$r$, is sensitive not merely to linearity in the original scores but to any kind of monotonicity. Easy-$r$, however, can have a value of $+1$ or $-1$ even if the relationship is not monotonic, while Spearman's rho can attain $+1$ or $-1$ *only* for a perfectly monotonic relationship.

Inferences about the population value of Spearman's rho can be made using special tables for the purpose, but for $n$ greater than 10 the values of Spearman's rho needed for a particular $p$ value are very close to the values given in the $r$ table. The nice thing about Spearman's rho is that we need make no assumptions about the shape of the population distribution of the original scores—we threw them away in favor of ranks, which always have the same distribution (one each of the numbers 1 through $n$, barring ties).

Spearman's rho for the TV/aggressiveness data of Section 7.2 is .611, which is quite close to the $r$ value of .605. Situations in which we might expect a big difference between $r$ and Spearman's rho are those in which some extreme points—*outliers* is the term usually used—are affecting the value of $r$. Spearman's rho, since it works only with ranks, is not affected by how far out a point is.

A shortcut formula for calculating Spearman's rho is as follows:

$$\text{Spearman's rho} = 1 - \frac{6\sum_{i=1}^{n} d_i^2}{n(n2 - 1)}$$

where $d_i$ represents the difference in ranks for individual $i$ on the two variables— that is, the rank for individual $i$ on the first variable minus the rank for individual $i$ on the second variable.

## Kendall's tau

*Kendall's tau* is another index of correlation that varies from $-1$ to $+1$. It, too, is sensitive to monotonicity of any kind, and like Spearman's rho, it works with ranks instead of the original scores. It differs from Spearman's rho in that Kendall's tau considers the degree to which pairs of points are ''concordant'' or ''discordant.''

A pair of points is said to be *concordant* if the ordering of the two points on the first variable is the same as that on the second variable. For example, consider two children in the TV/aggressiveness sample. Suppose that child 1 watches more TV than child 2. If child 1 also has a higher aggressiveness score than child 2, the pair of children is said to be concordant. If child 1 has a lower aggressiveness score than child 2, the pair is said to be *discordant*.

Kendall's tau is calculated by looking at all possible pairs of points and counting up how many are concordant and how many are discordant. The difference between the proportion of concordant and the proportion of discordant pairs of points, out of all possible pairs, is Kendall's tau. In symbols,

$$\text{Kendall's tau (with no ties)} = \frac{2C}{n(n-1)}$$

where $C$ is the number of concordant pairs of individuals. Various corrections have been introduced to deal with ties; see, for example, Siegel (1956).

Kendall's tau for the TV/aggressiveness data is .464. Kendall's tau is usually lower in absolute value than Spearman's rho. This does not mean that Kendall's tau is less sensitive, but rather that it must be judged on its own terms. We can compare Kendall's tau values with other Kendall's tau values.

Inferences using Kendall's tau can be performed using special tables, or using the following normal approximation (also from Siegel, 1956): Under the null hypothesis that the population value of Kendall's tau equals zero, the sample value of Kendall's tau has an approximately Normal distribution with mean zero and standard error

$$\sqrt{\frac{2(2n+5)}{9n(n-1)}}$$

## STUDY QUESTIONS—*Section 11.5*

**11.5.1** [Statpal exercise] Use Statpal's NOnparametric tests procedure to calculate Spearman's rho and Kendall's tau for the data of Exercise 11.2.1. In each case determine the $p$ value for a two-tailed test of the null hypothesis that age is unrelated to response.

**11.5.2** [Statpal exercise—Coronary Care data set (see Appendix A)] Calculate Kendall's tau and Spearman's rho between the variables AGE and NDAYS. Determine the $p$ value for each coefficient and write a sentence describing the results.

**11.5.3** [Statpal exercise—Electricity Consumer Questionnaire data set (see Appendix B)] Calculate Kendall's tau and Spearman's rho between the variables YRS-MEMBR and RATESRVC. Determine the $p$ value in each case, and compare these results to those for Exercise 11.4.7.

## 11.6 A nonzero $r$ does not imply causality, nor does a zero $r$ imply no correlation.

### Correlation versus causality

A teacher administers a test of arithmetic skills to 40 pupils. Scores on the test range from 4 to 17 (of a possible 20). The pupils' weights range from 16 to 45 kg (35 to 99 lb). Figure 11.10 shows a scatterplot of the weights versus the test scores. The scatterplot shows a very marked positive relationship between

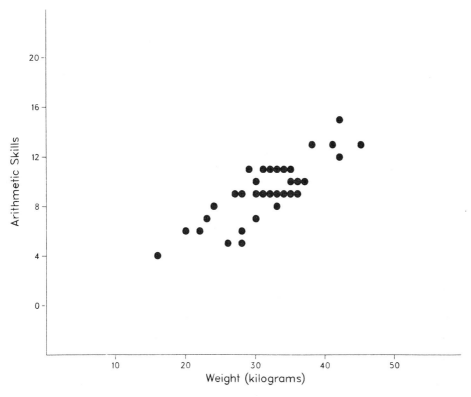

**Figure 11.10**   Weight versus arithmetic skills for a sample of 40 children ($r = .818$).

weight and test score. In fact, $r$ for the data equals .818. Does this relationship constitute evidence for a causal link between weight and arithmetic skills? Should arithmetic teachers learn more about nutrition? Or could it be that skill in arithmetic causes weight gain, or at least predisposes one to it?

Suppose we are now told that the 40 pupils ranged in age from 5 through 10 years old. On a test of arithmetic skills, who would tend to do better, all other things equal? The older children, of course. And which children are heavier? The older children. Hence the correlation between weight and test score. Perhaps arithmetic teachers *should* learn more about nutrition—but not based on this set of data!

The point is that the existence of a nonzero correlation between two variables does not necessarily imply any causal link between the variables. If both variables are related to a third variable—as, in this case, both variables were

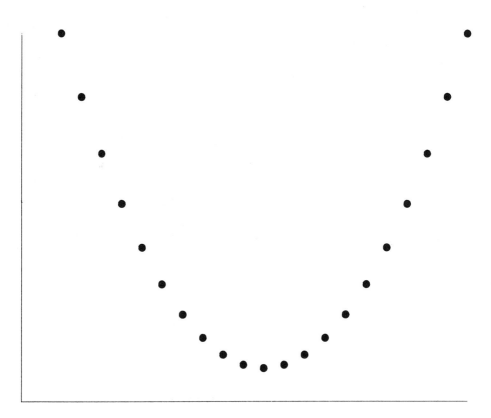

**Figure 11.11**  Curvilinear relationship.

related to age—there can be a very strong correlation between two variables that have nothing to do with each other in a deterministic sense.

This sort of thing often happens if we consider two variables over time. For example, the correlation of total cigarette consumption with average life expectancy over the years 1900 through 1960 is almost certainly strong and *positive*. Does this mean that increased cigarette consumption would increase life expectancy? Of course not—it means that both variables happened to increase over time, for different reasons.

### Correlation versus linear correlation

Just as a nonzero $r$ does not necessarily imply a causal link, a zero $r$ does not necessarily imply the lack of a correlation between two variables. We have noted previously that $r$ measures *linearity*. Two variables may be strongly related, even

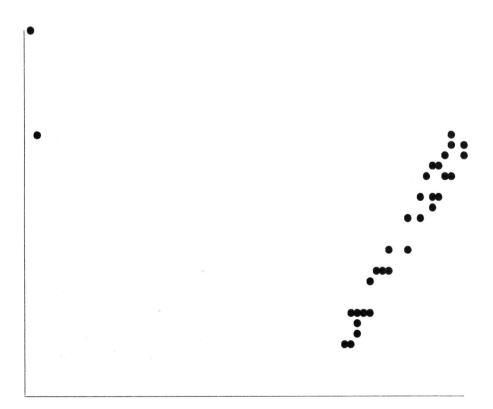

**Figure 11.12**   Scatterplot with outliers.

in a causal sense, but have a very low $r$ if the relationship is nonlinear. For example, Figure 11.11 shows a scatterplot of two variables with a very clear, but curvilinear relationship. Easy-$r$, $r$, Spearman's rho, and Kendall's tau are all close to zero for this set of data, in spite of the clear pattern.

This means that we must not rely solely on an index such as $r$ to tell us about the relationship between two variables. A scatterplot is indispensible for identifying nonlinear relationships. If we do encounter a nonlinear relationship, we can often transform one or both variables so that we do have a linear relationship (e.g., by using a polynomial in one of the variables).

## Outliers

Another situation in which a value of $r$ can be misleading is if there are extreme points in the data that greatly influence the value of $r$. Such points are called outliers. Figure 11.12 shows a scatterplot in which an apparently positive relationship results in a negative $r$ because of the influence of two outliers. The value of $r$ for this set of data is $-.129$. If we deleted the outliers, $r$ would be .977.

What should we do with outliers? It depends on the situation. As a minimum, we should look at the individual cases and try to determine if there was something special about them. Often, outliers are the result of blunders in data entry or mistakes in computer programs. Other times, outliers represent a set of individuals who are qualitatively different in some important respect. Sometimes, it is appropriate to describe the results both with and without the outliers and comment on how the outliers affect the results.

### STUDY QUESTIONS—*Section 11.6*

**11.6.1**  Consider again the data plotted in Figure 11.12. The medians for the two variables are both about 62. Using a ruler, divide up the figure into the four quadrants based on the medians, and calculate easy-$r$. Then exclude the outliers and calculate it again. (Excluding the two outliers does not affect the median very much, so do not bother changing the quadrant boundaries.)
   a.  How much does easy-$r$ change when you exclude the outliers?
   b.  Compare the effect of outliers on easy-$r$ to the effect of outliers on $r$.

**11.6.2**  [Continuation of Exercise 11.6.1] What about Spearman's rho and Kendall's tau? To what extent would they be affected by outliers? (*Note*: Including the outliers, Spearman's rho turns out to be .62 and Kendall's tau turns out to be .65. Excluding the outliers, the corresponding figures are .96 and .88.) Compare the effect of outliers on these measures to the effect of outliers on $r$.

# 12

# Introduction to Regression Methods

## 12.1 Often we wish to form a model to describe how one variable responds to others.

*Deterministic versus stochastic models*

The Internal Revenue Service (IRS) is responsible for administering the complex laws that govern the collection of taxes in the United States. Among the provisions of the tax code is a stipulation that business expenses may be deducted from taxable income. One major category of business expense for many businesses is expenses associated with travel by private car.

How much should a taxpayer be allowed to deduct for use of a private car for business purposes? If the car is used only for business purposes and the taxpayer keeps careful records, the problem is straightforward, in theory. If, however, the car is used for other purposes in addition to business, the problem is more complicated. The total expense of operating the car then needs to be apportioned between deductible and nondeductible parts, usually according to the number of miles the car was driven for each use.

Unfortunately, many taxpayers keep inadequate records of their car expenses, particularly if the business use of the car is secondary. Even when the taxpayer claims that good records have been kept, the IRS needs some way of determining whether a given deduction is reasonable for a given amount of driving, in order to be able to detect inflated deductions. For both of these purposes, the IRS needs to determine a reasonable figure for the average cost of owning a car as a function of the number of miles driven.

Note that the question here is quite different from the question of strength of correlation. We are not wondering to what extent a straight line, or any other function, describes the relationship between cost and mileage (although that may be an important question in its own right). Rather, we are wondering how much cost changes, on the average, in response to miles driven. As we saw in Chapter 11, two scatterplots could both show perfect linear relationships but have vastly different slopes—that is, they could differ in how much one variable responds to changes in the other. It is with the latter question we are concerned in this example.

Note also that the question is stated in terms of the *average* cost as a function of the number of miles driven. Cars vary in cost of ownership, and it would be unrealistic to expect to be able to state the exact cost of owning a car based solely on mileage. Two cars driven the same number of miles might differ considerably in how much it costs to own them.

If we were trying to give the exact cost of owning any car based solely on the number of miles driven, we would be seeking a *deterministic* model. A deter-

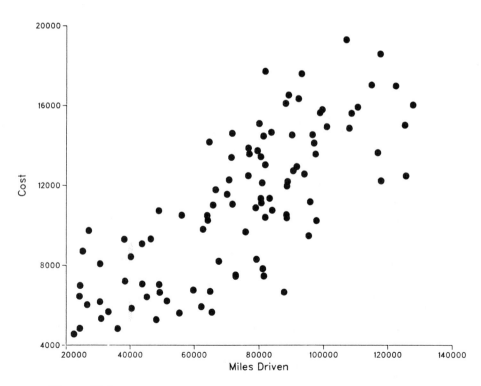

**Figure 12.1** Miles driven versus cost for a sample of 100 cars.

ministic model for giving cost as a function of mileage would not allow any room for variability in cost among cars with the same mileage. Such a model is clearly incorrect for this example.

What we need is a model that allows for the fact that mileage does not tell the entire story of cost. Such a model is called a *stochastic model*, from a Greek word meaning "to aim at." A stochastic model, in general, starts with a deterministic model giving the average of a variable as a function of others and then adds an element of randomness to account for the fact that the deterministic model does not work perfectly. We will look at how stochastic models work shortly. First, we need some special terminology.

### Dependent and independent variables

In our example we are trying to form a model that will allow us to give the average cost associated with a given number of miles. Thus we are trying to describe how cost depends on mileage—not the other way around. Cost is therefore called the *dependent variable* in this example. Mileage is considered an *independent variable*. In general terms, *the variable we are trying to predict is called the dependent variable. The variables we use to predict the dependent variable are called the independent variables.* By convention, when we draw scatterplots we put the dependent variable on the $y$ (vertical) axis and an independent variable on the $x$ (horizontal) axis.

## 12.2 Stochastic models may include one independent variable, or several.

### Some stochastic models with one independent variable

Let's look at a scatterplot of some data relating cost to miles driven and see how we might formulate a stochastic model to predict cost from mileage. Figure 12.1 shows a scatterplot of data on 100 cars. For each car the cost of owning the car over its entire life was determined, together with the number of miles driven. Each point on the plot represents a car.

One of the simplest stochastic models we might consider would predict cost as a direct multiple of mileage, plus some random variation to reflect the fact that the actual cost depends on things in addition to mileage. We could state the model like this:

Model 1:   cost = ($B$ × mileage) + random variation

where $B$ represents the cost per mile. The random variation is some added or subtracted cost, different for each car, that depends on factors unrelated to mileage. We assume that the random variation is as likely to be negative as positive—that is, we assume that the average random variation over all cars is zero. It is the random variation that makes this a stochastic rather than a deterministic model.

Since we assume that the random variation averages out to zero, we can restate model 1 in terms of the *average* cost as follows:

Model 1:    average cost = $B$ × mileage

How reasonable is model 1? Let's consider what kind of picture model 1 specifies on a scatterplot and compare it with the scatterplot of our data. As you can easily verify, model 1 specifies a straight line through the origin with slope $B$ (where $B$ is in dollars per mile). That is, for zero mileage, model 1 asserts that the average cost is zero, with a constant increase of $B$ dollars for every additional mile.

The idea that the average cost of a car with zero mileage should necessarily be zero is not particularly realistic. After all, there are some fixed costs associated with owning a car that are independent of mileage. To incorporate this idea, we can modify our model to include a constant cost associated with a car with zero mileage, giving us model 2:

Model 2:    cost = $B0$ + ($B1$ × mileage) + random variation

or, taking the average:

Model 2:    average cost = $B0$ + ($B1$ × mileage)

Like model 1, model 2 defines a straight line, but not through the origin unless $B0$ is zero. (Since $B0$ is the value of $y$ where the line intercepts the $y$ axis, it is often called the *y intercept*, or simply the *intercept*.)

Model 2 presumably does a better job than model 1, but we might still want to refine our model further. For example, we might postulate that the cost per mile of owning a car changes over the life of the car. Most of the depreciation in the value of the car occurs during the early miles, and if we are considering only mileage in predicting average cost, we would need to reflect that in our model. The apparent bend in the scatterplot suggests this sort of situation.

There are several ways to incorporate the idea of a changing slope into a model. For our example, one way could be to consider low-mileage cars separately from high-mileage cars, using a straight-line model for each. Model 3 shows how this could be done:

Model 3:

  cost = $C1$ + ($B1$ × mileage) + random variation
        for mileage up to 50,000 miles

  cost = $C2$ + ($B2$ × mileage) + random variation
        for mileage over 50,000 miles

where $C1$ and $C2$ are the constants and $B1$ and $B2$ are the slopes for the two lines.

Another way to allow for a changing slope is to use some function of mileage, rather than simply mileage, as the independent variable. For example, a quadratic model would look like this:

Model 4:

$$\text{cost} = B0 + (B1 \times \text{mileage})$$
$$+ (B2 \times \text{mileage squared})$$
$$+ \text{random variation}$$

This model defines a parabola rather than a straight line for the average cost. Other commonly used functions of the independent variable include the logarithm and square root, both of which serve to decrease the slope for larger values of the independent variable.

### Stochastic models involving more than one independent variable

In our cost/mileage example, we are looking for a model that will allow us to predict cost as a function of mileage alone, without any other information about the car. This approach, of course, ignores the fact that other information could allow us to make much more accurate predictions. In the models that use only mileage, we are still left with a substantial amount of variability in cost for which we cannot account (and hence assume to be random variation). But if we incorporated other information into the model, perhaps we could reduce the amount of unexplained variation still further.

Let's consider a model that takes into account not only mileage but also number of cylinders in the car's engine (typically, four, six, or eight, although there are a few exceptions):

Model 5:

$$\text{cost} = B0 + (B1 \times \text{mileage})$$
$$+ (B2 \times \text{number of cylinders})$$
$$+ \text{random variation}$$

or, in terms of the average cost:

average cost $= B0 + (B1$ times mileage$) + (B2$ times number of cylinders$)$

What do $B0$, $B1$, and $B2$ mean in model 5? $B0$ would be the average cost of a car with zero mileage and no cylinders—a nonexistent car, of course. We might be able to imagine a car that never gets driven throughout its life, but it is difficult to imagine a car with a zero-cylinder engine, provided that we are restricting ourselves to standard engines. The point is that $B0$ may or may not be meaningful, depending on whether zero is a meaningful value for the independent variables.

A simple way to see the correct interpretation of $B1$ and $B2$ is to consider what happens to the average of the dependent variable when each independent variable changes by 1 unit. Consider, for example, two cars with mileages that differ by 1 mile. Provided that the two cars have the same number of cylinders, model 5 asserts that the average costs of owning the two cars will differ by $B1$ dollars.

That is, a 1-unit change in mileage is associated, on the average, with a $B1$-unit change in cost, provided that the number of cylinders remains constant. Note the subtle but important difference between the interpretation of $B1$ in model 5 and $B1$ in model 2: Whereas in model 2 $B1$ represents the average change in cost for a unit change in mileage, in model 5 $B1$ represents the average change in cost for a unit change in mileage *provided that the other variable in the model remains constant.* Thus *the inclusion of another variable in the model changes the interpretation of the parameters for variables already in the model.*

Note, incidentally, that it is difficult to draw a picture of model 5 unless you are a very good artist. That is because we need three dimensions to represent the scatterplot. Model 5 specifies not a line but a two-dimensional surface located in the three-dimensional space of the scatterplot. If we were to add still another independent variable, we would have a model that specifies a three-dimensional "hypersurface" in a four-dimensional space, and so on for higher dimensions. As we see in Chapter 13, there are indeed graphical techniques that can be used to help evaluate models with more than one independent variable, but they certainly do not require drawing four-dimensional pictures!

### Choosing a model

How do we decide which model to use in a given situation? If we have a theoretical basis for choosing a particular model, the problem is relatively simple. If we can specify that the deterministic part of the model must be of a certain mathematical form, based on theory, then all we have to do is write down the form, making appropriate provision for random variation.

However, in many situations we have no strong theoretical basis for a particular model and want to use the data to help us formulate one. In such situations we can specify several candidate models and choose from among them the one that seems to be most in accord with the data. The "stepwise" techniques we cover in Chapter 13 represent one approach to this problem—and the cautions discussed there are relevant to any situation in which we are using the data to help us choose from among several competing models.

It is important to keep the techniques discussed in this chapter in perspective. The ultimate test of a particular model is not how well it fits the data with which it was derived, but how well it fits new data. The process of finding a model for a particular study and evaluating how well that model performs is only one part of a larger process through which models are refined and theories are formulated to account for the available data.

## STUDY QUESTIONS—*Sections 12.1 and 12.2*

**12.2.1**  It is important to understand what the parameters of a model mean. As we noted in Section 12.2, an easy way to figure out the interpretation of a particular parameter in a model is to see what happens to the dependent variable, on the average, when an independent variable changes by 1 unit.

Let's look at an example. Consider a model that predicts the number of votes cast in a given precinct for the Democratic candidate in a statewide election as a function of three independent variables: (1) the number of registered Democrats in the precinct, (2) the median household income in the precinct, and (3) the dollar amount spent on campaign materials and advertising within the precinct. We will assume that all precincts have roughly the same number of registered voters (although in practice we would need to make adjustments for this). The model we consider is as follows:

$$
\begin{aligned}
\text{Democratic votes} = B0 &+ (B1 \times \text{number of registered Democrats}) \\
&+ (B2 \times \text{median household income, in \$1000s}) \\
&+ (B3 \times \text{campaign spending, in dollars}) \\
&+ \text{random variation}
\end{aligned}
$$

a.  Campaign spending is measured, of course, in dollars. The unit of the dependent variable is votes. What is the unit of $B3$? What is the unit of $B0$?

b.  Why do we need to include random variation in the model? What does it represent?

c.  Use your knowledge (or prejudices) about American politics to guess what signs (plus or minus) $B1$, $B2$, and $B3$ ought to have.

d.  We have not yet discussed how to estimate the parameters (the topic of Section 12.3), but suppose that we find estimates for $B1$, $B2$, and $B3$ equal to .3, $-100$, and .05, respectively. $B0$ is estimated to be equal to 3500. Based on these estimates, what is the estimated average number of Democratic voters in a precinct with 5000 registered Democrats, a median household income of $20,000, and campaign expenditures of $2500?

e.  A newspaper columnist interprets $B3$ as indicating that the party spent 5 cents per Democratic vote. However, the actual amount spent in all precincts worked out to considerably more than 5 cents per Democratic vote, even using the same figures. Does this mean that $B3$ is wrong? How could this happen? (*Hint*: What is the complete interpretation of $B3$?)

**12.2.2**  Six preschool children are given a screening test consisting of 20 items, with each item testing mastery of a particular concept. For each child, the total number of items answered correctly is recorded. Three weeks later, the testing is repeated, so that each child has two scores on the test. The scores are as follows:

| Child | Test 1 | Test 2 |
|-------|--------|--------|
| 1     | 14     | 14     |
| 2     | 9      | 13     |
| 3     | 19     | 16     |
| 4     | 8      | 10     |
| 5     | 10     | 11     |
| 6     | 15     | 15     |

a. Draw a scatterplot of the data, with test 2 on the vertical axis.

b. Write down a straight line model to predict test 2 (the dependent variable) using test 1 (the independent variable).

c. Using a straightedge (a clear ruler works best, but anything will do), draw a straight line on the scatterplot that fits the points as closely as you can. This requires you to have some basis for judging *goodness of fit*. Try to write down, in specific terms, how you evaluated goodness of fit—that is, your basis for choosing one line over others as the "best" line. There are several ways of doing it—just write down your way.

d. Having drawn your line, figure out its slope and intercept and write out the equation of the line. (*Hint*: The slope of a line is the change in $y$ associated with a unit change in $x$. The intercept is the $y$ score associated with an $x$ score of zero.)

e. Suppose that all the test 2 scores were shifted upward by 4 points. Would that change the slope of the best line? Would it change the intercept?

f. Calculate easy-$r$ for the data set. (Refer to Chapter 11 for details.) What does this indicate? Does easy-$r$ say anything about the slope of the best line?

## 12.3 Regression methods are used to estimate the parameters of a stochastic model.

### The term "regression"

The term *regression* is due to Sir Francis Galton (1822–1911), whose work on heredity led to the development of many of the methods we discuss in this chapter. In studying measurements of the sizes of seeds in mother and daughter pea plants, Galton noticed that the sizes of seeds from daughter plants tended to "regress" to the mean. That is, mother plants with very large seeds would produce daughter plants with seeds that although larger than average were nevertheless closer to the mean. Galton found the same effect with respect to the heights of fathers and sons: Very tall fathers had tall sons, and very short fathers had short sons, but the sons were closer to the mean than the extreme fathers were, an effect that Galton termed "regression to mediocrity."

In modern usage, *regression methods* are methods for finding the parameters of some stochastic model relating a dependent variable to one or more independent variables. The term *simple regression* refers to regression procedures involving one independent variable, while *multiple regression* refers to procedures involving more than one independent variable.

## Estimating the parameters of a stochastic model

The problem of estimating the parameters of a stochastic model is really two separate problems. First, we have to consider how to find the parameters for a particular set of data—that is, how to calculate the values of the $B$'s that best fit the data we have. Second, we need to consider the problem of statistical inference—that is, how to draw inferences about the values of the parameters that best fit the *population*, given only *sample* data.

We consider the first problem in this section. How do we calculate the $B$'s for a stochastic model, given a set of data? The problem is to find those values of the $B$'s that yield the best fit to the data, in some sense. What do we mean by "best"?

We use model 2 from Section 12.2 as our example. Figure 12.2 shows the data (100 cars, with cost on the $y$ axis and mileage on the $x$ axis). Since we are considering only one independent variable, this is a simple regression problem. Recall our model:

$$\text{cost} = B0 + (B1 \times \text{mileage}) + \text{random variation}$$

As we have noted, this model defines a straight line with slope $B1$ and intercept equal to $B0$. What we want is the particular straight line, out of the infinite range of possible straight lines, that best fits the data. Figure 12.2 shows a few lines, all of which look pretty good. How should we determine which line is the best?

## Evaluating goodness of fit

The key to evaluating how well a particular line fits the points is to consider what we are trying to accomplish: We are trying to predict the dependent variable using the independent variable. Thus a line fits the points to the extent that the values of the dependent variable predicted by the line coincide with the actual values of the dependent variable. In terms of the model, we want to find the line that keeps the amount we have to attribute to *random variation* as small as possible. This means that we should evaluate how well a line fits the points *in the vertical direction*—that is, in terms of scores on the dependent variable (see Figure 12.3).

One way to evaluate how well a line fits the points is simply to add up the vertical distances of all the points to the line. This sum represents lack of fit—the bigger the sum, the less the line fits the points. We call this sum the *sum of*

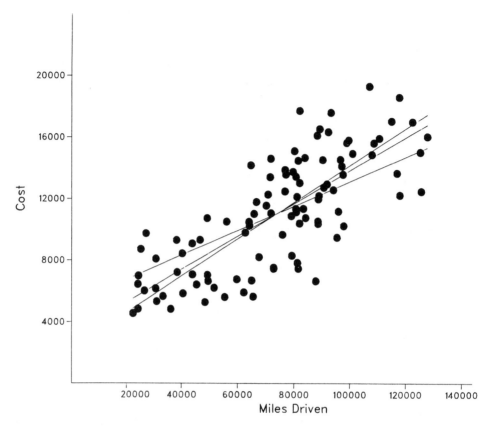

**Figure 12.2** Miles driven versus cost for a sample of 100 cars.

*absolute deviations*, since the vertical distances represent deviations from the value of the dependent variable predicted by the line.

## SSE

Although the sum of absolute deviations is simple to compute and easy to understand, we usually use a slightly different criterion to evaluate goodness of fit: We use the sum of *squared* deviations rather than absolute deviations. We use the sum of squared deviations for the same reason that the standard deviation is used in preference to the mean absolute deviation: It is mathematically more tractable. Since these deviations represent ''errors'' in predicting the dependent variable using the model, the sum of squared deviations is often abbreviated

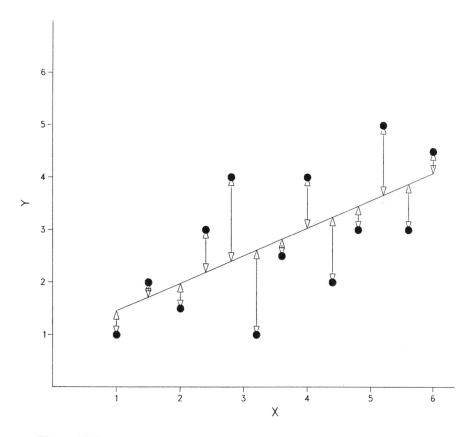

**Figure 12.3**  Vertical deviations from a line.

SSE, for *sum of squares, error*. Thus, we define the ''best'' line as that line which gives the smallest SSE. This line is called the *least-squares line*.

### Finding the least-squares line

Now that we know what we are looking for—the least-squares line—the problem is to find it. How do we find the one line that has the smallest SSE? How can we be sure that we have it? The answer will be familiar if you happen to know calculus: By using the mathematical procedure called differentiation, we can find precisely those values of $B0$ and $B1$ that yield the smallest SSE. Appendix C shows how it is done for the case of simple regression.

Fortunately, we do not need to know calculus to *use* (as opposed to derive) regression methods, since the necessary formulas are readily available and are incorporated into many good computer programs. All we need to do is apply the formulas—or, most often, let our computer program apply them for us. For the cost/mileage data, the least-squares line works out to be as follows:

predicted cost = 3125 + (.107 × mileage)

Thus for each additional mile a car is driven, we predict that the cost will increase by something over 10 cents.

Notice that the equation of the least-squares line gives the *predicted* cost. The *actual* cost of a car with a given mileage will generally differ from the predicted cost—indeed, it is the sum of the squares of these differences over all cars that constitutes SSE. What the least-squares method assures us is that no other line would give us a smaller SSE. We have achieved the smallest SSE possible with a straight line.

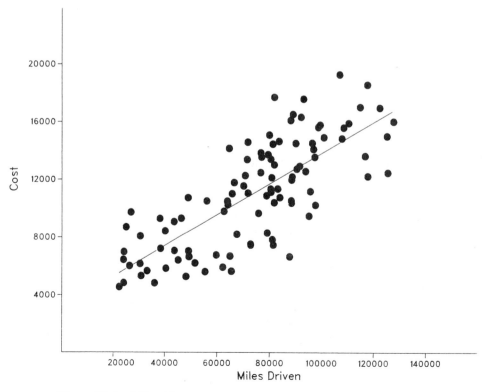

**Figure 12.4**   Miles driven versus cost for a sample of 100 cars.

Figure 12.4 shows the scatterplot of the data with the least-squares line drawn on it. SSE for the least-squares line works out to be approximately 573,057,272. (The unit of SSE in this case is "dollars squared.") By comparison, the SSEs for the three "eyeball" lines shown in Figure 12.2 were 647,793,584 for line A, 584,179,000 for line B, and 626,135,048 for line C—all greater than SSE for the least-squares line, as we know they must be.

## Notational conventions for sample estimates of the B's

At this point we have considered only the problem of finding the least-squares formula for a particular set of data. We deal in a later section with the problem of drawing inferences about the population from sample data. However, there are some notational conventions for distinguishing between population and sample regression formulas that we should cover now.

The symbols that we have been using for the regression coefficients—the $B$'s—will refer in this book to *population* values (parameters). We will use lowercase $b$'s ($b0$, $b1$, etc.) for *sample* values (statistics). We might have used the Greek letter beta ($\beta$) for the parameters, as many books do, but this symbol is also used for another regression term, and we can avoid some confusion by using $B$.

It is important in using computer programs to make sure that you know what the program is giving you when it labels something "$B$," "BETA," or "$b$." Different programs use different conventions.

## Calculating predicted value

To find the predicted value for a given car, we simply plug the car's mileage into the equation. For example, a car with 50,000 miles would have a predicted cost (over the life of the car) of $3125 + (.107 times 50,000), or $8475. Although the actual cost for any particular car with 50,000 miles will deviate from $8475, we are predicting that *on the average*, cars with 50,000 miles will have a cost of $8475.

STUDY QUESTIONS—*Section 12.3*

12.3.1   Using the least-squares formula given for the predicted cost of owning a car as a function of miles driven, answer the following questions. Recall that the formula was found to be

predicted cost = 3125 + (.107 $\times$ mileage)

a.   What is the predicted cost of owning a car driven 40,000 miles?
b.   What is the predicted cost of owning a car driven 100,000 miles?
c.   What is the predicted cost of owning a new car—that is, a car driven no miles? Does this prediction necessarily make sense?

d. What is the predicted cost of a car driven 138,247,922,016 miles—that is, about 138 billion miles?

e. Did you answer part d? Under what circumstances should a least-squares formula be used, or not used, to predict the dependent variable?

## 12.4 The coefficient of determination tells how well the model fits the data.

### SST, SSE, and SSR

We noted in Section 12.3 that the least-squares solution gives the smallest possible SSE compared to any other solution for the same model. So relative to other straight lines, our least-squares line for predicting cost from mileage is the best. But how good a fit is it in absolute terms? Does the figure of 573,057,272 dollars squared for SSE represent a good fit or a poor one? (What is a dollar squared, anyway?)

We can deal with this by considering how much we have gained by using the independent variable, in terms of accuracy of our predictions. (Note the similarity of this reasoning to that of the analysis of variance discussed in Chapter 10.) Without using the independent variable, our best prediction of a car's cost would simply be the mean cost over all cars. How well would such a prediction fit the data? We can evaluate it in terms of the sum of squared deviations of the actual scores from the mean (the predicted value, in this case). This sum is called SST (for "sum of squares, total"). For our data, SST works out to be 1,386,780,137 dollars squared.

How much do we gain, then, by using the independent variable to modify our prediction (via the least-squares equation)? Without it, the total lack of fit is SST. With it, the total lack of fit is SSE. Thus what we have gained, in terms of reducing lack of fit, is the difference between SST and SSE, called SSR (for "sum of squares, regression"). SSR in this case equals 1,386,780,137 minus 573,057,272, or 813,722,865 dollars squared.

### The coefficient of determination, R-squared

Of course, we are still left wondering whether this number should be considered large or not. However, if we consider SSR as a proportion of SST, we will have a measure with a simple interpretation: We will have the proportion of the total variation in the dependent variable that is explainable by the independent variable via the least squares equation.

This measure, SSR divided by SST, is called the *coefficient of determination* and is usually denoted *R-squared*. For our data, *R*-squared equals .587, or 58.7%. Thus 58.7% of the variability in cost can be explained by the least-squares line using mileage.

To recap:

SST equals the sum of squared deviations from the overall mean.
SSE equals the sum of squared deviations from the predicted values using the least-squares equation.
SSR equals SST minus SSE.
$R$-squared (the coefficient of determination) equals SSR divided by SST.

## R-squared versus r squared

Recall from Chapter 11 that the square of $r$, the correlation coefficient, has a very similar interpretation to that of $R$-squared: The square of $r$ is the proportion of the variability of one variable that can be explained by a straight-line relationship with the other variable. The similarity of the terms "$r$ squared" and "$R$-squared" is, of course, more than merely coincidental. If we are using a straight-line model with one independent variable, $R$-squared is the same as the square of the correlation between the dependent and independent variables—that is, $r$ squared.

## Multiple regression

We have introduced the basic ideas of regression in the context of simple regression (i.e., regression involving only one independent variable) for ease of presentation. In particular, it is considerably easier to show pictures of scatterplots in two dimensions than in three, four, or 17 dimensions. (Can you draw on two-dimensional paper a projection of, say, a six-dimensional hypersurface in a seven-dimensional space?)

Nevertheless, when we move from simple to multiple regression, the basic concepts remain the same. In particular, our discussion of SST, SSE, and SSR, and $R$-squared applies the same way to multiple regression as to simple regression. $R$-squared in multiple regression still represents the proportion of the variability in the dependent variable that can be accounted for by the least-squares equation. The only difference is that the least-squares equation now includes a set of variables rather than just one.

## STUDY QUESTIONS—*Section 12.4*

**12.4.1**  Consider the following simple regression formula:

predicted systolic blood pressure = $120 + (6 \times \text{indicator})$

where "indicator" equals 0 for females, 1 for males.
    a.  Based on the formula, what is the predicted blood pressure for a male? For a female?
    b.  Suppose that the formula is the least-squares line from a sample of 60 adults

(30 men and 30 women). Sketch a scatterplot showing what the data might have looked like. Note that the independent variable has only two possible scores.

c. How would you interpret the intercept in the formula? (*Hint*: What does a score of zero on the independent variable indicate?) How about the slope?

d. Now let's consider the problem of drawing inferences about a population based on sample regression results. Based on your answers to part c, what sort of test do you suppose we could do to determine if the sample evidence is strong enough to conclude that the population slope is greater than zero? What assumptions might be required?

## 12.5 The *F* test is used to test the model as a whole for statistical significance.

### *Statistical inference in regression problems*

We have discussed the first part of the overall problem of estimating the parameters of a stochastic model: finding the best values, in terms of the least-squares criterion, for a given set of data. We have also discussed a way of evaluating how well our best regression formula fits the data—that is, the coefficient of determination.

Now we consider the second part of the problem: drawing inferences about a population regression formula based on sample data. This problem can in turn be broken into two parts: drawing inferences about the model as a whole and drawing inferences about each individual parameter (i.e., the constant and the *B*'s) in a model.

### *Drawing inferences about the model as a whole*

Let's return to our car cost example. Suppose that our group of 100 cars is a random sample from the population of all cars currently in service in the United States. Using the simple regression model with mileage as the only independent variable, we found the least-squares formula and found that $R$-squared was equal to .587. Thus we were able to account for 58.7% of the variability in cost in the sample data by a straight-line relationship with mileage. But what about the population? It is theoretically possible that the population coefficient of determination is actually zero, and we were simply unlucky in drawing an unrepresentative random sample. Is this plausible?

This calls, of course, for a hypothesis test. The null hypothesis is that the population coefficient of determination is zero—that is, that mileage contributes nothing to the prediction of cost. The alternative hypothesis is that the population coefficient of determination is greater than zero—that is, that mileage does contribute to the prediction of cost.

## Assumptions for the F test

We introduced the $F$ test in Chapter 10 in the context of one-way analysis of variance. The $F$ test used in regression is not merely similar to that used in ANOVA, but mathematically equivalent. Indeed, as we saw in a study question, problems involving group differences can be cast in regression terms by a suitable choice of dummy variables. Thus it should come as no surprise that the assumptions required for the $F$ test in regression are the same as those required for ANOVA, merely stated in other terms. Recall that these assumptions are also the assumptions needed for a two-sample $t$ test.

To use the $F$ test to determine if we can conclude that the population coefficient of determination differs from zero, we need three assumptions:

1. *Independence.* Individual scores on the dependent variable must be mutually independent. That is, there must be no tendency for one individual's score to be affected by another individual's score.
2. *Normality.* Consider all the individuals in the population that have a particular set of scores on the independent variables (e.g., all cars with exactly 50,000 miles). We require that the scores of those individuals on the dependent variable have a normal distribution about their mean. This must be true for all possible settings of the independent variables.
3. *Homogeneity.* Consider again all the individuals that have a particular set of scores on the independent variables. Call the variance of their scores on the dependent variable $\sigma$. We require that $\sigma$ be the same for all possible settings of the independent variable—that is, that the variance of the normal distributions mentioned above be homogeneous.

We discuss some ways of checking the validity of these assumptions in Chapter 13.

## The F test in regression

Given that our assumptions are valid, we can proceed to test our hypotheses. Just as for one-way ANOVA, our test statistic is $F$, which is computed from the sums of squares and their respective degrees of freedom. In regression terminology, $F$ is defined as MSR divided by MSE, where MSR (*mean square, regression*) is SSR divided by its *df*, and MSE (*mean square, error*) is SSE divided by its *df*. You might wish to refer back to Section 10.2 for a discussion of the rationale behind the $F$ statistic.

All we need to know now are the respective *df* for SSR and SSE. The *df* for SSR is equal to the number of independent variables in the model. Thus for a simple regression model, *df* for SSR equals 1. For SSE, the *df* equals the number of scores ($n$), minus the number of parameters, *including B0*. Thus for our example, we have $100 - 2$, or 98, *df* for SSE. Note that, as in one-way ANOVA, the *df* for the two sums of squares add up to $n - 1$, which is the *df* for SST.

Just as in one-way ANOVA, these results are generally presented in the form of an ANOVA table. Appendix C gives the general formulas for each of the terms in the table. Here are the results for our example:

| Source | *df* | SS | MS | *F* |
|--------|------|-----|-----|-----|
| Regression | 1 | 813,722,865 | 813,722,865 | 139.16 |
| Error | 98 | 573,057,272 | 5,847,523 | |
| Total | 99 | 1,386,780,137 | | |

Recall that if the null hypothesis is true, both MSR and MSE are estimates of the same quantity—the common variance σ referred to in the homogeneity assumption. Since *F* equals MSR/MSE, we would expect *F* to be about 1, on the average, if the null hypothesis were true. Larger values of *F* support the alternative hypothesis.

How likely is it, then, that we would observe an *F* of over 139 if the null hypothesis were true? The *F* table (Table D.6) indicates that we can expect an *F* of 6.93 or greater (for 1 and 90 *df*—close enough to 1 and 98) to occur less than 1% of the time by chance. Our *F* is much larger than 6.93, so the *p* value for our data is less than 1%. We therefore reject the null hypothesis and conclude that the population coefficient of variation is indeed greater than zero. (This would have been a very unrealistic example if we had not concluded that mileage is helpful in predicting the cost of a car!)

## STUDY QUESTIONS—*Section 12.5*

**12.5.1**  [Statpal exercise—Coronary Care data set (see Appendix A)]
   a.  Go into the REgression procedure. Specify NDAYS as the dependent variable and AGE as the independent variable. (Recall that NDAYS is the number of days hospitalized.)
   b.  When Statpal prompts you for a command, specify DISplay. This will give you the least-squares formula for the model predicting NDAYS using no independent variables. What does the intercept represent?
   c.  Next, select the ADD command, specifying the variable AGE to be added to the model. Then select DISplay to view the results.
   d.  Interpret the value of *B* for AGE, in specific terms.
   e.  Interpret the value of the intercept. Does this interpretation have any practical meaning in this context?
   f.  To what extent does AGE predict NDAYS in this set of data?

g.  Does the model using AGE predict NDAYS significantly better than chance? (You may accept the validity of the necessary assumptions for now.)

**12.5.2**  [Statpal exercise—Coronary Care data set (see Appendix A)]

a.  Use the REgression procedure to obtain the least-squares solution for the $b$'s in the formula

predicted NDAYS $= b0 + (b1 \times$ HADMI)

Recall that HADMI indicates whether or not the patient had had a myocardial infarction, with $1 =$ yes and $2 =$ no. Note that in Statpal's terminology, $B$ for the intercept is $b0$ and $B$ for HADMI is $b1$.

b.  Interpret the coefficient $b1$. Be specific!

c.  Interpret the coefficient $b0$. Is this a meaningful number in this context?

d.  Test the null hypothesis that the model performs no better than chance in predicting NDAYS.

**12.5.3**  [Statpal exercise—Coronary Care data set (see Appendix A)]

a.  Use the REgression procedure to obtain the coefficients for the following model:

predicted NDAYS $= b0 + (b1 \times$ AGE$) + (b2 \times$ HADMI)

To do this, specify the three variables NDAYS, AGE, and HADMI. Then use the ADD command twice to add the two variables AGE and HADMI to the model. (Save your results for future exercises.)

b.  Interpret the coefficients $b1$ and $b2$. Be specific!

c.  Compare the results for the coefficients $b1$ and $b2$ to the coefficients obtained in Exercises 12.5.1 and 12.5.2. Why do the coefficients differ between the simple and multiple regression models?

## 12.6  Individual regression coefficients can be tested using a form of $t$ test.

### Testing individual regression coefficients for significance

Let's recap the interpretations of the $B$'s in simple and multiple regression models. Recall that the $B$'s (as opposed to the $b$'s) are *parameters*. In both simple and multiple regression, $B0$ represents the average of the dependent variable for individuals with scores of zero on all independent variables in the model. This is meaningful only if such individuals are possible—which, as we have seen, is often not the case. Usually, $B0$ is not of much interest by itself (although there are occasional exceptions).

In simple regression, $B1$ represents the average change in the dependent variable for each unit change in the independent variable. A value of $B1$ equal to zero, therefore, would indicate that changes in the independent variable make no difference in the average of the dependent variable. This is equivalent to saying that the population coefficient of determination equals zero.

In multiple regression, there are two or more $B$'s of interest, and the interpretation of a particular $B$ reflects the fact that there are other independent variables in the model. To say that the $B$ for a particular independent variable equals zero means that changes in that variable make no difference, on the average, in the dependent variable—*provided* that the other independent variables in the model remain constant.

In either case we are often interested in testing whether the $B$ for a particular independent variable is different from zero. To do this, it turns out that we can use a form of a $t$ test.

## The t test for a specific B

Once again, since a particular $B$ is a type of average (referring to average change in the dependent variable associated with a unit change in the independent variable), we should not be surprised that we can use a $t$ test to test hypotheses concerning it. Recall the basic procedure: We evaluate how many standard errors our statistic is away from the null hypothesis value of the parameter. This number is the $t$ score. We then consult the $t$ table with the appropriate number of $df$ to determine the $p$ value. In the case of regression, the parameter of interest is one of the $B$'s—say, $B2$. The statistic is the corresponding $b$ (from sample data)—for $B2$, it would be $b2$. The number of degrees of freedom is the number of $df$ for SSE: the number of cases ($n$) minus the number of parameters in the model including $B0$.

To use the $t$ table, we require the same assumptions as for the $F$ test (independence, normality, and homogeneity). Again, we'll consider ways of checking the validity of these assumptions in a later section.

All we need now is the standard error of the statistic. The formula for this in the case of simple regression is given in Appendix C. We simply accept the fact that nearly all computer programs calculate it for you. Nevertheless, it is important to understand what the standard error indicates: Just as for any other statistic, the standard error of a sample regression coefficient indicates how variable the statistic tends to be over all possible samples. The larger the standard error, the more variable the statistic—and the less certain we can be about our inferences using the statistic.

Let's look at our car cost example again, this time using a multiple regression model with two independent variables: (1) mileage and (2) number of cylinders. The model is as follows:

$$\text{cost} = B0 + (B1 \times \text{mileage}) + (B2 \times \text{number of cylinders}) + \text{random variation}$$

We have already looked at the results of a simple regression using only mileage.

Using both mileage and number of cylinders, we get the following sample results:

predicted cost = 290 + (0.096 × mileage)
+ (728.2 × number of cylinders)

ANOVA table:

| Source | df | SS | MS | F |
|--------|----|----|----|---|
| Regression | 2 | 898,609,356 | 449,304,678 | 89.28 |
| Error | 97 | 488,170,727 | 5,032,688 | |
| Total | 99 | 1,386,780,083 | | |

Thus $b0$ equals 290, $b1$ equals 0.096, and $b2$ equals 728.2. The standard errors for $b1$ and $b2$ work out to be .0089 and 177.3, respectively.

Before we consider the individual regression coefficients, let's compare these results with those from the simple regression model using mileage only. First, notice that the coefficient for mileage is not the same as it was in the simple regression model. This reflects the fact that the coefficient no longer has the same meaning as it did in simple regression: In multiple regression, the coefficient for mileage represents the average change in cost for a unit increase in mileage *provided that the number of cylinders remains constant.*

That is, if we compare two randomly picked cars whose mileage differs by, say, 1000 miles, we would expect their cost to differ by $96 (that is 1000 × .096)—provided that the two cars have the same number of cylinders. If in our data set mileage is at all correlated with number of cylinders—if, for example, the larger cars have been driven farther—then the coefficient for mileage in *simple* regression will reflect the effect of number of cylinders to some extent, as well as the effect of mileage itself.

Now let's do the *t* tests to determine if the sample data allows us to claim that the population coefficients for mileage and number of cylinders are different from zero. For mileage, we have $b1$ = .096, with a standard error of .0089. Thus the *t* score for $b1$ is .096 ÷ .0089, or 10.79. Consulting the *t* table with 100 *df* (the closest tabled *df* to 97), we find that for a two-tailed *p* value of .05, we need a *t* score of 1.98. Since our *t* score is much greater than 1.98, we can reject the null hypothesis and conclude that $B1$ is indeed different from zero. For number of cylinders we have $t$ = 728.2 ÷ 177.2, or 4.11. This, too, exceeds 1.98, so we can conclude that $B2$ is also different from zero.

## The partial F test

We noted in our discussion of one-way ANOVA that in the particular case of two groups, the *F* test was equivalent to the two-sample *t* test, with the *F* score equal to the square of the *t* score. The same correspondence holds in the case of testing individual regression coefficients. Specifically, the test of an individual regression coefficient that we have just described as a *t* test can also be performed as an *F* test, where the *F* score is simply the square of the *t* score. The *df* for the *F* score are 1 and *k*, where *k* is the number of *df* for SSE.

To distinguish the *F* test for an individual *B* from the *F* test for the model as a whole, the former is sometimes called a *partial F test*. The word "partial" reflects the fact that the test is evaluating the effect of one part of the model—that is, one particular independent variable—with the effects of the other parts held constant. Similarly, the *B* for a particular independent variable in a multiple regression model is sometimes called a *partial regression coefficient*.

It is very important to remember that the interpretation of a *B* in a multiple regression model explicitly refers to the presence of the other terms in the model. *Adding or deleting an independent variable from the model changes the interpretations of the coefficients for the remaining variables.* If the variable being added or deleted happens to be correlated with the variables left in the model, the numerical values of the coefficients will also change. The word "partial" reminds us of this fact.

### STUDY QUESTIONS—*Section 12.6*

**12.6.1**   [Statpal exercise—Coronary Care data set (see Appendix A)]
  a.   Refer to your output for Exercise 12.5.3. Perform a *t* test for each of the regression coefficients for AGE and HADMI. Interpret the results.
  b.   Now refer to your output for Exercises 12.5.1 and 12.5.2. Why are the *t* scores for the two coefficients in the multiple regression model different from the corresponding coefficients in the simple regression models?

**12.6.2**   [Statpal exercise—Electricity Consumer Questionnaire data set (see Appendix B)]
  a.   Use the REgression procedure to find the coefficients of the following regression formula:

RATESRVC = $b0$ + ($b1$ × LIKENUKE) + ($b2$ × YRSMEMBR)

   Recall that RATESRVC is a rating of member satisfaction with the electric cooperative's service; LIKENUKE indicates the member's orientation toward nuclear power, with 1 = proponent and 2 = opponent; and YRS-MEMBR is the number of years the respondent has been a member of the cooperative.
  b.   Interpret the regression coefficient for YRSMEMBR.
  c.   Test the coefficient for YRSMEMBR for statistical significance.

d.  Interpret the regression coefficient for LIKENUKE. Be specific!
e.  Test the coefficient for LIKENUKE for statistical significance.
f.  Now use the REMove command to remove YRSMEMBR from the model and display the results for the remaining model (i.e., the model including LIKENUKE only). Test the coefficient for LIKENUKE for statistical significance.
g.  What other procedure could you have used to perform an equivalent test to that in part f?

# 13

## Further Topics in Regression

### 13.1 Stepwise methods can be used to select independent variables for a model.

*Building a model*

Up to now, we have been considering situations in which we know which independent variable or variables we wanted to use, according to a specified model, to predict the dependent variable. But often researchers have available data for a large number of independent variables and wish to find a model that will do a good job of predicting the dependent variable with as few independent variables as possible.

Why would researchers wish to reduce the number of independent variables involved in a model? There are two main reasons. First, scientific investigation seeks to explain the observed facts as parsimoniously as possible; that is, the simpler the explanation, the better in scientific terms. A model that involves a few independent variables is generally simpler to interpret than one that involves many independent variables. Second, the fewer independent variables used in the model, the cheaper the data collection and analysis (usually). So researchers have both a scientific and an economic interest in trying to keep the number of independent variables used in a model as small as possible.

Let's consider an example. The personnel department of a large company wishes to develop a battery of tests to predict success in sales positions to help evaluate applicants for such positions. Consulting with a personnel psychologist, the department selects eight tests believed to measure various constructs relevant to success in sales occupations and administers the tests to a large group of applicants. A randomly selected subset, numbering 120 applicants, of this group is then hired for a 6-month trial period, after which each new employee's total sales performance (in dollars) is recorded. [Studies of this kind, called *validation studies*, involve important methodological issues which we are glossing over in this example; see, for example, Guion (1965) for a discussion of these issues.]

Thus, after the 6-month trial period, the personnel department has available nine pieces of data for each of the 120 new employees: total sales (the dependent variable) and scores on each of the eight tests administered at the time of application. We label the eight tests TEST1 through TEST8. The department wants to find some combination of the test scores that does a good job of predicting total sales. Moreover, the department has an interest in using as few tests as possible, to keep down the cost of testing each applicant.

### Finding the best model

We note at the outset that we will be considering only "linear" models at this point: that is, models which specify the predicted scores on the dependent variable as a weighted sum of the independent variables. We will not be concerned with models that involve, for example, products of independent variables (although nothing prevents us from defining, say, TEST9 as TEST1 times TEST2 and including TEST9 in our model); nor will we be concerned with models that involve products of the weights. Linear models are by far the most widely used models for this kind of study, largely because of their simplicity. All the models we discuss in this chapter are linear models.

What, then, would be the best linear model for the department to use? The answer, of course, depends on what we mean by "best." If we define best in terms of $R$-squared, the problem is simple: The model with the most variables must necessarily achieve the highest $R$-squared! This is because the addition of a variable to a model can only increase (or at worst, leave unchanged) the value of SSR and hence $R$-squared. Supplying more information cannot possibly make our predictions worse than they were.

But even though $R$-squared cannot decrease when a variable is added, the amount of increase could be very small. Indeed, it could be so small that it could easily be attributed to chance variation. How might we test this? One way is to use the $t$ test we discussed in Section 12.6. If we cannot claim based on the sample evidence that the $B$ for a variable is different from zero, we cannot claim that the increase in $R$-squared associated with the addition of that variable to the model is anything but a chance occurrence.

Moreover, an increase in $R$-squared does not necessarily imply that the $F$ test of the model as a whole will yield a higher $F$ score. This is because the addition of another variable to a model adds 1 to the *df* for SSR and subtracts 1 to the *df* for SSE. If the increase in SSR (which is equal to the decrease in SSE) is very small, MSR could actually go down and MSE could go up. This would mean that *F* would go down—perhaps far enough so that it would no longer be significant. (You might want to look at the formulas for the mean squares and *df* given in Appendix C to convince yourself that this is possible.)

So *R*-squared by itself is not an appropriate criterion for judging which model is best, but we do have other criteria available. Let's consider a very popular approach to building a model that makes use of these criteria.

## Stepwise procedures

The idea of stepwise procedures is to start with some model involving one or more of the available independent variables and then add or delete variables step by step, one at a time, to make the model better. When no appreciable gain is achieved by taking another step, the procedure stops. The resulting model is not necessarily the best of all possible models, but it is generally one of the best.

There are a variety of different stepwise procedures, depending on what model the procedure starts with, what criteria are used to determine which variables should be added or deleted, and what criteria are used to determine when the procedure should stop. Some procedures start with one independent variable and add additional variables at each step. These are called *forward selection procedures*. Others start with all the independent variables in the model, and delete variables at each step. These are called *backward elimination procedures*.

The procedure we discuss, by example, is a combination of forward and backward procedures and is a very popular method. When the term *stepwise regression* is used without explanation, it usually refers to this procedure.

Stepwise regression starts by finding the one independent variable of the set that, by itself, explains the greatest amount of variability in the dependent variable. This variable can be found, of course, by looking at the correlation of each independent variable with the dependent variable: The one with the highest correlation (in absolute value) wins.

## The correlation matrix

It is useful to have available not only the correlations of each independent variable with the dependent variable, but also the correlations of the independent variables with each other. An easy way to present all these correlations is in the form of a *correlation matrix*. Here, then, is the correlation matrix for the nine variables (one dependent variable and eight independent variables) in our personnel department example:

|        | SALES | TEST1 | TEST2 | TEST3 | TEST4 | TEST5 | TEST6 | TEST7 | TEST8 |
|--------|-------|-------|-------|-------|-------|-------|-------|-------|-------|
| SALES  | 1.000 | .146  | .031  | .396  | .263  | .171  | −.091 | .021  | −.032 |
| TEST1  | .146  | 1.000 | .464  | .178  | .109  | .123  | .003  | −.082 | .099  |
| TEST2  | .031  | .464  | 1.000 | .042  | −.036 | .022  | −.098 | −.090 | .125  |
| TEST3  | .396  | .178  | .042  | 1.000 | .932  | .135  | .740  | −.176 | −.115 |
| TEST4  | .263  | .109  | −.036 | .932  | 1.000 | .059  | .834  | −.148 | −.164 |
| TEST5  | .171  | .123  | .022  | .135  | .059  | 1.000 | .015  | .071  | −.075 |
| TEST6  | −.091 | .003  | −.098 | .740  | .834  | .015  | 1.000 | −.156 | −.092 |
| TEST7  | .021  | −.082 | −.090 | −.176 | −.148 | .071  | −.156 | 1.000 | −.103 |
| TEST8  | −.032 | .099  | .125  | −.115 | −.164 | −.075 | −.092 | −.103 | 1.000 |

Each number in the correlation matrix shows the correlation ($r$) between the row variable and the column variable. For example, the correlation between SALES and TEST4 is .263, and the correlation between TEST3 and TEST4 is .932. The diagonal contains the correlations of each variable with itself—which is, of course, perfect. Note that we are actually presenting each correlation twice (except for the diagonal), because the correlation of, say, TEST1 with TEST2 is the same as the correlation between TEST2 and TEST1. Many authors present correlation matrices in the form of a triangle, with each correlation shown only once.

### The process of stepwise regression

Now we are ready to begin building a model using stepwise regression. Let's follow the process step by step.

*Step 1*:  *Find the independent variable that has the largest correlation with the dependent variable. If this correlation is significant, add the independent variable to the model.* From the correlation matrix, we see that the highest correlation with SALES is .396, for the independent variable TEST3.

Testing this correlation for significance involves the $t$ test described in Section 11.4. As you can easily verify from the $r$ table, an $r$ of .396 with 118 $df$ gives a $p$ value less than .001. Using (as is customary) an $\alpha$ of .05, we conclude that the population correlation is indeed different from zero—or, in other words, that the sample correlation is significant.

Adding TEST3 to a model and finding the least-squares formula, we get the results shown in Figure 13.1. (*Note*: The results for each step are as printed by Statpal.)

Note that $B$ for the "intercept" refers to $b0$. So we could write our equation at step 1 as follows:

predicted SALES = 26,016.33 + (182.23 × TEST3)

```
Dependent variable:  SALES
  Independent variables in the model:
    TEST3
  Independent variables not in the model:
    TEST1     TEST2     TEST4     TEST5     TEST6     TEST7     TEST8

Variable            B          Std Error    t Score    2-tail Sig.

Intercept      26016.3329     2587.0882     10.0562      0.0000
TEST3            182.2261       38.8766       4.6873      0.0000

Valid cases:  120    Missing cases:  0


Analysis of Variance

Source            SS          DF         MS          F      Sig.
Regression 702373547.22        1 702373547.22    21.9707 0.0000
Residual    3772294710.6     118 31968599.242

Total       4474668257.8     119 37602254.267

R-squared =   0.1570
R-squared adjusted for DF =   0.1498
```

**Figure 13.1**   Regression results from Statpal.

The printed results for the *p* values (labeled ''sig.,'' for ''significance'') for both
the *F* test (for *R*-squared) and the *t* test (for the *b* for TEST3) are .000. Of course,
no *p* value can really be zero; when a computer program reports a *p* value of
''.000,'' it means that the *p* value was zero to three decimal places—in other
words, less than .001.

Note that at this stage, with only one variable in the model, the overall *F* test
and *t* test for the *b* for TEST3 are equivalent and necessarily yield the same *p*
value (provided that we are doing a two-tailed *t* test).

*Step 2*:   *Find the independent variable, of those not yet in the model, that
would give the greatest increase in R-squared if we added it to the model. If this
increase would be significant, and the tolerance of the variable is not too small,
add the variable to the model.* We accomplish this by calculating (or letting our
computer calculate) what the *t* score would be for the coefficient for each
variable if that variable were added to the model. The variable that gives the
biggest *t* score (in absolute value) is the variable that would give the greatest
increase in *R*-squared. (Some computer programs use the partial *F* test instead,
which, as we have noted, is an equivalent procedure.)

The term ''tolerance'' relates to the degree to which the independent variable being considered is itself related to the other independent variable or variables already in the model. We discuss it when we discuss the problem of multi-collinearity in Section 13.4. For now, we simply accept that the tolerance is adequate.

For our example data, Statpal reports the information shown in Figure 13.2 for the variables not yet in the model after step 1. While two of the $t$ scores have $p$ values less than .05, the $t$ score for TEST6 is the largest (in absolute value). So TEST6 is added to the model, yielding the results shown in Figure 13.3.

*Step 3*: *Reevaluate the independent variables already in the model to deter-mine if any should be removed.* This step is what distinguishes stepwise regres-sion, as the term is usually used, from forward selection. We have already seen that the addition of a variable to a regression model changes the coefficients for the variables already in the model if the newly added variable is correlated with the variables already in. Under some conditions, the addition of a new variable may render the coefficient for a variable already in the model nonsignificant.

Thus when a variable is added, we reevaluate the significance of the variables already in the model. If any have become nonsignificant, we remove the one with the smallest $t$ score (in absolute value). For our example, the addition of TEST6 to the model did not render the coefficient for TEST3 nonsignificant (as can be seen in the previous display), so we need not remove TEST3.

*Step 4*: *Repeat steps 2 and 3 until there are no more variables that meet the criteria for addition or deletion.* For our example, Statpal gives the information shown in Figure 13.4 about the variables not in the model after the addition of TEST6. Since none of the variables would produce a significant increase in $R$-squared (as evidenced by the nonsignificant $t$ scores), the process stops at this

```
Dependent variable:  SALES
  Independent variables in the model:
    TEST3
  Independent variables not in the model:
    TEST1    TEST2    TEST4    TEST5    TEST6    TEST7    TEST8
For variables not in the model:
```

| Variable | B-in | Std Error | Tolerance | t Score | 2-tail Sig. |
|---|---|---|---|---|---|
| TEST1 | 33.5101 | 36.8687 | 0.9682 | 0.9089 | 0.3653 |
| TEST2 | 2.7574 | 16.3735 | 0.9982 | 0.1684 | 0.8667 |
| TEST4 | -250.3720 | 68.5668 | 0.1313 | -3.6515 | 0.0003 |
| TEST5 | 39.0063 | 27.7260 | 0.9817 | 1.4068 | 0.1610 |
| TEST6 | -381.9236 | 44.4089 | 0.4517 | -8.6002 | 0.0000 |
| TEST7 | 28.7776 | 26.3425 | 0.9690 | 1.0924 | 0.2766 |
| TEST8 | 4.1113 | 25.7956 | 0.9867 | 0.1594 | 0.8738 |

**Figure 13.2**  Regression results from Statpal.

```
Dependent variable:  SALES
  Independent variables in the model:
      TEST3     TEST6
  Independent variables not in the model:
      TEST1     TEST2     TEST4     TEST5     TEST7     TEST8


Variable              B          Std Error    t Score    2-tail Sig.

Intercept       26934.7862      2036.4594    13.2263      0.0000
TEST3             471.7765        45.4693    10.3757      0.0000
TEST6            -381.9236        44.4089    -8.6002      0.0000

Valid cases:  120   Missing cases:  0


Analysis of Variance

Source            SS         DF       MS            F        Sig.
Regression 2163441834.6       2 1081720917.3    54.7594 0.0000
Residual    2311226423.2    117 19754071.993

Total       4474668257.8    119 37602254.267

R-squared  =  0.4835
R-squared adjusted for DF =  0.4747
```

**Figure 13.3**  Regression results from Statpal.

```
Dependent variable:  SALES
  Independent variables in the model:
      TEST3     TEST6
  Independent variables not in the model:
      TEST1     TEST2     TEST4     TEST5     TEST7     TEST8
For variables not in the model:

Variable       B-in      Std Error  Tolerance  t Score    2-tail Sig.

TEST1        -15.6641     29.6187     0.9314    -0.5289      0.5983
TEST2        -19.0583     12.9959     0.9616    -1.4665      0.1440
TEST4         56.6308     70.2475     0.0856     0.8062      0.4220
TEST5         15.2777     22.1143     0.9657     0.6909      0.4914
TEST7         21.8905     20.7297     0.9675     1.0560      0.2929
TEST8          2.3465     20.2802     0.9866     0.1157      0.9082
```

**Figure 13.4**  Regression results from Statpal.

point, with TEST3 and TEST6 in the model. Our final least-squares formula is therefore

predicted SALES $= 26,934.79 + 471.78(\text{TEST3}) - 381.92(\text{TEST6})$

$R$-squared for this formula was .484.

So the department could investigate a test battery composed of TEST3 and TEST6. Note that the coefficient for TEST6 is negative—that is, lower scores on TEST6 are associated with higher sales performance, holding TEST3 constant. Note too that the final result is not what we would have expected from a cursory look at the SALES column of the correlation matrix. The two variables that had the greatest correlation, *taken singly*, with SALES were TEST3 and TEST4. The correlation of SALES with TEST6 was quite small—nonsignificant, in fact. This points up yet again the fact that the regression coefficient for a particular independent variable depends very much on the other variables in the model.

## 13.2   Beware of the misuses of stepwise regression!

### Misuses of stepwise regression

Stepwise regression can be a useful tool for building a model. However, it can also be misused and its results misinterpreted. We will consider two very popular misuses of stepwise regression. (With a little experience, you can soon invent your own!)

### Misuse 1: The fishing expedition

Given a large set of independent variables, stepwise regression is a very handy way to find a subset of independent variables that together predict the dependent variable. If the researcher has no particular theoretical basis for expecting certain independent variables to be useful as predictors, it is tempting to simply plug all the variables into a stepwise regression routine and let the procedure find a good-fitting model.

Here is a real example. A researcher has data for 20 individuals on 15 variables (the dependent variable, called DEPVAR, and 14 independent variables, called INDVAR1 through INDVAR14). The assumptions needed for the validity of the overall $F$ test and the individual $t$ tests were met: The individual scores were independent and normally distributed, with homogeneous variance. (How the researcher could be sure about this will be made clear shortly.)

Since the researcher had no particular theory available to help construct a model, he used stepwise regression to build a model for him. The results were as follows:

$$\text{predicted DEPVAR} = 15.76 - .34(\text{INDVAR5}) + .57(\text{INDVAR6})$$

| | | |
|---|---|---|
| $t$ scores: | $-2.25$ | $2.94$ |
| $p$ values: | $.04$ | $.01$ |

$R$-squared $= .418$      $F = 6.12$ with 2 and 12 $df$      $p = .010$

So a regression formula using two independent variables was able to account for over 40% of the variability of the dependent variable. The overall $F$ test was significant, as were both of the $t$ tests for the individual regression coefficients for the independent variables. Nevertheless, the researcher was not particularly impressed.

Why was the researcher not impressed? Because the researcher (who happens to have been the author) generated all the data by computer *using a random number generator*! In other words, there were no true relationships whatever among any of the 15 variables. Any observed relationships, including the "significant" $R$-squared, were due entirely to chance. (This explains how the researcher could be so sure about the assumptions. The data were generated as independent random selections from normal distributions. Thus he could be sure that the dependent variable had a normal population distribution, with homogeneous variance over all settings of the independent variables.)

How could this have happened? Doesn't a significant $R$-squared mean that the probability of observing an $R$-squared this large purely by chance is small? Was the researcher simply observing a rare event?

The answer is that the researcher did observe a rare event. In any one test, the chance of observing a statistic as great as the one observed is quite small. But in doing the stepwise procedure the computer did so many tests that the chance of finding one or two significant results was not particularly small at all, and, of course, the significant results were the only ones the stepwise procedure bothered to tell us about. If you do enough tests, you are almost certain to find something that looks significant, even if in reality the data are random numbers!

This illustrates the danger of the fishing expedition—fishing for significant results in a sea of tests. When multiple tests are performed, even if we control the chance of a type I error in any one test, the chance of making a type I error in the study as a whole is considerably greater than the nominal value of $\alpha$.

*This means that stepwise regression is dangerous*, *particularly with a large number of variables and a small sample size*. A handy rule of thumb is to require that the sample size be at least 10 times the number of variables. (So, for our example above, the researcher should not have proceded unless he had had at least 150 individuals.) Even then, results of stepwise regression should be considered preliminary and need to be confirmed on additional data.

## Misuse 2: Finding the important variables

Let's return to our personnel department example of Section 13.1. Recall that the personnel department was trying to find a combination of tests that together would predict sales performance. Using stepwise regression (with an adequate sample size, by the 10-to-1 rule), the department arrived at a regression formula that included two of the original set of eight independent variables.

The regression formula provides a combination of two variables that accounts for 48.3% of the variability in SALES. It does *not*, however, tell us that the two variables that were included in the model are the two best predictors of SALES. Why not? Because the regression procedure was not seeking the two best predictors—it was seeking the best *combination* of predictors. As we noted in Section 13.1, the two best predictors, *taken individually*, were TEST3 and TEST4. The reason TEST4 was not included in the model is that TEST3 and TEST4 are highly correlated with each other. (You should confirm this in the correlation matrix.) Thus after TEST3 was included in the model, TEST4 made little *additional* contribution to the prediction of SALES. It was another variable—TEST6—that made the greatest additional contribution to predicting SALES, given that TEST3 was already in the model.

What this means is that stepwise regression results should not be used to find that subset of a set of independent variables that are *individually* important as predictors of the dependent variable. If that sort of information is desired, it may be obtained simply by looking at the individual correlations of each independent variable with the dependent variable.

On the other hand, stepwise regression is useful in finding out which variables make *additional* contributions to the prediction of the dependent variable, given that one or more independent variables are already in the model. As we saw in our example, it is entirely possible for a variable to have a low correlation with the dependent variable on its own, but nevertheless make a substantial additional contribution to prediction after a variable is in the model.

## STUDY QUESTIONS—*Sections 13.1 and 13.2*

**13.2.1** The administration of a large university wished to identify those assistant professors who could be considered underpaid or overpaid, in the sense that their salaries seem out of line with what would be predicted from their years of experience, formal qualifications (i.e., degrees), number of publications, and general level of salaries prevalent in their particular field. There was also concern about whether there were sex differences in salary that could not be accounted for by any of the other factors. The consulting statistician wrote down the following model:

$$\text{SALARY} = B0 + B1(\text{YEARS}) + B2(\text{DEGREE}) + B3(\text{PUBS}) + B4(\text{FIELD}) + B5(\text{SEX}) + \text{random variation}$$

where
YEARS = number of years employed at the assistant professor level
DEGREE = 1 if the faculty member had earned the terminal degree in his or her field; 0 if not

PUBS      = number of publications listed for the faculty member in the past 5 years

FIELD     = national median salary for university faculty in the particular field

SEX       = 0 for males, 1 for females

The statistician obtained data on these variables for a total of 130 assistant professors at the university and obtained the least-squares regression formula. The results were as follows:

predicted SALARY = −787 + 924(YEARS) + 148(DEGREE)
                          + 109(PUBS) + 1.06(FIELD) − 2300(SEX)

| Variable | $t$ Score | $p$ |
|----------|-----------|-----|
| YEARS    | 9.52      | .000 |
| DEGREE   | .34       | .735 |
| PUBS     | 1.52      | .130 |
| FIELD    | 18.79     | .000 |
| SEX      | −5.91     | .000 |

$R$-squared = .848      $F$ = 138.28 with 5 and 124 $df$      $p$ = .000

Correlation matrix:

|        | SALARY | YEARS | DEGREE | PUBS | FIELD | SEX |
|--------|--------|-------|--------|------|-------|-----|
| SALARY | 1.000  | .593  | .377   | .504 | .649  | −.177 |
| YEARS  | .593   | 1.000 | .608   | .759 | −.037 | .096 |
| DEGREE | .377   | .608  | 1.000  | .487 | −.032 | .006 |
| PUBS   | .504   | .759  | .487   | 1.000 | .006 | .090 |
| FIELD  | .649   | −.037 | −.032  | .006 | 1.000 | −.046 |
| SEX    | −.177  | .096  | .006   | .090 | −.046 | 1.000 |

a.   What would be the predicted salary of a female assistant professor with 6 years' experience, a terminal degree, and seven publications, whose field has a median salary of $30,000?

b.   What would be the predicted salary of a male assistant professor in the same field with the same experience, degree, and number of publications? What is the difference in predicted salary between the female and male assistant professors?

c. An administrator looked at the regression results and commented that neither number of publications nor the having of a terminal degree had a significant effect on salary. Criticize the administrator's reasoning and point out evidence to the contrary.

d. How should the administration identify those assistant professors who might be considered underpaid? One way would be to find those whose actual salary was substantially below the salary predicted by the model. What would be the ramifications of using the model for this purpose? Can you suggest a modification that might be more equitable?

**13.2.2** [Continuation of Exercise 13.2.1] Now let's consider the problem of checking the validity of the assumptions of independence, normality, and homogeneity. We continue with the salary example.

a. State the three assumptions clearly.

b. Up to now, we have been stating the assumptions with respect to scores on the dependent variable. However, we could also state them in terms of the "random variation" component in the model. To see why, think of the model as being composed of two parts: the deterministic part and the random variation part. The model asserts that an individual's salary can be considered a sum of the two parts, where the deterministic part (as the name implies) is entirely determined by the independent variables. Given individuals with the same scores on the independent variables, there is no variation in their predicted scores—these are fixed, according to the model. The only reason such individuals vary, according to the model, is because of random variation. Thus when the assumptions refer to the distribution of salaries for a given setting of the independent variables, they could equivalently refer to the distribution of the random variation part of salaries, since that is where all the variation occurs. With this in mind, restate the assumptions in terms of random variation.

c. The assumption of independence requires that random variation be truly random—that is, that one individual's deviation from the model be unrelated to any other individual's deviation. Can you think of any problems with this assumption in our salary example? How might the fact that salaries for all the members of a single department are assigned by the same individual affect this assumption?

d. The sample provides an estimate of each individual's random variation. How can that estimate be calculated? (*Hint*: Random variation is the difference between the individual's actual score on the dependent variable and that individual's score as predicted by the model.)

e. What should the mean of all the random variation scores be?

f. Given that each individual's random variation can be estimated, how might the assumption of normality be checked? What sort of picture could you draw?

g. How could the assumption of homogeneity of variance be checked? Suppose that we were to plot our estimates of random variation against the predicted scores. If the homogeneity assumption is valid, what should the scatterplot look like?

### 13.3  Plotting the residuals can help identify violations of assumptions.

*Residuals and predicted values*

As we noted in the previous set of study questions, the assumptions of independence, normality, and heterogeneity can be stated either in terms of the scores on the dependent variable (as we have been doing up to now) or in terms of the random variation part of the model. If we can estimate each individual's random variation, we can look at the distribution of all such estimated random variation scores in the sample. This should give us an idea of what the population distribution of random variation looks like.

How should we estimate an individual's random variation score? All we have to do is take the difference between the individual's actual score on the dependent variable and the individual's *predicted* score—predicted, that is, using the sample regression formula. Since the sample regression formula is itself an estimate, this will not give us the individual's actual random variation score, but it will give us our best sample estimate.

This difference is called a *residual*. Thus for a given individual, *the residual is the actual score minus the predicted score*. The residual is what is left over—the residue—in the actual score after the "predictable" part is taken out.

Note that the residuals always have a mean of zero as a consequence of the way the least squares regression formula is calculated. Note also that the residuals are the sample version of the "vertical deviations" we were talking about in Section 12.3 in the context of simple regression. Thus the sum of squared residuals is none other than SSE.

*Using residuals to check assumptions*

By examining the residuals, we can get an idea of how reasonable the three assumptions of independence, normality, and homogeneity are in a given situation. There are three kinds of scatterplots we can use for this purpose, each useful for checking a particular assumption. We consider each in turn.

*Checking for independence: the casewise plot of residuals*

The independence assumption requires that each individual's score be unrelated to that for all other individuals. In particular, we require that the occurrence of, say, a large positive random variation score for one individual should not have any effect on the random variation score for any other individual.

Usually, we rely on the design of the study to ensure that the scores are really independent. Sometimes, however, dependencies can occur among individuals in subtle ways. This is particularly apt to occur if the data consist of observations taken over time—for example, daily stock prices, or monthly temperatures, or hourly blood pressures. If there are time trends in the data, the scores for individuals close together in time will tend to be similar.

How can we check for such lack of independence? One simple way is to plot the individual residuals versus time (or more generally, versus the order the scores appear in the data set). Such a plot is called a *casewise* plot of residuals. If the scores are independent, there should be no particular pattern in the plot—the residuals should be evenly scattered about their mean of zero. Figure 13.5 shows an example of such a plot (as plotted by Statpal) for the car cost data set (which, we suppose, was entered into the computer in the order the data were collected). There is no apparent pattern in the plot.

Now consider Figure 13.6, which is a plot of residuals from a regression of computer response on number of users over a 3-day period. The residuals do not fluctuate randomly; instead, they meander around the zero line. This indicates a lack of independence among the residuals.

Another use of the casewise plot is to detect outliers—that is, data points with unusual or extreme scores. Such points can have an inordinately large impact on the regression results and often indicate some substantive problem with the data collection. Figure 13.7 shows a casewise plot in which two outliers are readily detectable (cases 7 and 19). Outliers must be examined individually to determine how to handle them. If they seem to represent points that are "marching to a different drummer," they should be excluded from the analysis and treated separately.

Note, incidentally, that under some circumstances, an outlier can have such a large impact on the least squares solution that the regression surface will be forced to come close to the outlier. When this happens, the outlier will have a small residual, which means that a casewise plot of residuals will fail to show the outlier as anything unusual. This type of problem is usually the result of faulty data entry—which again emphasizes the importance of checking the data before starting the analysis!

### Checking for normality: the normal probability plot

The normality assumption requires that the random variation scores have a normal distribution whatever the settings of the independent variables. If this assumption holds, the sample estimates of the random variation scores—that is, the residuals—should also have a normal-looking distribution.

A simple way to check for normality, then, is to look at a histogram of the residuals. Figure 13.8 shows such a histogram for our car cost data, using the model with both mileage and number of cylinders (from Section 12.6). Note that the histogram looks at least mound-shaped, which is generally sufficiently close to normality.

A slightly fancier way to check for normality is to prepare a *normal proba-*

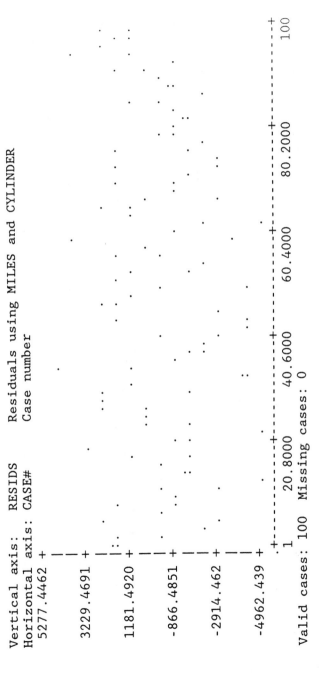

**Figure 13.5**   Scatterplot from Statpal.

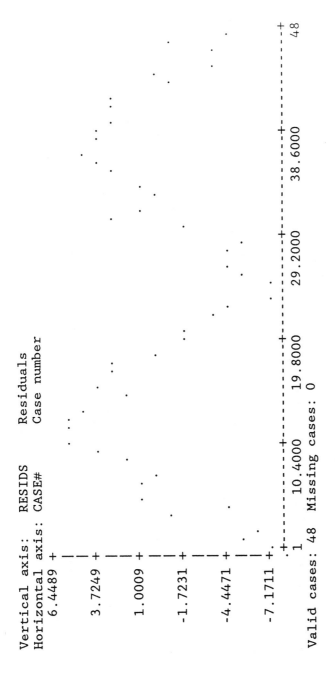

**Figure 13.6** Scatterplot from Statpal.

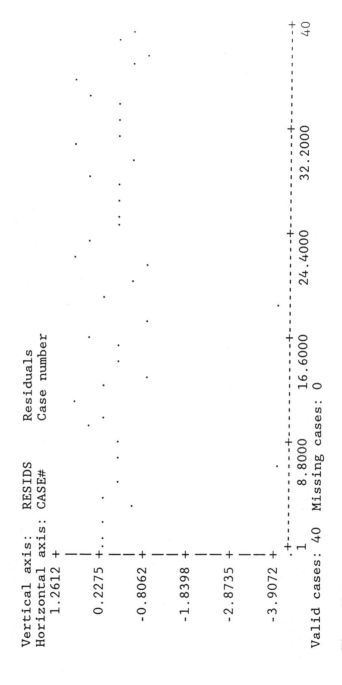

**Figure 13.7**  Scatterplot from Statpal.

```
For variable: RESIDS       Residuals using MILES and CYLINDER
      Lower          Upper            Frequency
      bound          bound         N       prop.

[ -5000.00      -4000.00  )        5      0.050    *****
[ -4000.00      -3000.00  )        4      0.040    ****
[ -3000.00      -2000.00  )       11      0.110    ***********
[ -2000.00      -1000.00  )       13      0.130    *************
[ -1000.00        0.0000  )       17      0.170    *****************
[    0.0000     1000.0000 )       12      0.120    ************
[ 1000.0000     2000.0000 )       18      0.180    ******************
[ 2000.0000     3000.0000 )       12      0.120    ************
[ 3000.0000     4000.0000 )        6      0.060    ******
[ 4000.0000     5000.0000 )        1      0.010    *
[ 5000.0000     6000.0000 ]        1      0.010    *

Valid cases: 100    Missing cases: 0
```

**Figure 13.8**    Histogram from Statpal.

*bility plot* of the residuals. A normal probability plot is a scatterplot of the residuals (on the vertical axis) versus the scores that would be expected from a normal distribution with the same mean and standard deviation. If the residuals are (approximately) normally distributed, the points in the normal probability plot should form a straight line. Appendix C describes how to find the scores for a normal probability plot. Figure 13.9 is the normal probability plot for the same data as in Figure 13.8. Note the generally straight line.

## Checking for homogeneity: the plot of residuals versus predicted scores

The homogeneity assumption requires that for all settings of the independent variables, the variance of the random variation scores be the same. A handy way to check this is to look at a scatterplot of residuals (on the vertical axis) versus predicted scores (on the horizontal axis). The predicted scores represent all observed settings of the independent variables, so if the homogeneity assumption is valid, we would expect to see the points have about the same degree of (vertical) spread over the entire range of predicted values.

Figure 13.10 shows the plot of residuals versus predicted scores for the car cost data using the model with two independent variables. Note that as we move across the plot from left to right, the degree of variability in the residuals seems fairly consistent.

Now consider Figure 13.11, which shows a plot of residuals versus predicted

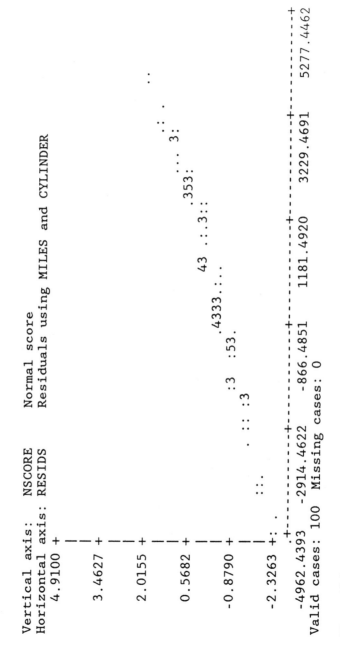

**Figure 13.9**   Scatterplot from Statpal.

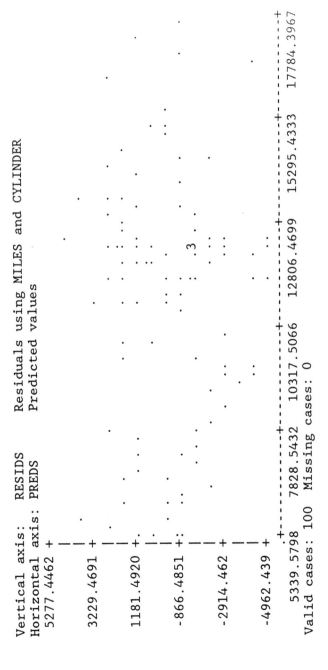

**Figure 13.10** Scatterplot from Statpal.

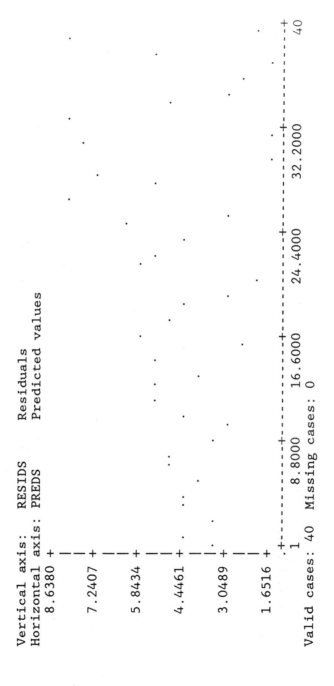

**Figure 13.11**  Scatterplot from Statpal.

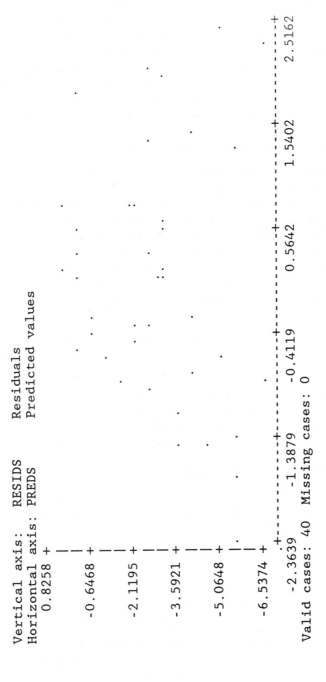

**Figure 13.12** Scatterplot from Statpal.

scores for a different data set (and regression formula). Notice the distinct tendency for individuals with lower predicted scores to vary less in their residuals. This indicates a problem with the homogeneity assumption, which can be addressed using one of the methods described in Section 13.4.

Incidentally, homogeneity of variance has a special name: *homoscedasticity*, from a Greek word meaning "scatter." The absence of homoscedasticity is, of course, *heteroscedasticity*. (Heteroscedasticity is tied with multicollinearity for the title of "common statistical term with the most syllables.")

## *Checking the adequacy of a model*

Another use of the scatterplot of residuals versus predicted scores is to help identify problems in the specification of the model. Consider Figure 13.12. Although the degree of spread of the residuals seems fairly constant across predicted scores, there is a distinct bend in the scatter.

This type of pattern indicates that a nonlinear relationship exists between the dependent variable and the independent variables, and that a different model may be more appropriate. One way to approach this problem is to use *polynomial regression*—that is, use a polynomial in the independent variables as the model. For example, if we are using only one independent variable, we could use the model:

dependent variable $= B0 + B1$(independent variable)
$+ B2$(independent variable squared)
$+$ random variation

To use a standard regression program to fit this model, all we need to do is calculate the square of the independent variable and include it in a multiple regression together with the (unsquared) independent variable. A significant coefficient for the squared term indicates a departure from linearity.

## STUDY QUESTIONS—*Section 13.3*

13.3.1  [Statpal exercise—Coronary Care data set (see Appendix A)]

    a.  Use the REgression procedure to obtain the coefficients for the regression formula

        predicted NDAYS $= b0 + (b1 \times \text{AGE}) + (b2 \times \text{HADMI})$

        Recall that NDAYS is the number of days hospitalized, and HADMI equals 1 for patients who had had a myocardial infarction, 2 for those who had not.

    b.  Interpret the coefficients for AGE and HADMI.

    c.  Determine if the coefficients for AGE and HADMI are significantly different from zero.

    d.  After you have displayed the results, choose the RESiduals command. Specify that you wish to compute residuals and give the name RESIDS to the variable to hold the results.

    e.  Also have predicted values calculated, specifying the variable PREDS to hold the results.

    f.  Now go back to the Main Menu and use the HIstogram procedure to obtain a histogram for RESIDS. Comment on the histogram in terms of the assumptions you made in doing the tests in part c.

    g.  Use the SCatterplot procedure to plot PREDS on the horizontal axis versus RESIDS on the vertical axis. Again, comment on the resulting plot in terms of the assumptions you made in doing the tests in part c.

**13.3.2**  [Statpal exercise—Coronary Care data set (see Appendix A)]

    a.  Use the REgression procedure to obtain the coefficients for the regression formula

predicted PAINTIME $= b0 + (b1 \times$ AGE$) + (b2 \times$ HADMI$)$

The variable PAINTIME is the length of time (in hours) the patient reported having experienced the symptoms (generally chest pain) that led to their coming into the emergency room. Recall that HADMI equals 1 for patients who had had a myocardial infarction, 2 for those who had not.

    b.  Request that residuals be calculated, naming the resulting variable RESIDS. (If you already have a variable called RESIDS, you may overwrite it.)

    c.  Request that predicted values be calculated, naming the resulting variable PREDS. (If you already have a variable called PREDS, you may overwrite it.)

    d.  Go back to the Main Menu and use the HIstogram procedure to obtain a histogram of the variable RESIDS. Comment on the results.

    e.  Use SCatterplot to plot PREDS on the horizontal axis against RESIDS on the vertical axis. Comment on the results.

    f.  How might you deal with the problems evident in the plots?

## 13.4  Violations of assumptions can be treated by a variety of methods.

Our discussion in this section is necessarily a brief treatment of a complex topic. If you are in doubt, talk to a statistician.

### Effects of violations of the assumptions

The reason we need to be concerned about the validity of the assumptions of independence, normality, and homogeneity is that violations of the assumptions can lead us to make incorrect inferences about the parameters of the regression model. We might assert, for example, that a particular regression coefficient is different from zero with 95% certainty when our actual level of certainty should be much less. Our estimates of the parameters will generally not be biased but we will tend to give them more credence than we should. In particular, the $F$ test and $t$ test we discussed earlier will give incorrect $p$ values if the assumptions do not hold.

Some violations are more serious than others, and in some situations even large deviations from the ideal have little effect on the results. Some situations call for a greater degree of fussiness than others. Indeed, the purpose of the study is an important factor in deciding whether remedial measures should be taken.

### Lack of independence

If the individual scores on the dependent variable (or equivalently, the random variation scores) are not mutually independent, the standard errors (and therefore the $t$ tests) for the regression coefficients can be seriously incorrect. As we mentioned in Section 13.3, observations taken over time are often related to other observations taken around the same time. Thus we are particularly concerned with this problem with time-series data.

One way to handle this problem is to use a type of model that explicitly takes account of the relationships among individual scores. The general category of techniques called time-series analysis, discussed in many books on the subject, includes methods that make use of such models.

### Heteroscedasticity

The problem of nonconstant variance in the random variation scores can be manifest in several ways. Most frequently, the problem is that the residuals become more spread out as the predicted scores increase (as in Figure 13.11). When this occurs, one way to reduce the severity of the problem is to use some type of transformation on the data so as to change the large numbers more than the small numbers. For example, taking the square root of the scores on the dependent variable (provided that they are all positive) will tend to reduce the large numbers more than the small numbers and may therefore ''stabilize'' the variance. Taking the logarithm is another popular transformation.

Another approach to the problem of heteroscedasticity, called *weighted least squares*, is appropriate when the standard deviation of the random variation scores increases in some systematic fashion. The idea of weighted least squares is to weight the observations in inverse proportion to their variance. Information on weighted least squares can be found in more advanced books on regression.

### Non-normality

In general, nonnormality is a less severe problem than heteroscedasticity. The $F$ test and $t$ tests we have described are ''robust'' with respect to violations of the normality assumption—that is, they tend to give correct $p$ values even when the random variation scores are not normally distributed. Thus special treatment of the problem of nonnormality is usually not needed. If the problem occurs in conjunction with heteroscedasticity, the same transformations that reduce the latter usually reduce the former.

## Multicollinearity

One additional problem that sometimes arises in multiple regression does not represent a violation of any assumption, but nevertheless can cause serious difficulties. We briefly alluded to this problem, called *multicollinearity*, in our discussion of stepwise regression. Multicollinearity refers to the existence of correlation among independent variables in a regression model. Generally, when authors refer to multicollinearity, they mean that there is a high degree of correlation among independent variables, since low or moderate correlation among independent variables does not present much of a problem.

Why should high correlation among independent variables in a multiple regression model cause a problem? To see why, think about the interpretation of a regression coefficient in a multiple regression model. For example, in our stepwise regression example about personnel selection tests, the coefficient for TEST3, in a formula that also included TEST6, was 471.78. The interpretation of the coefficient is as follows: Each 1-point increase in an applicant's score on TEST3 is associated on the average with a $471.78 increase in SALES, provided that the applicant's score on TEST6 remains constant.

It is the last phrase—the proviso that other variables in the model remain constant—that makes clear why a high degree of correlation among independent variables leads to problems. It makes no sense to talk about the effect of a change in a variable holding another variable constant if the two variables change together to a great extent. As an extreme example, suppose that we had included TEST3 in the model twice. Then any combination of coefficients that add up to 471.78 would do equally well for the two occurrences of TEST3 in the model. That is, there would be an infinite number of possible solutions for each of the two coefficients, all doing equally well!

Although including the same variable twice would be rather silly, this situation is simply an extreme example of a very common problem. It is not unusual to have available several measures of similar constructs, all intended to be used as independent variables to predict some dependent variable using regression. To the extent that the independent variables are correlated among themselves, the regression coefficients become "unstable," in the sense that they are subject to change greatly from sample to sample. The overall fit is not impaired and the validity of the $F$ test is not affected, but the standard errors of the coefficients are apt to be so large that it is difficult to draw conclusions about them.

How can we tell if multicollinearity is a problem before adding a variable to a regression model? One way is to use the measure called *tolerance* (which we mentioned as one of the criteria involved in selecting variables in stepwise regression). Tolerance measures the degree to which a particular variable is independent of the other variables in a regression model. The formula is

tolerance of variable $j = 1 - R_j^2$ on $1, 2, \ldots, j - 1, j + 1, \ldots, k$

where the $R^2$ term indicates the $R^2$ obtained using the other $k - 1$ independent variables in the model to predict variable $j$. Thus tolerance represents the proportion of the variance of variable $j$ that is not explainable by a linear combination of the other independent variables in the model. A tolerance of zero means that the variable is perfectly correlated with some linear combination of the variables already in the model, while a tolerance of 1 means that the variable is independent of the other variables. Thus a low tolerance for a variable indicates multicollinearity.

There are several ways to handle multicollinearity. The simplest is to choose variables so as to avoid the problem, by excluding variables with low tolerance. Another approach is to combine highly correlated variables into a composite measure and then use the composite measure in the regression. (One way to obtain such composite measures from a set of variables is called *principal components analysis*, described in books on multivariate analysis.)

### STUDY QUESTIONS—*Section 13.4*

**13.4.1**  A researcher in recreation management collected data on the number of bathers at a public beach during 40 weekdays in a summer season, together with information on the average temperature and relative humidity of each day. Using temperature and humidity as independent variables to predict numbers of bathers, the researcher obtained the following regression results:

predicted number of bathers $= 41.70 + 4.81(\text{temperature}) + 2.78(\text{humidity})$

$R$-squared $= .394$     $F = 12.00$     $df = 2, 37$     $p < .001$

   a.  Use the regression formula to predict the number of bathers on the beach for a day with a temperature of 82°F and a humidity of 80%.
   b.  What would you predict the number of bathers to be for a day with a temperature of zero degrees and a humidity of zero?
   c.  The term *extrapolation* refers to the practice of trying to apply regression results beyond the range of values actually observed. If you used the regression formula to answer the previous question, you have extrapolated the results. Let's consider an extrapolation in the other direction. What would you predict the number of bathers to be for a day with a temperature of 120°F and a humidity of 90%?
   d.  Under what conditions would it make sense to extrapolate regression results? Under what conditions might extrapolation yield nonsense?

**13.4.2**  We have pointed out that violations of assumptions can lead us to state our $p$ value or level of confidence incorrectly. So what? Why is it important to have correct $p$ values? You might want to go back to Chapter 2 and review the logic of statistical inference at this point.

# 14

# Further Topics in Analysis of Variance

## 14.1 Analysis of variance is used to compare group means.

### Situations calling for ANOVA

Recall from Chapter 10 that analysis of variance (ANOVA for short) is used to test for differences among the means of two or more groups. (You might wish to review Sections 10.5 and 10.6 at this point.) One-way ANOVA is used when each individual is grouped only one way. For example, we could use one-way ANOVA to analyze differences in mean activity levels among three groups of rats, where each group was injected with a different drug. The grouping is on a single basis— that is, drug—with each rat in one and only one group. The basis on which the individuals are grouped is often called a *factor*, with each group defining a *level*. Thus our rat study would have one factor (drug) with three levels (types of drug). We discussed the statistical treatment of one-way ANOVA in Chapter 10.

Two-way ANOVA is used when each individual is grouped two ways. In our rat study, we might also be interested in the differences between two different strains of rat and design the study so that each drug is administered to some rats of each strain. Now we have two factors in the design: strain and drug. Strain has two levels and drug has three. Three-way and higher-way ANOVAs are defined in analogous fashion.

### The scope of our discussion

Our discussion of ANOVA methods will concentrate on applications that are easily performed with common statistical computer packages, to give an idea of the types of questions that can be addressed. We will not cover the methodological issues involved in designing studies.

In particular, we concentrate on those situations in which (1) each individual is measured only once on the dependent variable; (2) levels of each factor are fixed rather than random; and (3) factors are "crossed," which means that each level of a factor occurs in combination with each level of every other factor in the data. This covers many common situations and will serve to introduce the topic. If you are in doubt as to whether the methods discussed in this chapter apply to a particular situation, you should consult a statistician. More complete discussions of experimental designs and related topics can be found in many books on the subject; see, for example, Montgomery (1976).

## The basic ANOVA model

The basic ANOVA model asserts that an individual's score on the variable of interest can be considered a sum of three parts: (1) the overall mean, (2) a constant associated with the group (or groups) to which the individual belongs, and (3) some random variation assumed to have a mean of zero over all individuals. The second part may in turn be broken down into several different terms, depending on how many factors there are in the model.

For example, for our study of drug effects on activity levels in rats, a one-way ANOVA model could be written as follows:

activity level for rat number $i$ (receiving drug type $j$)
 = overall mean activity level for all rats
  + constant effect of drug type $j$
  + random variation specific to rat $i$

Just as in regression, the model can be divided into a deterministic part (consisting of the overall mean and the constant drug effect) and a stochastic part (the random variation). Note that the deterministic part in the model above is simply the mean activity level for the rats receiving drug type $j$. To say that drug $j$ has a certain constant "effect"—let's call it EJ—means that the activity levels for the rats in group $j$ deviate from the overall mean, on the average, by EJ units. Thus the mean activity level for the rats receiving drug $j$ is equal to the overall mean plus EJ. (Of course, EJ can be positive or negative.)

## Additive model for two-way ANOVA

For the two-way ANOVA including both drug type and strain, a model can be written as follows:

activity level for rat number $i$ (receiving drug type $j$, of strain $k$)
 = overall mean activity level for all rats
  + constant effect of drug type $j$

+ constant effect of strain $k$

+ random variation specific to rat $i$

This model represents the effects of receiving a certain drug and being of a certain strain as entirely *additive*. In particular, the effect of getting a certain drug is represented as constant regardless of which strain the rat is. Equivalently, the effect of being of a particular strain is represented as constant regardless of which drug the rat is receiving.

## ANOVA as a regression method

Although most textbooks and computer programs treat analysis of variance and regression as separate topics, ANOVA is itself a form of regression. In a situation that calls for ANOVA, the basis for choosing one technique over the other is simply convenience. Sometimes it is more convenient to state the problem in regression terms, whereas other times the ANOVA terminology is more useful. In fact, most computer programs perform ANOVA work by changing the problem into a regression problem, producing the regression results, and then reporting the results in ANOVA terms.

How are ANOVA and regression connected? How can you translate an ANOVA problem into regression terms? The trick, as we have seen in several study questions, is to use dummy variables. By defining dummy variables appropriately, it is possible to change an ANOVA model into a regression model, so that standard regression procedures can be used. We look at an example of how this is done in a study question.

Although the details of defining dummy variables are beyond the scope of this introductory discussion, we can state the essential equivalence between the ANOVA and regression models quite simply. Both forms specify that an individual score is made up of a deterministic part and a stochastic part. The only difference between the ANOVA and regression forms is in the way the deterministic part is written down.

## STUDY QUESTIONS—Section 14.1

**14.1.1**   Forty cigarette smokers, including 20 men and 20 women, are recruited to participate in a study of the effectiveness of various methods for modifying smoking behavior. Each smoker is randomly assigned to one of four groups, with five men and five women in each group. Individuals in group A were shown a series of ads emphasizing the health effects of smoking, and met together as a support group to help each other find ways to stop or cut down on smoking. Individuals in group B were shown the ads but did not meet as a support group. Individuals in group C met as a support group but did not see the ads. Individuals in group D received neither the ads nor the support group. After 4 weeks, all individuals were asked to record the number of cigarettes smoked during one 24-hour period. This number was used as the dependent variable.

a. This study can be viewed as a design with three factors (ads, support group, and sex), each with two levels. Write an ANOVA model for this study.

b. Now define three dummy variables, called ADS, SUPPORT, and SEX, as follows:

ADS       = 1 if the individual saw the ads, 0 if not
SUPPORT = 1 if the individual was in a support group, 0 if not
SEX       = 1 for women, 0 for men

Consider the following multiple regression model:

number of cigarettes smoked
$$= B0 + B1(\text{ADS}) + B2(\text{SUPPORT}) + B3(\text{SEX})$$
$$+ \text{random variation}$$

In specific terms, what is the meaning of $B0$ in this model? (*Hint*: $B0$ is the average number of cigarettes smoked for the population of people for whom all three independent variables are zero. Who are those people?)

c. What is the meaning of $B1$? Of $B2$? Of $B3$?

d. Suppose you wished to test the null hypothesis that the ads had an effect. How could you do this using the regression model? Be specific.

e. Note that $B0$ in the regression model is not the same as the "overall mean" in the ANOVA model. Nevertheless, each of the parameters in the regression model can be written in terms of the ANOVA parameters. Explain how each of the following identities can be derived:

$B0$ = overall mean + effect of not seeing ads
                        + effect of not being in support group
                        + effect of being male
      = mean of males in group D

$B1$ = effect of seeing ads − effect of not seeing ads
      = mean of groups A and B − mean of groups C and D

(similarly for $B2$ and $B3$)

f. If the effect of seeing the ads is the same as the effect of not seeing the ads, we can simply say that the ads had no effect one way or the other. In view of the correspondences between ANOVA and regression parameters, how could we perform a hypothesis test of the null hypothesis that ads have no effect?

14.1.2 [Continuation of Exercise 14.1.1] Suppose that the ads had the effect of making women smoke less but did not affect men at all (on the average). How could you modify the ANOVA model to allow for this possibility?

## 14.2 Interaction means that the effect of one factor depends on another factor.

### Interaction between factors

The additive model for two-way ANOVA that we discussed in the preceding section made no provision for the possibility that a drug might have a different effect on one strain than on the other. The effect of a particular drug and the effect of being of a particular strain were represented separately, with the drug effect constant regardless of strain and the strain effect constant regardless of drug.

But suppose that the effect of a drug differs between strains of rats. For example, suppose that the two strains of rats are (1) Wistar-Kyoto (WKY) rats, a normal strain, and (2) spontaneously hypertensive (SH) rats, a strain that tends to be hyperactive. Certain drugs could have the effect of making the activity levels of WKY rats higher but the activity levels of SH rats lower. The additive model has no way of representing this sort of situation.

A situation in which the effects of one factor depend on another factor is said to exhibit *interaction* between the two factors. If the effect of a drug depends on which strain we are talking about, then drug and strain interact.

## ANOVA models with interaction terms

Let's modify our two-way ANOVA model to allow for the possibility of interaction between drug and strain:

activity level for rat number *i*
  (receiving drug type *j*, of strain *k*)

  = overall mean activity level for all rats
    + constant effect of drug type *j*
    + constant effect of strain *k*
    + constant effect of drug type *j* AND strain *k*
    + random variation specific to rat *i*

The term we have added represents some constant effect specific to a particular *combination* of drug and strain. Since this effect can be different for each of the six (three times two) combinations of drug and strain, the addition of the new term to the model provides for different drug effects for the two strains—that is, for interaction between drug and strain.

## Main effects

The two terms representing the separate effects of the factors—one for drug and one for strain—are called *main effects*. More generally, when we say that a factor has a nonzero main effect, we mean that at least one of the levels of the factor has some effect different from the other levels of that factor.

What does a main effect represent? Let's first consider our two-way ANOVA model without interaction. If, for example, we have a main effect of drug, then the mean activity level of the rats in at least one of the drug groups differs from the mean of the rats in the other drug groups. Similarly, a main effect of strain indicates that the strains differ in their mean activity levels.

When an interaction is present, the interpretation of main effects gets a bit muddy. To see why, consider what the presence of interaction indicates: An interaction between two factors means that the effect of one factor depends on the other. Thus if an interaction is present, it may not be meaningful to talk about a constant main effect for a factor.

Therefore, when an interaction term is included in a model, a reasonable approach is to consider the interaction first. If we find (through the hypothesis test to be described in Section 14.3) that the interaction term is not needed in the model, we can proceed to interpret the main effects.

## 14.3   The *F* test is used to test for significant main effects and interactions.

### Inferences about main effects and interactions

Having noted that the ANOVA models we have discussed are simply regression models in disguise, we should not find it surprising that we can draw inferences about the parameters of those models using the same tests as in regression. In particular, we can use an *F* test for inferences concerning the model as a whole as well as for tests on main effects or interactions.

Once again, let's review the logic of hypothesis testing, this time in the context of our rat activity-level example. Suppose that we have 15 rats in each of the two strains (WKY and SH), with five rats in each strain receiving each of the three drugs (A, B, or C). The rats of each strain are randomly allocated among the three drug types. Each rat's activity level is measured (using a special device that counts movements during a 5-minute period) before and after the administration of the drug. The change in activity level (positive scores indicating more activity, negative indicating less activity) is the dependent variable of interest. The scores for the 30 rats are as follows:

| Strain | Drug type | | |
| | A | B | C |
| --- | --- | --- | --- |
| WKY | 18   22   26 | 16    9   10 | −8        3   5 |
|      | 35   35 | 15  −6 | −6   −5 |
| SH | −23  −17  −9 | −6   14    5 | 3        3   1 |
|    | −17  −26 | −1   13 | 6  −13 |

It will also be handy to have the mean and standard deviation of each of the six strain-by-drug groups:

| Strain | Drug | | | |
|---|---|---|---|---|
| | A | B | C | All |
| WKY | | | | |
| Mean | 27.200 | 8.800 | −2.200 | 11.267 |
| Std. dev. | 7.662 | 8.815 | 5.805 | 14.360 |
| SH | | | | |
| Mean | −18.400 | 5.000 | .000 | −4.467 |
| Std. dev. | 6.542 | 8.689 | 7.483 | 12.580 |
| Both | | | | |
| Mean | 4.400 | 6.900 | −1.100 | 3.400 |
| Std. dev. | 24.954 | 8.491 | 6.420 | 15.491 |

We note first that the group means are not all the same. Does this mean, then, that drug and strain necessarily have an effect on activity level? Of course not—indeed, the groups would not have exactly the same means even if all six groups were from the same strain and received the same drug, because the activity levels will vary by chance. The question (as always) is whether the sample differences are large enough to conclude that the populations differ.

### Graphing the means

Before proceding with the analysis of the data, it is good practice to prepare a graph of the means of the groups. Figure 14.1 shows how it can be done for a two-way design. Notice that we have placed the dependent variable on the vertical axis, with one of the factors on the horizontal axis (drug type, in this case). The other factor is indicated by the symbol used on the plot: W for the WKY rats and S for the SH rats. For a three-way or higher-way design, we could use symbols representing various combinations of levels on two or more factors.

The graph of the means is especially useful for analyzing interaction between factors. For our two-way design, if drug and strain do not interact, the drug effects would be (roughly) constant for both strains. This means that the two lines representing the two strains would be roughly parallel. Lines that are not parallel suggests that the drug effects differ between the two strains—that is, that drug and strain interact.

In Figure 14.1 it appears that the two factors do interact, since the lines actually cross. Drug A seems to work in opposite directions for the two strains, producing an increase in activity levels for WKY rats but a decrease for SH rats. But, of course, we have not yet determined whether the effects are greater than what we might expect by chance. To do this, we use the $F$ test.

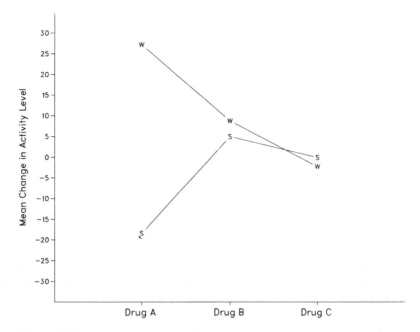

**Figure 14.1**    Mean change in activity level for each strain by drug route in rat example.

## Assumptions required for the F test

Since ANOVA is a form of regression, the assumptions required for the $F$ test are identical to those we discussed in Chapter 8: independence, normality, and homogeneity.

The independence assumption requires that the random variation for each individual be independent of the random variation for every other individual. This assumption requires special methods for situations in which repeated measurements for the same individuals are being analyzed. In designs such as the one we have been discussing in our activity-level example, we rely on careful randomization to ensure us that the independence assumption is met.

The normality and homogeneity assumptions refer to the distributions of random variation for individuals in each combination of levels: We require that the population defined by each combination of levels have a normal distribution with a common standard deviation (usually denoted $\sigma$). Of the two, the homogeneity assumption is generally more critical. Using equal sample sizes in all groups is one way to minimize the impact of heterogeneous variances. The transformations discussed in Chapter 13 are also useful in some situations.

## The F test in ANOVA

The $F$ test used in ANOVA to test for significant main effects and interactions is, of course, the same as that in regression. Just as in regression, we partition the total sum of squares of the dependent variable, with each part representing the variability that can be explained by a particular main effect or interaction. As in regression, we divide each sum of squares by its degrees of freedom to produce a mean square. The $F$ statistic for a particular main effect or interaction is computed as its mean square divided by MSE.

We will not go into the computational details (which can be found in more advanced presentations), but instead we show how to interpret results as calculated by a standard computer program. For our activity-level example, the ANOVA table looks like this:

| Source | SS | df | MS | F | Sig. |
|---|---|---|---|---|---|
| Main effects | | | | | |
| STRAIN | 1856.533 | 1 | 1856.533 | 32.344 | .000 |
| DRUG | 335.000 | 2 | 167.500 | 2.918 | .073 |
| Interaction: STRAIN by DRUG | 3390.067 | 2 | 1695.033 | 29.530 | .000 |
| Error | 1377.600 | 24 | 57.400 | | |
| Total | 6959.200 | 29 | 239.972 | | |

Notice that the sums of squares for the main effects, interaction, and error all add up to the total. This will be the case, in general, only for orthogonal designs. Notice also that the $df$ add up in similar fashion (which will always be the case, whether or not the design is orthogonal).

How can we interpret the ANOVA table? As we noted in Section 14.2, we should start with the interaction term. The $F$ statistic for the interaction term is 29.530 with 2 and 29 $df$, which gives a $p$ value (in the column headed "sig.") of less than .001. Therefore, we can conclude that there is an interaction between strain and drug in our study—as we expected from the graph of the means. So the effect of drug on activity level does indeed depend on which strain is being considered, as we suspected from our examination of the graph of the means.

How about the main effects? Although the main effect of strain was highly significant, and that for drug nearly so, the fact that the two factors interact renders the main effects less meaningful. We can say that there is a significant difference among drugs, ignoring strain—but the interaction means that we should not ignore strain.

## STUDY QUESTIONS—*Sections 14.2 and 14.3*

**14.3.1**  The group means and standard deviations and ANOVA table for the smoking study described in Exercise 14.2.1 are shown below. Recall that the dependent variable was the number of cigarettes smoked during a particular 24-hour period at the conclusion of the study.

| Support group: | Women | | Men | |
| --- | --- | --- | --- | --- |
| | In | Not in | In | Not in |
| Saw ads | | | | |
|   Mean | 7.200 | 7.600 | 7.200 | 14.400 |
|   Std. dev. | 2.775 | 3.912 | 3.114 | 5.128 |
| No ads | | | | |
|   Mean | 7.600 | 9.600 | 11.400 | 15.100 |
|   Std. dev. | 3.286 | 3.209 | 4.270 | 4.667 |

Analysis of variance:

| Source | Sum of squares | df | Mean square | F | Sig. |
| --- | --- | --- | --- | --- | --- |
| Main effects | | | | | |
|   ADS | 40.000 | 1 | 40.000 | 2.621 | .115 |
|   SUPPORT | 122.500 | 1 | 122.500 | 8.026 | .008 |
|   SEX | 176.400 | 1 | 176.400 | 11.558 | .002 |
| Two-way interactions | | | | | |
|   ADS by SUPPORT | .900 | 1 | .900 | .059 | .810 |
|   ADS by SEX | 6.400 | 1 | 6.400 | .419 | .522 |
|   SUPPORT by SEX | 52.900 | 1 | 52.900 | 3.466 | .072 |
| Three-way interaction: | | | | | |
|   ADS by SUPPORT by SEX | 12.100 | 1 | 12.100 | .793 | .380 |
| Error | 488.400 | 32 | 15.262 | | |
|     Total | 899.600 | 39 | 23.067 | | |

a.  Prepare a graph of the means for this set of data.

b.  Using the information above, does the homogeneity assumption seem tenable? What information should you use to examine this?

c. A three-way interaction among ADS, SUPPORT, and SEX would indicate that everything depends on everything else. Was the three-way interaction significant in this case?

d. Based on the ANOVA, which factors have a significant effect?

**14.3.2** [Continuation of Exercise 14.3.1] Some potentially important variables have been left out of the design. For example, consider the amount each individual smoked prior to the study. Certainly we would expect amount smoked prior to the study to bear some relationship to amount smoked after the study, and if our study groups happen to differ in the amount they smoked prior to the study, we might falsely attribute the related differences among the groups in after-study smoking to the effects of our ads or support groups. Therefore, we want to control for prestudy smoking. Another factor we might want to control for is age, which could also be related to amount smoked.

Variables such as these, representing factors we want to control for in our analysis, are called *covariates*. What we want to do, in essence, is perform our ANOVA not on the original data, but rather on data that have been adjusted to remove the effects of the covariates. (Such an analysis is often called an *analysis of covariance*.)

Suppose that we obtained information from each participant on (1) the average number of cigarettes smoked per day during the month prior to the start of the study, and (2) the participant's age. How might we control for these variables in our analysis? (*Hint*: The residuals from a regression of the dependent variable on the covariates represent that part of the dependent variable that is unrelated to the covariates. So if we want to make sure we are analyzing data on smoking that is unrelated to the covariates, how should we proceed?)

**14.3.3** [Statpal exercise]

a. Create a Statpal system file for the rat activity-level example data of Section 14.3. To do this, go into the CReate facility and specify keyboard data entry. Define three variables: (1) ACTLEVEL, the activity-level score for the rat; (2) DRUG, with 1 = A, 2 = B, and 3 = C; and (3) STRAIN, with 1 = WKY and 2 = SH. Then input the three variables for each rat.

b. Use the BReakdown procedure to reproduce the means and standard deviations given on page 261.

c. Use the ANalysis of variance procedure to reproduce the ANOVA table shown in page 261.

**14.3.4** [Statpal exercise—Coronary Care data set (see Appendix A)] Recall that the variable RISk indicates which of three risk groups the patient belonged to, based on the patient's ECG (electrocardiogram) at the time of emergency room treatment. A score of 1 indicates low-risk, as evidenced by normal or unchanged ECG; a score of 2 indicates high-risk, as evidenced by abnormal or changed ECG; and a score of 3 indicates indeterminate risk. Recall also that the variable HADMI indicates whether the patient had had a myocardial infarction, with 1 = yes and 2 = no.

a. Use the BReakdown procedure to find the means and standard deviations on the variable NDAYS (number of days hospitalized) for the six groups defined by the two grouping variables RISK and HADMI.

b.  Draw a graph of the means, similar to Figure 14.1. Put NDAYS on the vertical axis, HADMI on the horizontal axis, and draw one line for each risk group.

c.  Looking at the graph of the means, what does the effect of risk group seem to be on number of days hospitalized? What does the effect of having an MI seem to be? Do the two factors seem to interact with each other?

d.  Use the ANalysis of variance procedure to obtain the ANOVA table. Include the interaction term.

e.  Interpret the results.

# 15

# Analyzing Categorical Data

## 15.1 In many situations, our data are not measurements, but simply counts.

### Examples of categorical data

A sample of birth records for 200 births reveals that 136 occurred in a hospital attended by a physician, 38 occurred in a hospital attended by a certified nonphysician attendant (such as a nurse-midwife), 12 occurred outside of a hospital attended by a physician, 8 occurred outside of a hospital attended by a certified nonphysician attendant, and the remaining 6 occurred outside of a hospital attended by a noncertified attendant. During the previous year the percentage of births in each of the five categories had been 80%, 4%, 10%, 4%, and 2%, respectively. Does the sample evidence indicate that a change has occurred in the relative frequencies of births in these categories? How can we test the null hypothesis that the relative frequencies have not changed?

Here is another example, this time involving two categorical variables. A researcher for a state Department of Corrections wishes to determine whether a first-offense diversion program is effective in reducing recidivism. Data are available on a total of 70 convicted first-time offenders, of whom 35 had been randomly assigned to the diversion program while the other 35 had gone through standard court procedures. For each offender, the researcher consulted court records to determine if the individual had been convicted of a later offense within 2 years of the first offense. The results can be summarized as follows:

|                       | Later conviction? |      |       |
|-----------------------|-------------------|------|-------|
|                       | Yes               | No   | Total |
| Diversion program     | 9                 | 26   | 35    |
| Standard procedures   | 17                | 18   | 35    |
| Total                 | 26                | 44   | 70    |

This type of layout is called a *contingency table*, since it gives the distribution of one variable contingent on the values of the other. As we can see in the table, the offenders in the diversion program had a lower rate of subsequent convictions than those who went through standard procedures: 9/35 versus 17/35. But these 70 offenders represent only a sample from a population. Do the observed data support the conclusion that offenders in the diversion program experience a lower recidivism rate? To answer this question, we need some type of hypothesis test procedure.

Let's consider another example. A survey of 1800 recent retirees from employment at universities asked (among other things) whether the respondent had retired voluntarily or not, and whether the respondent had been in a managerial, technical, administrative, or academic position. The results were as follows:

|                  | Retired voluntarily? |      |       |
|------------------|----------------------|------|-------|
| Type of position | Yes                  | No   | Total |
| Managerial       | 205                  | 80   | 285   |
| Technical        | 327                  | 318  | 645   |
| Administrative   | 420                  | 119  | 539   |
| Academic         | 170                  | 161  | 331   |
| Total            | 1122                 | 678  | 1800  |

Is type of position related to whether or not people in the population from which this sample came retire voluntarily? Again, we need a procedure for hypothesis testing.

These situations differ from what we have discussed in previous chapters in one important respect: The data do not consist of measurements or scores on some quantitative variable, but rather consist of counts of the numbers of individuals in various categories. In this chapter we consider some techniques for analyzing this type of data.

## STUDY QUESTIONS—*Section 15.1*

**15.1.1**  A six-sided die, if it is fair, should yield each of the numbers from 1 through 6 with about equal frequency over a series of throws. Suppose that a die is thrown 60 times, giving the following results:

| Result | Frequency (number of throws) | Relative frequency |
|--------|------------------------------|--------------------|
| 1 | 7 | .117 |
| 2 | 4 | .067 |
| 3 | 12 | .200 |
| 4 | 10 | .167 |
| 5 | 15 | .250 |
| 6 | 12 | .200 |

a.  If the die were fair, what would the distribution of results look like on the average? Any particular run of 60 throws will deviate somewhat from the ideal average run, but what would the ideal average run of 60 throws produce in the frequency column if the die is fair?

b.  Our data, of course, deviate from the ideal average for a fair die. But do they deviate far enough to suggest that the die is not fair? How might you evaluate the degree to which the observed frequencies deviate from the ideal average frequencies?

c.  Suppose that the die is deliberately manufactured to be off-balance such that on the average, the results 1 through 6 come up with relative frequency 10%, 10%, 10%, 20%, 25%, and 25%, respectively. What would be the ideal average frequencies out of 60 throws for such a die, and how might you evaluate the degree to which the observed frequencies deviate from those ideal frequencies?

**15.1.2**  Now let's consider the same types of questions in the context of a contingency table. Suppose that 30 residents of urban areas and 20 residents of rural areas are asked whether or not they approve of the performance of an elected official. Among the 50 respondents, 34 said they approve and 16 said they disapprove.

a.  If approval had nothing to do with location of residence (urban or rural), what would the contingency table look like, on the average? Note that the *average* table need not be restricted to integer counts. (An average need not be a possible value—consider, for example, the famous average family with 2.4 children.) Make up a contingency table showing your answer.

b.  Suppose the survey revealed that 10 urban residents approved of the official. Fill in the rest of the table. Why is it sufficient to report the count for only one cell to be able to fill in the rest of the table?

c.  How might you evaluate the degree to which the observed table you just filled in deviates from the average table you prepared in part a?

d.  Now suppose that a third category of response, "no opinion," was added to a new survey. Also, suppose that three categories of residence location were used: urban, suburban, and rural. A total of 200 people are surveyed. Of the 200, 125 approved, 40 expressed no opinion and the remaining 35 disapproved. Also, of the 200, 60 were urban, 110 were suburban, and the remaining 30 were rural. Prepare the table of average counts that would be expected if location and approval are not related.

e.  For the 3-by-3 situation of part d, suppose that rural residents are really more likely to approve than either suburban or urban residents. Make up a contingency table that would reflect such a difference. As you make up the numbers, note that you are not completely free to make up numbers for all nine cells of the table. How many numbers could you make up before the others were determined?

f.  In a general case of a table with $r$ rows and $c$ columns, with fixed row and column totals, how many numbers can you pick freely before the rest are determined? Try to figure out a general formula. (*Hint*: Note that in the case of a single variable with $k$ categories, with a fixed number of cases, you are free to pick $k - 1$ of the $k$ numbers. Now extend that fact to a two-variable contingency table.)

## 15.2  Expected counts are what we would observe on the average if the null is true.

### Expected frequencies

Let's consider again the first example from Section 15.1. For the study of 200 births, we were interested in determining whether the sample counts of births in the various categories indicate a change from the previous year's known percentages in the population as a whole. In other words, we wish to test the null hypothesis that the population distribution of births among the categories is unchanged. We realize that the sample results will not exactly match the population percentages in each category. But if the sample results are far from what we would expect on the average if the null hypothesis were true, we will reject the null hypothesis.

So to test the null hypothesis that the population percentages are unchanged, a three-step procedure is involved. First, we need to consider what the average frequencies would look like if the null hypothesis were true. Such average frequencies are called *expected* frequencies (or counts)—not because we particularly expect to see exactly those frequencies, but because they are our best guess as to what we might expect. Second, we need to devise some way of evaluating how far our sample frequencies are from the expected frequencies, yielding a test statistic. Finally, we need to consider the sampling distribution of the test statistic in order to find the $p$ value.

### Finding expected frequencies

How do we find expected frequencies? In the birth example, the null hypothesis was that the percentages in the various categories had not changed from the previous year, during which they were 80%, 4%, 10%, 4%, and 2%, respectively, for the five categories. Therefore, if we have sampled 200 births, we would expect the respective counts to be 160, 8, 20, 8, and 4 births. Notice, incidentally, that since the five counts must add up to 200, specifying four of the counts determines the remaining count.

Now let's consider the court diversion example, which involved a 2-by-2 contingency table. We have data on whether or not a later conviction occurred for 35 offenders in each of the two groups (diversion program and standard procedures). The null hypothesis of interest is that group is unrelated to the occurrence of a later conviction. What would the expected frequencies be for each of the four cells of the table if the null hypothesis were true?

The answer is implicit in the meaning of the term "unrelated." If group is unrelated to occurrence of later conviction, we mean that the proportion of offenders with a later conviction was the same for both groups. Since 26/70 of the offenders had a later conviction, we would expect that same proportion to hold for both groups if the null hypothesis is true. Thus we would expect 26/70 out of 35, or 13, offenders in each group to have a later conviction. We can therefore present the expected frequencies as follows:

Expected Frequencies if Group and Later Conviction Are Unrelated

|  | Later conviction? | | |
|---|---|---|---|
|  | Yes | No | Total |
| Diversion program | 13 | 22 | 35 |
| Standard procedures | 13 | 22 | 35 |
| Total | 26 | 44 | 70 |

Note that specifying one expected frequency was sufficient to determine the entire table, since both row and column totals are fixed.

For the retiree example, the null hypothesis of interest is that type of position is unrelated to whether or not retirement was voluntary. To say that the two variables are unrelated means that the proportion of retirees who retired voluntarily is the same regardless of type of position. Overall, the proportion who retired voluntarily was 1122/1800. So, for example, the expected frequency of retirees from managerial positions who retired voluntarily is 1122/1800 times 285, or 177.65. Note that the expected frequency need not be an integer, even though actual counts must be, because an expected frequency is an average, not an actual observation. The remaining expected frequencies are shown below.

Expected Frequencies if Type of Position and Voluntary Retirement Are Unrelated

|  | Retired voluntarily? | | |
| --- | --- | --- | --- |
| Type of position | Yes | No | Total |
| Managerial | 177.65 | 107.35 | 285 |
| Technical | 402.05 | 242.95 | 645 |
| Administrative | 335.98 | 203.02 | 539 |
| Academic | 206.32 | 124.68 | 331 |
| Total | 1122 | 678 | 1800 |

## A general formula for expected frequencies

We noted that the expected proportion of people in managerial positions (of a total of 285) who retired voluntarily is 1122/1800. We could have achieved the same result by noting that the expected proportion of people who retired voluntarily (of a total of 1122) who were in managerial positions was 285/1800. This leads us to a simple, general formula for expected frequencies in a contingency table: *The expected frequency in a given cell is the row total times the column total divided by the total sample size.*

## Degrees of freedom in a contingency table

In the last instance, we needed to specify three of the expected counts before the row and column totals determined the remaining numbers: Specifically, we needed to specify one of the two numbers in each of three of the four rows.

This can be extended, of course, to a general table with $r$ rows and $c$ columns: If the row and column totals are fixed, specifying all but one of the frequencies in each of all but one of the rows (or columns) specifies the entire table. Therefore, the number of cell frequencies needed to specify the entire table is $(r - 1)$ times

$(c - 1)$. This number is referred to as the number of degrees of freedom $(df)$ for the table. So a 4-by-2 table (such as in our retiree example) has 3 times 1, or 3, $df$. A 2-by-2 table (as in our diversion program example) has 1 $df$.

## 15.3 The chi-square test evaluates the difference between observed and expected frequencies.

### The test statistic used for the chi-square test

As we have noted, to test a null hypothesis, we need to evaluate how far our observed data deviates from what we would expect, on the average, if the null hypothesis is true. In the case of categorical data, we have discussed how to find expected frequencies. Now we need to consider how to evaluate how far the observed frequencies are from the expected frequencies.

We begin with the births example. How might we evaluate how far the observed frequencies in the five categories (136, 38, 12, 8, and 6) deviate from the expected frequencies under the null hypothesis that the population percentages have not changed (160, 8, 20, 8, and 2)?

One way would be to add up the absolute deviation in each cell. However, as we have seen previously, the statistic generally used is not the sum of absolute deviations but a different quantity with a similar rationale. That quantity is called the *chi-square* $(\chi^2)$ *statistic*. To calculate $\chi^2$, we first calculate, for each cell, the squared difference between observed and expected frequencies, divided by the expected frequency. The sum of those quantities over all cells is $\chi^2$. For the births example, the calculations look like this:

| Category | Observed frequency, $O$ | Expected frequency, $E$ | $O - E$ | $(O - E)^2$ | $(O - E)^2/E$ |
|---|---|---|---|---|---|
| Hospital–physician | 136 | 160 | $-24$ | 576 | 3.6 |
| Hospital–certified attendant | 38 | 8 | 30 | 900 | 112.5 |
| Out of hospital– physician | 12 | 20 | $-8$ | 64 | 3.2 |
| Out of hospital– certified attendant | 8 | 8 | 0 | 0 | 0.0 |
| Out of hospital–not certified attendant | 6 | 4 | 2 | 4 | 1.0 |
| | | | | Total $(\chi^2)$: | 120.3 |

## The distribution of the test statistic

Does a $\chi^2$ statistic of 120.3 for the births example represent a large enough deviation from the expected frequencies to lead us to reject the null hypothesis? To find out, we need to consider the sampling distribution of the $\chi^2$ statistic under the null hypothesis.

If the null hypothesis is true, and provided that one assumption is valid, we can use the same chi-square distribution that we first encountered in Chapter 10 (in the context of the Kruskal-Wallis test) as an approximation of the sampling distribution of the $\chi^2$ statistic. (That, of course, is why the statistic is generally referred to as $\chi^2$.) Not surprisingly, the degrees of freedom used in consulting the chi-square table is the *df* for the table itself. For the birth example, with five categories, we have 4 *df*.

Consulting Table D.7 with 4 *df*, we find that a $\chi^2$ value of 9.488 gives a *p* value of .05. Our $\chi^2$ value of 120.3 is much larger than 9.488, which means that the chance of getting a $\chi^2$ value that large by chance, if the null hypothesis is true, is much smaller than .05. Thus we reject the null hypothesis. It is clear from the table that our conclusion is based largely on the category of hospital births with a certified (nonphysician) attendant, which had a considerably higher frequency than expected.

## Assumption and conditions needed for the chi-square test

We noted above that one assumption is required for the chi-square test. That assumption is that the individuals in each category are independent of those in all other categories. If, for example, it were possible to classify a single birth into more than one category—say, if both a physician and a nurse-midwife attended the birth and it was counted in both categories—then the independence assumption is violated and this form of the chi-square test should not be used. Similarly, if the result of one birth influenced the result of another, the independence assumption is violated.

We noted also that the chi-square distribution is *approximately* the sampling distribution of the $\chi^2$ statistic under the null hypothesis. The approximation will be a good one as long as the expected frequencies are not too small. A good rule of thumb is to require that (1) all expected frequencies be at least 1, and (2) not more than 20% of the categories have expected frequencies less than 5. If these requirements cannot be met with the original categories, it is often possible to meet them by collapsing several categories into one, as long as this is meaningful.

## The chi-square test for contingency tables

For contingency tables, the chi-square test works exactly the same way as for data on a single categorical variable: we calculate $(O - E)^2/E$ for each cell and then sum these quantities over all cells to produce the $\chi^2$ statistic. We then consult the chi-square table with $(r - 1)$ times $(c - 1)$ $df$, where $r$ is the number of row categories and $c$ the number of column categories.

The calculations for the diversion program example are as follows:

|  | Later conviction? | |
| --- | --- | --- |
|  | Yes | No |
| Diversion program | $O = 9$ <br> $E = 13$ | $O = 26$ <br> $E = 22$ |
| Standard procedures | $O = 17$ <br> $E = 13$ | $O = 18$ <br> $E = 22$ |
| $\chi^2 =$ | 16/13 $\quad +$ <br> 16/13 $\quad +$ | 16/22 $\quad +$ <br> 16/22 |
|  | $= 3.92$ | |

With 1 $df$, we require $\chi^2$ value of 3.84 for a $p$ value of .05. Thus with a $\chi^2$ value of 3.92, we can reject the null hypothesis. Since 3.92 is between the tabled values for $p$ values of .05 and .025, we can state that the $p$ value for this table is less than .05 but greater than .025. From the table we can see that the offenders in the diversion program were indeed less likely to have been convicted of a later offense than were offenders who received standard court procedures.

We leave the calculations for the retirees example as an exercise. You should verify that the $\chi^2$ value works out to be 121.14, giving a $p$ value (with 3 $df$) less than .01. From the table it appears that the technical and academic retirees were less likely to have retired voluntarily than were the managerial and administrative retirees.

## Use frequencies, not relative frequencies

Note that the $\chi^2$ statistic must be calculated using counts (i.e., frequencies), *not* percentages or proportions (relative frequencies). To see why, consider two different studies, each obtaining the same proportions in all cells, but with one having 10 times the sample size of the other. If proportions were used to calculate $\chi^2$, both studies would produce exactly the same $p$ value. This would not reflect the fact that a small deviation from expected frequency in *percentage* terms might be quite large in *absolute* terms for a large sample size. Clearly, a large sample size should lead us to reject the null hypothesis based on a smaller percentage deviation from expected frequencies than we would require from a small sample. So in situations in which your expected values are expressed as percentages, it is important to convert them to frequencies before calculating $\chi^2$.

## STUDY QUESTIONS—*Sections 15.2 and 15.3*

**15.3.1**   a.   Use the chi-square test to determine if you can reject the null hypothesis that the die used to produce the results in Exercise 15.1.1 was fair.
 b.   Now test the null hypothesis that the results of the die throws were from an unbalanced die with probabilities as specified in Exercise 15.1.1, part c.

**15.3.2**   Test the null hypothesis described in Exercise 15.1.2, part c.

**15.3.3**   [Statpal exercise—Coronary Care data set (see Appendix A)] One of the major questions of the study was whether the emergency room ECG (electrocardiogram) could be used to predict the likelihood of major complications developing later on. Patients for whom major complications could be ruled out could be monitored in a standard hospital room rather than the coronary care unit, with a dramatic cost savings. The variable SERCOMP indicates whether the patient subsequently developed a serious complication, with 1 = yes and 2 = no. Recall that the variable RISK indicates which of three groups the patient is in: 1 for normal or unchanged ECG (hence low-risk); 2 for abnormal or changed ECG (hence high-risk); and 3 for indeterminate ECG. Using the TAble procedure, cross-tabulate SERCOMP with RISK, and calculate $\chi^2$. Interpret the results.

**15.3.4**   [Statpal exercise—Coronary Care data set (see Appendix A)]
 a.   The variable DEATH indicates whether or not the patient subsequently died during the same hospital admission, with 1 = yes and 2 = no. Use the TAble procedure to cross-tabulate DEATH with RISK. What do the results seem to indicate?
 b.   Should you calculate $\chi^2$ for this table? Why or why not?
 c.   Regardless of your answer to part b, instruct Statpal to calculate $\chi^2$ and comment on the results.

**15.3.5**   Suppose that you have obtained test scores for 100 students on a standardized exam and you wish to determine whether the scores come from a normal distribution. The mean of the scores was 220, with a standard deviation of 40. Suppose that you divide the range of the exam scores into eight categories and count the number of students in each category. The results were as follows:

| Scores | Frequency |
|---|---|
| Below 100 | 0 |
| 100–139.99 | 1 |
| 140–179.99 | 14 |
| 180–219.99 | 37 |
| 220–259.99 | 29 |
| 260–299.99 | 16 |
| 300–339.99 | 3 |
| 340 and above | 0 |

a. If the null hypothesis that the scores come from a normal distribution with mean 220 and standard deviation 40 were true, what would the expected frequency be for each category? Note that the categories are conveniently set up to be 1 standard deviation wide and are symmetric about the mean. (Refer back to Chapter 4 if you do not remember how to find normal curve proportions.)

b. Note that of the eight categories, four (two on each end of the distribution) have expected frequencies less than 5. By combining categories, we can avoid this problem. Calculate new observed and expected tables using the following four categories: (1) below 180, (2) 180–219.99, (3) 220–259.99, and (4) 260 and above.

c. Use the chi-square test to determine if the null hypothesis can be rejected. (With the collapsed table, you now have 3 *df*.)

d. In collapsing the table, we lost some information. Specifically, our test would not be able to detect deviations from normality in the tails of the distribution. What would be required for us to test for normality in the tails as well as the middle of the distribution using the chi-square test?

**15.3.6** [Continuation of Exercise 15.3.5] The chi-square test you performed in Exercise 15.3.5 is called a *goodness-of-fit test*, since it tests how well a normal distribution fits the observed distribution of scores. This test can be used to test the goodness of fit of any distribution, not just the normal. Suppose that we wanted to test not for normality but for uniformity in the original eight categories. That is, we wish to test the null hypothesis that the eight categories of scores occur with equal frequency in the population. Find the expected frequencies for our sample size of 100, and use the chi-square test to determine if the null hypothesis of uniformity can be rejected. Note that there is no need to collapse categories in this instance, since all expected frequencies are large enough.

# References

Brush, J., Chalmer, B., and Wackers, F. Normal or near-normal admission electrocardiogram identifies low risk group of patients with suspected acute myocardial infarction. Presented at the meeting of the American Federation of Clinical Research, Washington, May, 1984.

Cochran, W. G. *Sampling Techniques*. New York: Wiley, 1953.

Eckland, B. K., and Wisenbaker, J. M. *National Longitudinal Study: A capsule description of young adults four and one-half years after high school.* National Center for Educational Statistics, Department of Health, Education, and Welfare, 1979.

Filliben, J. J. The probability plot correlation coefficient test for normality. *Technometrics*, *17*, 1975, p. 111.

Guion, R. M. *Personnel Testing*. New York: McGraw-Hill, 1965.

Heimendinger, J. *Acta Helvetica paediatrica, 19,* 1964.

Howell, D. C. *Statistical Methods for Psychology*. Boston: Duxbury Press, 1982.

Kruskal, J. B. Transformations of data, in W. H. Kruskal and J. M. Tanur, eds. *International Encyclopedia of Statistics*, vol. 2. New York: Free Press, 1978.

Montgomery, D. C. *Design and Analysis of Experiments*. New York: Wiley, 1976.

Ott, L. *An Introduction to Statistical Methods and Data Analysis*. Boston: Duxbury Press, 1977.

Schaeffer, R. L. and Mendenhall, W. *Introduction to Probability: Theory and Applications*. Boston: Duxbury Press, 1975.

Siegel, S. *Nonparametric Statistics for the Behavioral Sciences*. New York: McGraw-Hill, 1956.

Snedecor, G. W., and Cochran, W. G. *Statistical Methods*. Ames, Iowa: The Iowa State University Press, 1967.

Tukey, J. *Exploratory Data Analysis*. Reading, Ma.: Addison-Wesley, 1977.

Wonnacott, R. J., and Wonnacott, T. H. *Introductory Statistics*. New York: John Wiley, 1985.

# Appendix A: The Coronary Care Data Set

The Coronary Care data set consists of data on 120 patients who were admitted to a medical center in 1982. Each patient had come to the emergency room (ER) complaining of chest pain, had been examined by the ER staff, and had been admitted to the coronary care unit (CCU) for observation. Part of the routine ER examination was to perform an electrocardiogram (ECG).

Subsequent to admission to the CCU, tests were done to determine if the patient had had a myocardial infarction [(MI) commonly known as a heart attack]. Various complications were recorded as they occurred.

The data were collected retrospectively—that is, by inspecting medical records some time after the admission. The main purpose of collecting the data was to determine if information obtained in the initial ER examination, particularly the ECG, could be used to predict which patients really needed the special CCU facilities as opposed to observation on a standard medical/surgical floor. Since the cost of CCU care is approximately $475 per day more than the cost of care on a standard medical/surgical floor, a finding that some patients could be safely assigned to the less expensive care would have potentially enormous economic implications. (In fact, extrapolating the results from this study yielded an estimated cost savings nationally of about $600 million!)

The data used in this book are taken from a larger set and were slightly modified for pedagogical purposes. Information on the study can be found in Brush, Chalmer and Wackers (1984). Variables listed below are as follows:

AGE                    Patient's age in years

SEX                    1 = male
                       2 = female

RISK                   1 = normal or unchanged ECG
                       2 = abnormal or changed ECG
                       3 = indeterminate ECG

NDAYS                  Number of days in hospital (total for this
                         admission)

HADMI                  0 = did not have MI
                       1 = did have MI

NCCU                   Number of days in CCU (this admission)

SERCOMP                1 = had serious complication
                       2 = did not have serious complication

DEATH                  1 = died
                       2 = survived

PAINTIME               Hours since patient began experiencing pain
                         (0 = no pain)


Coronary Care Data Set

| AGE | SEX | RISK | NDAYS | HADMI | NCCU | SERCOMP | DEATH | PAINTIME |
|-----|-----|------|-------|-------|------|---------|-------|----------|
| 66  | 2   | 2    | 17    | 1     | 16   | 1       | 1     | 0        |
| 45  | 2   | 1    | 5     | 0     | 2    | 2       | 2     | 0        |
| 40  | 2   | 1    | 2     | 0     | 2    | 2       | 2     | 24       |
| 73  | 1   | 2    | 9     | 1     | 4    | 1       | 2     | 0        |
| 51  | 1   | 2    | 13    | 1     | 4    | 1       | 2     | 0        |
| 77  | 1   | 1    | 3     | 0     | 2    | 2       | 2     | 1        |
| 66  | 1   | 1    | 2     | 0     | 1    | 2       | 2     | 2        |
| 42  | 1   | 2    | 1     | 0     | 1    | 2       | 2     | 0        |
| 74  | 1   | 2    | 10    | 1     | 4    | 2       | 2     | 0        |
| 46  | 1   | 2    | 11    | 1     | 4    | 2       | 2     | 9        |
| 62  | 1   | 2    | 75    | 1     | 5    | 2       | 2     | 0        |
| 72  | 1   | 3    | 28    | 1     | 3    | 2       | 2     | 1        |
| 64  | 1   | 1    | 3     | 0     | 3    | 2       | 2     | 1        |
| 37  | 2   | 1    | 2     | 0     | 2    | 2       | 2     | 0        |
| 45  | 2   | 1    | 2     | 0     | 2    | 2       | 2     | 24       |
| 36  | 1   | 1    | 5     | 0     | 2    | 2       | 2     | 1        |
| 52  | 1   | 1    | 3     | 0     | 1    | 2       | 2     | 2        |
| 76  | 2   | 2    | 44    | 1     | 11   | 2       | 2     | 0        |
| 43  | 1   | 2    | 12    | 1     | 5    | 2       | 2     | 1        |

Appendix A (*Continued*)

| AGE | SEX | RISK | NDAYS | HADMI | NCCU | SERCOMP | DEATH | PAINTIME |
|-----|-----|------|-------|-------|------|---------|-------|----------|
| 73 | 1 | 2 | 2 | 0 | 2 | 2 | 2 | 0 |
| 63 | 1 | 2 | 2 | 0 | 1 | 2 | 2 | 0 |
| 81 | 1 | 2 | 13 | 0 | 3 | 2 | 2 | 3 |
| 44 | 1 | 1 | 4 | 0 | 4 | 2 | 2 | 0 |
| 69 | 1 | 2 | 16 | 0 | 11 | 1 | 2 | 3 |
| 69 | 2 | 2 | 13 | 1 | 4 | 2 | 2 | 0 |
| 52 | 1 | 2 | 11 | 1 | 4 | 2 | 2 | 3 |
| 36 | 1 | 2 | 7 | 1 | 5 | 2 | 2 | 0 |
| 53 | 2 | 1 | 2 | 0 | 2 | 2 | 2 | 1 |
| 89 | 2 | 2 | 8 | 1 | 3 | 2 | 2 | 0 |
| 49 | 1 | 1 | 2 | 0 | 2 | 2 | 2 | 1 |
| 60 | 1 | 2 | 3 | 0 | 2 | 2 | 2 | 0 |
| 57 | 2 | 2 | 11 | 1 | 5 | 1 | 2 | 0 |
| 39 | 1 | 1 | 7 | 0 | 3 | 2 | 2 | 2 |
| 73 | 1 | 2 | 7 | 1 | 7 | 1 | 1 | 3 |
| 51 | 2 | 2 | 12 | 1 | 6 | 2 | 2 | 0 |
| 65 | 2 | 2 | 11 | 1 | 3 | 2 | 2 | 0 |
| 49 | 2 | 2 | 4 | 1 | 4 | 1 | 1 | 0 |
| 79 | 1 | 3 | 3 | 0 | 2 | 2 | 2 | 3 |
| 41 | 1 | 1 | 1 | 0 | 1 | 2 | 2 | 0 |
| 66 | 2 | 3 | 4 | 0 | 1 | 2 | 2 | 0 |
| 42 | 1 | 2 | 2 | 0 | 2 | 2 | 2 | 0 |
| 49 | 1 | 1 | 5 | 0 | 2 | 2 | 2 | 3 |
| 66 | 1 | 2 | 14 | 1 | 6 | 2 | 2 | 0 |
| 59 | 2 | 2 | 4 | 0 | 1 | 2 | 2 | 6 |
| 85 | 2 | 3 | 5 | 0 | 1 | 2 | 2 | 0 |
| 46 | 1 | 2 | 48 | 1 | 5 | 1 | 2 | 0 |
| 53 | 1 | 1 | 12 | 1 | 3 | 2 | 2 | 2 |
| 62 | 2 | 3 | 24 | 1 | 2 | 1 | 2 | 0 |
| 42 | 1 | 3 | 1 | 0 | 1 | 2 | 2 | 0 |
| 43 | 1 | 1 | 2 | 0 | 2 | 2 | 2 | 3 |
| 70 | 2 | 3 | 2 | 0 | 1 | 2 | 2 | 0 |
| 50 | 1 | 1 | 8 | 0 | 1 | 2 | 2 | 1 |
| 41 | 2 | 1 | 6 | 0 | 1 | 2 | 2 | 0 |
| 56 | 2 | 2 | 4 | 0 | 2 | 2 | 2 | 1 |
| 47 | 1 | 2 | 28 | 1 | 8 | 2 | 2 | 1 |
| 47 | 2 | 1 | 2 | 0 | 2 | 2 | 2 | 3 |
| 78 | 2 | 2 | 11 | 1 | 3 | 2 | 2 | 0 |
| 78 | 2 | 3 | 4 | 0 | 3 | 2 | 2 | 0 |
| 52 | 1 | 1 | 2 | 0 | 1 | 2 | 2 | 24 |
| 52 | 1 | 2 | 11 | 1 | 4 | 2 | 2 | 1 |

Appendix A (*Continued*)

| AGE | SEX | RISK | NDAYS | HADMI | NCCU | SERCOMP | DEATH | PAINTIME |
|-----|-----|------|-------|-------|------|---------|-------|----------|
| 58 | 2 | 1 | 2 | 0 | 2 | 2 | 2 | 1 |
| 76 | 1 | 3 | 4 | 0 | 4 | 2 | 2 | 0 |
| 72 | 1 | 2 | 5 | 0 | 5 | 1 | 1 | 0 |
| 48 | 1 | 2 | 15 | 1 | 3 | 1 | 2 | 1 |
| 75 | 2 | 2 | 3 | 0 | 3 | 2 | 2 | 0 |
| 49 | 1 | 1 | 9 | 0 | 3 | 1 | 2 | 72 |
| 62 | 1 | 2 | 10 | 1 | 4 | 2 | 2 | 12 |
| 85 | 2 | 1 | 4 | 0 | 1 | 2 | 2 | 1 |
| 75 | 1 | 3 | 11 | 1 | 3 | 1 | 2 | 1 |
| 55 | 1 | 3 | 8 | 0 | 1 | 2 | 2 | 12 |
| 57 | 2 | 1 | 6 | 0 | 1 | 2 | 2 | 12 |
| 48 | 2 | 1 | 4 | 0 | 2 | 2 | 2 | 0 |
| 52 | 1 | 1 | 15 | 1 | 6 | 2 | 2 | 1 |
| 88 | 2 | 3 | 13 | 1 | 3 | 1 | 2 | 0 |
| 51 | 1 | 1 | 1 | 0 | 1 | 2 | 2 | 2 |
| 57 | 1 | 2 | 12 | 1 | 2 | 2 | 2 | 12 |
| 63 | 1 | 2 | 13 | 1 | 4 | 2 | 2 | 0 |
| 64 | 2 | 2 | 12 | 1 | 5 | 1 | 2 | 1 |
| 35 | 1 | 1 | 2 | 0 | 1 | 2 | 2 | 1 |
| 65 | 1 | 2 | 9 | 1 | 4 | 2 | 2 | 0 |
| 61 | 1 | 2 | 11 | 1 | 4 | 1 | 2 | 0 |
| 54 | 1 | 1 | 4 | 0 | 4 | 2 | 2 | 2 |
| 69 | 1 | 2 | 11 | 1 | 3 | 2 | 2 | 0 |
| 73 | 2 | 2 | 12 | 1 | 1 | 2 | 2 | 24 |
| 57 | 1 | 2 | 16 | 1 | 12 | 1 | 1 | 0 |
| 72 | 1 | 2 | 3 | 0 | 1 | 2 | 2 | 1 |
| 80 | 1 | 2 | 2 | 0 | 2 | 2 | 2 | 9 |
| 54 | 1 | 1 | 9 | 1 | 2 | 2 | 2 | 1 |
| 56 | 1 | 2 | 12 | 1 | 5 | 2 | 2 | 0 |
| 57 | 1 | 2 | 7 | 0 | 3 | 2 | 2 | 0 |
| 63 | 1 | 2 | 4 | 0 | 2 | 1 | 2 | 0 |
| 62 | 1 | 1 | 1 | 0 | 1 | 2 | 2 | 24 |
| 70 | 1 | 1 | 1 | 0 | 1 | 2 | 2 | 5 |
| 75 | 1 | 3 | 2 | 1 | 2 | 1 | 1 | 0 |
| 62 | 2 | 1 | 3 | 0 | 2 | 2 | 2 | 0 |
| 52 | 1 | 1 | 16 | 0 | 2 | 2 | 2 | 0 |
| 72 | 1 | 3 | 13 | 1 | 3 | 2 | 2 | 0 |
| 54 | 1 | 1 | 5 | 0 | 1 | 2 | 2 | 1 |
| 70 | 2 | 2 | 12 | 1 | 7 | 1 | 2 | 6 |
| 33 | 1 | 1 | 2 | 0 | 2 | 2 | 2 | 0 |
| 58 | 1 | 2 | 6 | 0 | 3 | 2 | 2 | 0 |

Appendix A (*Continued*)

| AGE | SEX | RISK | NDAYS | HADMI | NCCU | SERCOMP | DEATH | PAINTIME |
|-----|-----|------|-------|-------|------|---------|-------|----------|
| 71 | 2 | 2 | 3 | 1 | 3 | 1 | 1 | 0 |
| 52 | 1 | 1 | 18 | 1 | 8 | 2 | 2 | 9 |
| 73 | 1 | 3 | 6 | 1 | 3 | 1 | 1 | 24 |
| 77 | 1 | 1 | 6 | 0 | 5 | 1 | 2 | 0 |
| 74 | 1 | 2 | 1 | 1 | 1 | 1 | 1 | 1 |
| 45 | 1 | 1 | 10 | 0 | 2 | 2 | 2 | 5 |
| 41 | 1 | 3 | 16 | 1 | 2 | 2 | 2 | 0 |
| 54 | 1 | 2 | 5 | 0 | 2 | 2 | 2 | 0 |
| 50 | 1 | 2 | 17 | 1 | 11 | 2 | 2 | 0 |
| 83 | 2 | 2 | 10 | 1 | 8 | 1 | 1 | 0 |
| 42 | 1 | 2 | 3 | 0 | 1 | 2 | 2 | 0 |
| 59 | 1 | 1 | 3 | 0 | 2 | 2 | 2 | 0 |
| 55 | 1 | 1 | 13 | 1 | 3 | 2 | 2 | 3 |
| 39 | 1 | 1 | 1 | 0 | 1 | 2 | 2 | 0 |
| 61 | 1 | 1 | 9 | 0 | 7 | 2 | 2 | 6 |
| 68 | 1 | 2 | 5 | 1 | 5 | 1 | 1 | 0 |
| 63 | 1 | 1 | 14 | 1 | 4 | 2 | 2 | 2 |
| 74 | 1 | 2 | 10 | 1 | 5 | 1 | 2 | 2 |
| 71 | 2 | 2 | 11 | 1 | 4 | 2 | 2 | 2 |

# Appendix B: The Electricity Consumer Questionnaire Data Set

An electric cooperative in Vermont sent out questionnaires to a sample of its members in an effort to determine members' attitudes on a number of issues of concern to the organization. The data set used in this book comes from a larger set and was modified for pedagogical purposes. Variables listed below are as follows:

BESTMNTH          Month considered best for annual meeting
                                   1 = January
                                   2 = February, etc.
                               −1 = no response

WDATTEND          1 = would attend annual meeting
                                   2 = would not attend annual meeting
                               −1 = no response

YRSMEMBR          Number of years of membership in the cooperative

LIKENUKE           1 = approve of cooperative's investment in nuclear power
                                   2 = disapprove
                               −1 = no response

CONSTYPE          Consumer type
                                   1 = residence
                                   2 = farm
                                   3 = industrial

RATESRVC          Rating of cooperative's service on a 5-point scale
                                   1 = highest
                                   5 = lowest
                               −1 = no response

Electricity Consumer Questionnaire Data Set

| BESTMNTH | WDATTEND | YRSMEMBR | LIKENUKE | CONSTYPE | RATESRVC |
|---|---|---|---|---|---|
| 3 | 2 | 17 | 2 | 3 | 5 |
| 7 | 2 | 6 | 1 | 2 | 3 |
| 5 | −1 | 18 | −1 | 2 | −1 |
| 5 | 1 | 20 | 2 | 2 | 2 |
| 4 | 2 | 16 | 2 | 1 | 5 |
| 5 | 2 | 1 | 2 | 1 | 4 |
| 5 | 1 | 18 | 2 | 1 | 3 |
| 5 | 1 | 16 | 2 | 1 | 2 |
| 5 | 2 | 6 | 2 | 2 | 3 |
| 5 | 1 | 13 | −1 | 1 | 2 |
| 11 | −1 | 11 | 2 | 2 | 3 |
| 5 | 2 | 17 | 2 | 1 | 3 |
| 5 | 1 | 6 | −1 | 2 | 4 |
| 5 | −1 | 9 | 2 | 2 | 1 |
| 5 | 1 | 19 | −1 | 1 | 3 |
| 5 | 1 | 7 | 1 | 1 | 4 |
| −1 | 1 | 2 | 2 | 1 | 2 |
| 5 | 2 | 12 | −1 | 3 | 3 |
| −1 | 2 | 6 | 1 | 2 | 2 |
| 5 | 2 | 15 | 2 | 2 | 4 |
| 5 | −1 | 17 | 1 | 1 | 4 |
| 5 | −1 | 4 | 2 | 1 | 3 |
| 5 | 2 | 4 | 2 | 2 | 2 |
| 5 | 1 | 13 | −1 | 1 | 3 |
| 5 | −1 | 4 | 2 | 3 | 2 |
| 5 | 1 | 13 | 1 | 1 | 4 |
| 5 | 1 | 17 | 1 | 1 | 2 |
| 5 | 2 | 15 | 2 | 2 | 3 |
| −1 | −1 | 5 | 2 | 3 | 4 |
| 5 | −1 | 3 | −1 | 2 | 3 |
| 5 | 2 | 5 | 2 | 2 | 3 |
| −1 | −1 | 17 | −1 | 1 | 2 |
| −1 | −1 | 9 | 2 | 2 | 4 |
| 5 | 2 | 7 | −1 | 2 | 3 |
| 5 | −1 | 13 | 2 | 2 | 2 |
| 5 | 1 | 12 | −1 | 2 | −1 |
| 4 | 1 | 18 | 1 | 1 | 5 |
| 5 | 1 | 13 | −1 | 1 | 3 |
| −1 | 2 | 12 | 2 | 2 | 3 |
| 5 | 2 | 15 | 2 | 1 | 4 |
| 9 | 1 | 3 | 2 | 1 | −1 |
| −1 | 1 | 19 | −1 | 3 | −1 |

Appendix B (*Continued*)

| BESTMNTH | WDATTEND | YRSMEMBR | LIKENUKE | CONSTYPE | RATESRVC |
|---:|---:|---:|---:|---:|---:|
| 7 | −1 | 18 | −1 | 1 | −1 |
| 5 | 2 | 19 | 2 | 1 | 3 |
| 5 | 1 | 5 | 1 | 3 | 1 |
| 5 | 2 | 13 | 2 | 2 | 3 |
| −1 | 1 | 17 | 1 | 1 | 3 |
| −1 | 2 | 7 | 1 | 2 | 1 |
| 5 | 2 | 10 | 2 | 3 | 2 |
| 5 | 2 | 20 | 2 | 1 | 1 |
| 7 | 1 | 16 | 2 | 2 | −1 |
| 1 | 1 | 13 | 1 | 1 | 3 |
| 5 | −1 | 2 | −1 | 1 | 3 |
| 5 | 1 | 13 | −1 | 3 | 2 |
| 5 | 1 | 1 | −1 | 1 | 2 |
| 2 | 2 | 14 | 2 | 1 | 4 |
| −1 | −1 | 10 | 1 | 1 | −1 |
| 7 | 2 | 18 | 2 | 3 | 1 |
| 5 | 2 | 15 | 2 | 3 | −1 |
| 5 | 2 | 14 | −1 | 2 | 2 |
| 4 | 2 | 15 | 1 | 2 | 2 |
| 5 | 1 | 3 | 1 | 2 | 1 |
| 5 | −1 | 6 | 2 | 2 | 2 |
| 5 | 1 | 11 | −1 | 3 | 3 |
| −1 | 1 | 18 | 1 | 1 | 5 |
| 5 | 1 | 3 | 2 | 1 | 2 |
| 5 | 2 | 5 | 2 | 1 | 5 |
| −1 | −1 | 2 | 2 | 3 | 3 |
| 10 | −1 | 15 | 2 | 3 | 2 |
| 11 | 1 | 2 | 2 | 2 | 3 |
| −1 | −1 | 16 | −1 | 3 | 2 |
| 5 | 1 | 5 | 2 | 3 | 3 |
| 5 | −1 | 15 | 1 | 1 | 5 |
| 10 | −1 | 5 | 2 | 1 | 3 |
| 5 | 1 | 16 | 2 | 1 | 2 |
| 8 | 1 | 3 | 2 | 1 | −1 |
| 5 | 1 | 19 | 2 | 2 | 3 |
| 1 | 1 | 7 | 1 | 1 | 3 |
| 4 | −1 | 4 | 2 | 2 | −1 |
| 5 | 1 | 15 | 2 | 1 | 3 |

# Appendix C:
# Selected Advanced Topics

This appendix contains brief treatments of various advanced topics referred to in the text. For convenience, here is a list of the topics covered arranged by chapter reference:

| Chapters | Topic |
|----------|-------|
| 9, 13 | Checking for normality and homogeneity |
| 9, 10 | Dealing with lack of homogeneity and normality |
| 10 | Comparing individual groups in ANOVA |
| 11 | Fisher's $z$ transformation for testing correlation coefficients |
| 12 | Derivation of least-squares formulas for simple regression coefficients |
| 12 | Formula for standard error of $b1$ in simple regression |
| 13 | Formulas for mean squares and $df$ in regression |

## C.1   Checking for normality and homogeneity

### Chi-square goodness-of-fit test for normality

This test is covered in Chapter 15 (Study Questions).

## Normal probability plot

This is a graphical technique for examining the degree to which a set of scores are consistent with normality. The procedure involves three steps:

1. Order the scores from smallest to largest and determine the percentile of each score. For tied scores, average the ranks to determine the percentile for the scores.
2. Determine the $z$ score from the standard normal distribution corresponding to each score's percentile. (For example, the $z$ score corresponding to the 90th percentile in the standard normal distribution is 1.28.) These $z$ scores are generally referred to as *normal scores*. A shortcut formula for finding the $z$ score from the percentile is

$$z = 4.91[p^{.14} - (1 - p)^{.14}]$$

   where $p$ is the percentile expressed as a proportion.
3. Draw a scatterplot (see Chapter 11) of the original scores versus the normal scores. If the scores come from a normal distribution, the points on the scatterplot should fall at least approximately) in a straight line. Substantial departures from a straight line indicate lack of normality.

A test is available (based on the correlation between the scores and the normal scores) to test the null hypothesis that the scores come from a normal distribution; for details, see, for example, Filliben (1975). Since the normality assumption is usually not critical unless departures are substantial, this test is generally not relevant to practical problems.

## F test for the equality of two variances

In the case of two samples, the null hypothesis that the variances of the two populations are equal, versus the alternative hypothesis that the variances are unequal, can be tested using an $F$ test, where the test statistic is calculated as follows:

$$F = \frac{s_1^2}{s_2^2}$$

where the numerator is the *larger* of the two variances. If the value of $F$ exceeds the tabled $F$ value for a given significance level with degrees of freedom $n_1 - 1$ and $n_2 - 1$, the null hypothesis is rejected. Like the two-sample $t$ test for comparing means, this test assumes that the two populations have normal distributions.

Many computer programs (e.g., Statpal, SPSSX) automatically perform this $F$ test whenever a two-sample $t$ test is requested, together with two different versions of the two-sample $t$ test. One of the $t$ tests, described in this book, makes the assumption of equal population variances, pooling the two sample variances in the calculation of the standard error of the difference between means; the other is called the *Welch test*, and does not assume equal population variances but rather keeps the two variance estimates separate. The idea is that the analyst can look first at the $F$ test and decide which of the two $t$ tests to apply. If the $F$ test indicates that the homogeneity assumption is not violated, the analyst can use the results of the pooled-variance $t$ test; otherwise, the analyst can use the results of the Welch test.

### Testing for homogeneity of variance in more than two populations

One test often used for testing the null hypothesis that several populations have equal variances is called *Bartlett's test*. Details of this test can be found in, for example, Snedecor and Cochran (1967); some packaged computer programs (e.g., SPSSX, BMDP) will do the computations. Unfortunately, Bartlett's test tends to be quite sensitive to departures from the normality assumption, producing too many significant results if the population distributions are nonnormal. Determining the practical importance of heterogeneity of variance is still as much art as science; in doubtful situations, consult a statistician.

### C.2  Dealing with lack of homogeneity and normality

One way to handle violations of the normality assumption, the homogeneity assumption, or both, is to use one of the transformations outlined below. Note that working with transformed data changes the question being asked; however, in many cases, the question stated in terms of transformed data is at least as meaningful as the original question.

Note also that in some cases the violation of the homogeneity assumption leads not only to mathematical difficulties (which transformations may or may not overcome), but also to conceptual difficulties. In many circumstances, the finding that populations being compared have substantially different variances makes the issue of differences in central tendency less meaningful.

Some of the more frequently used transformations are as follows:

1. *Taking the square root*: effective when the variances of the populations are proportional to their means; often used when the data are counts of relatively rare events
2. *Taking the logarithm*: effective when the standard deviations of the populations are proportional to their means; often used when the differences among populations are likely to be proportional rather than additive

3. *Taking the arcsine of the square root*: effective when the data consist of proportions
4. *Probit (also called normit) transformation for proportions*: that is, replacing each score (assumed to be a proportion between 0 and 1) with the $z$ score from the standard normal distribution whose percentile is the score being replaced
5. *Logistic (also called logit) transformation for proportions*: that is, replacing each score $p$ (assumed to be a proportion between 0 and 1) with $(1/2)\log[p/(1-p)]$
6. *Replacing the data with ranks*: especially effective when the data include a small number of extreme values

A complete discussion of transformations is beyond the scope of this book. See Kruskal (1978) for a more comprehensive discussion.

## C.3  Comparing individual groups in ANOVA

As noted in the text, the $F$ test provides only a test of the null hypothesis that all population means are equal. It gives no indication of which means are significantly different from which. Methods for testing hypotheses about subsets of group means are generally known as *multiple comparison* techniques. We offer only brief descriptions of some widely used methods; see, for example, Montgomery (1976) and Howell (1982) for discussions.

### A priori versus a posteriori (post hoc) techniques

If the analyst can specify in advance (before examining the data) which groups are to be compared, *a priori* techniques can be used. If the determination of which groups are to be compared depends on an examination of the data, or if all possible comparisons are to be made, *a posteriori* (also called *post hoc*) techniques must be used. The difference between the two types involves the way in which the probability of making an erroneous conclusion is controlled.

To see why examining the data makes a difference, consider a situation with a large number of groups—say, 100. Even if the null hypothesis were true, the most extreme groups would probably have quite different sample means just by chance, because there are so many groups. Thus, deciding which groups to compare based on an examination of the data is very likely to lead to an erroneous conclusion that the extreme groups are significantly different. To protect against this tendency, a posteriori techniques increase the amount of difference required for significance depending (generally) on how many groups are being compared.

## A priori techniques

One common method of testing for differences among particular groups specified a priori is the use of *linear contrasts*. For simplicity, we will consider only the case in which all sample sizes are the same, although the method can also be used with minor adjustments when the sample sizes differ; see, for example, Montgomery (1976) for details.

A linear contrast (or simply, a contrast) is a weighted sum of the group means, with the weights chosen so as to add up to zero. By choosing the appropriate weights, contrasts can be defined so as to implicitly compare any combinations of groups. For example, with four groups we could define the following contrast:

$$C = (0)\bar{X}_1 + (1)\bar{X}_2 + (-1)\bar{X}_3 + (0)\bar{X}_4$$

Notice that the four weights (0, 1, $-1$, 0) add up to zero. This contrast is simply the difference between the sample means for groups 2 and 3. Using weights (.5, .5, 0, $-1$) would define a contrast equal to the difference between the mean of group 4 and the mean of groups 1 and 2.

Thus a test of the null hypothesis that the population contrast equals zero implicitly constitutes a particular comparison among groups. The contrast can be tested using an $F$ test, with the following formulas:

MSW = same MSW defined for one-way ANOVA (see the text)

$$\text{SS(contrast)} = \frac{(\sum_{j=1}^{k} c_j T_j)^2}{n \sum_{j=1}^{k} c_j^2}$$

where

$T_j$ = total (rather than the mean) for the $j$th sample
$c_j$ = contrast weight for the $j$th group
$k$ = number of groups
$n$ = number of scores in each group (there are thus $kn$ scores altogether)

MS(contrast) = SS(contrast)        [i.e., SS(contrast) has 1 $df$]

$$F = \frac{\text{MS(contrast)}}{\text{MSW}}$$

If $F$ exceeds the tabled value for a given significance level, the contrast is significant.

## A posteriori techniques

In general, a posteriori techniques work on the principle that the greater the number of groups being compared, the greater the difference between extreme groups must be in order to be considered significant. Among the techniques commonly used are the Newman-Keuls procedure, Duncan's multiple range test, Tukey's honestly significant difference and wholly significant difference tests, and Scheffé's procedure. Some computer programs carry out one or more of these procedures, producing as output a description of which groups are significantly different from which. A full description of these procedures is beyond the scope of this book; see, for example, Howell (1982) for details.

## C.4   Fisher's z transformation for testing correlation coefficients

The $t$ test described in Chapter 11 can be used only for testing the null hypothesis that the population correlation coefficient equals zero (not any other number). However, by transforming the value of $r$ with a special transformation due to Sir Ronald Fisher, tests of other hypotheses can be performed. The transformation is as follows:

$$z_r = \frac{1}{2} \ln \frac{1 + r}{1 - r}$$

where $r$ is the sample correlation coefficient and ln indicates taking the natural logarithm. Fisher showed that $z_r$ is approximately normally distributed with

mean of $z_r = z_\rho$

(i.e., the transformed population correlation) and

standard error of $z_r = \dfrac{1}{\sqrt{n - 3}}$

Since $z_r$ can be treated as a normally distributed variable, we can use the normal distribution to test hypotheses about the population correlation coefficient ($\rho$), to form confidence intervals for $\rho$, or to test hypotheses about the equality of two population correlation coefficients. A brief description of these methods is given below; see, for example, Howell (1982) or Snedecor and Cochran (1967) for fuller discussions.

### Testing the null hypothesis that $\rho$ equals some specified constant k

The test statistic is

$$z = \frac{z_r - z_k}{\sqrt{1/(n - 3)}}$$

This $z$ value is compared to the $z$ score from the standard normal distribution needed for a given significance level; for example, if the absolute value of $z$ exceeds 1.96, the null hypothesis is rejected at $p < .05$ for a two-tailed alternative hypothesis.

## A confidence interval for ρ

To form a confidence interval for ρ, we first observe that a $Q\%$ confidence interval for $z_\rho$ is

$$z_r \pm z_{Q/2} \frac{1}{\sqrt{n-3}}$$

where $z_{Q/2}$ is the $z$ score that cuts off $Q/2$ percent of the standard normal distribution (thereby bracketing $Q\%$ between plus and minus $z_{Q/2}$).

This gives a confidence interval for *transformed* ρ; to turn it into a confidence interval for ρ itself, we reverse the transformation according to the following relationship between some number $k$ and the transformed value $z_k$:

$$k = \tanh(z_k) = \frac{\exp(2z_k) - 1}{\exp(2z_k) + 1}$$

(The symbol "tanh" is read "hyperbolic tangent.") For example, a sample size of 28 with a sample correlation coefficient of .40 leads to the following calculations for a 95% confidence interval for ρ: $z_{.40} = .424$; thus a 95% confidence interval for $z_\rho$ is $.424 \pm 1.96(.2)$, or the interval $[.032, .816]$. Reversing the transformation, we have $\tanh(.032) = .032$ and $\tanh(.816) = .673$. Thus we are 95% confident that ρ is between .032 and .673.

## Testing the null hypothesis that two population correlations are equal

Given two independent samples, the respective sample correlation coefficients $r_1$ and $r_2$ can be used to test the null hypothesis that the two population correlation coefficients are equal, using the following test statistic:

$$z = \frac{z_{r1} - z_{r2}}{\sqrt{[1/(n_1 - 3)] + [1/(n_2 - 3)]}}$$

where $z_{r1}$ and $z_{r2}$ are the transformed sample correlation coefficients, and $n_1$ and $n_2$ are the sample sizes. The test statistic is compared to the $z$ score from the standard normal distribution needed for a given significance level.

## C.5 Derivation of least-squares formulas for simple regression

The model under consideration is

$$\hat{y}_i = b_0 + b_1 x_i$$

We wish to find those values of $b_0$ and $b_1$ that jointly produce the smallest possible SSE; that is, we wish to minimize the quantity

$$\text{SSE} = \sum_{i=1}^{n} (y_i - \hat{y}_i)^2 = \sum_{i=1}^{n} [y_i - (b_0 + b_1 x_i)]^2$$

Taking partial derivatives with respect to $b_0$ and $b_1$ and setting equal to zero, we obtain

$$\sum_{i=1}^{n} (-2)[y_i - (b_0 + b_1 x_i)] = 0$$

and

$$\sum_{i=1}^{n} (-2)x_i[y_i - (b_0 + b_1 x_i)] = 0$$

Rearranging terms in the first equation gives

$$\sum_{i=1}^{n} y_i - \sum_{i=1}^{n} b_0 - b_1 \sum_{i=1}^{n} x_i = 0$$

Since $\sum_{i=1}^{n} b_0 = nb_0$, we have $nb_0 = \sum_{i=1}^{n} y_i - b_1 \sum_{i=1}^{n} x_i$. Solving, we get
$$b_0 = \bar{y} - b_1 \bar{x}$$

Substituting this result into the second equation above and solving for $b_1$ gives

$$b_1 = \frac{n \sum_{i=1}^{n} x_i y_i - \sum_{i=1}^{n} x_i \sum_{i=1}^{n} y_i}{n \sum_{i=1}^{n} x_i^2 - (\sum_{i=1}^{n} x_i)^2}$$

which can be shown to be equivalent to the covariance of $x$ and $y$ divided by the variance of $x$.

## C.6 Formula for the standard error of $b1$ in simple regression

$$\text{standard error of } b1 = \sqrt{\frac{\text{MSE}}{\sum (x_i - \bar{x})^2}}$$

### C.7 Formulas for mean squares and *df* in regression

$$MSR = \frac{SSR}{k}$$

where $k$ is the number of independent variables in the model. MSR thus has $k$ *df*.

$$MSE = \frac{SSE}{n - k - 1}$$

and has $n - k - 1$ *df*.

# Appendix D: Statistical Tables

Table D.1   Standard Normal Distribution. Each entry gives the area between zero (the mean) and $z$.

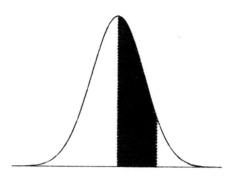

First
decimal
of z

Second decimal of z

|  | 0 | 1 | 2 | 3 | 4 | 5 | 6 | 7 | 8 | 9 |
|---|---|---|---|---|---|---|---|---|---|---|
| 0.0 | .0000 | .0040 | .0080 | .0120 | .0160 | .0199 | .0239 | .0279 | .0319 | .0359 |
| 0.1 | .0398 | .0438 | .0478 | .0517 | .0557 | .0596 | .0636 | .0675 | .0714 | .0753 |
| 0.2 | .0793 | .0832 | .0871 | .0910 | .0948 | .0987 | .1026 | .1064 | .1103 | .1141 |
| 0.3 | .1179 | .1217 | .1255 | .1293 | .1331 | .1368 | .1406 | .1443 | .1480 | .1517 |
| 0.4 | .1554 | .1591 | .1628 | .1664 | .1700 | .1736 | .1772 | .1808 | .1844 | .1879 |
| 0.5 | .1915 | .1950 | .1985 | .2019 | .2054 | .2088 | .2123 | .2157 | .2190 | .2224 |
| 0.6 | .2257 | .2291 | .2324 | .2357 | .2389 | .2422 | .2454 | .2486 | .2517 | .2549 |
| 0.7 | .2580 | .2611 | .2642 | .2673 | .2703 | .2734 | .2764 | .2793 | .2823 | .2852 |
| 0.8 | .2881 | .2910 | .2939 | .2967 | .2995 | .3023 | .3051 | .3078 | .3106 | .3133 |
| 0.9 | .3159 | .3186 | .3212 | .3238 | .3264 | .3289 | .3315 | .3340 | .3365 | .3389 |
| 1.0 | .3413 | .3438 | .3461 | .3485 | .3508 | .3531 | .3554 | .3577 | .3599 | .3621 |
| 1.1 | .3643 | .3665 | .3686 | .3708 | .3729 | .3749 | .3770 | .3790 | .3810 | .3830 |
| 1.2 | .3849 | .3869 | .3888 | .3907 | .3925 | .3944 | .3962 | .3980 | .3997 | .4015 |
| 1.3 | .4032 | .4049 | .4066 | .4082 | .4099 | .4115 | .4131 | .4147 | .4162 | .4177 |
| 1.4 | .4192 | .4207 | .4222 | .4236 | .4251 | .4265 | .4279 | .4292 | .4306 | .4319 |
| 1.5 | .4332 | .4345 | .4357 | .4370 | .4382 | .4394 | .4406 | .4418 | .4429 | .4441 |
| 1.6 | .4452 | .4463 | .4474 | .4484 | .4495 | .4505 | .4515 | .4525 | .4535 | .4545 |
| 1.7 | .4554 | .4564 | .4573 | .4582 | .4591 | .4599 | .4608 | .4616 | .4625 | .4633 |
| 1.8 | .4641 | .4649 | .4656 | .4664 | .4671 | .4678 | .4686 | .4693 | .4699 | .4706 |
| 1.9 | .4713 | .4719 | .4726 | .4732 | .4738 | .4744 | .4750 | .4756 | .4761 | .4767 |
| 2.0 | .4772 | .4778 | .4783 | .4788 | .4793 | .4798 | .4803 | .4808 | .4812 | .4817 |
| 2.1 | .4821 | .4826 | .4830 | .4834 | .4838 | .4842 | .4846 | .4850 | .4854 | .4857 |
| 2.2 | .4861 | .4864 | .4868 | .4871 | .4875 | .4878 | .4881 | .4884 | .4887 | .4890 |
| 2.3 | .4893 | .4896 | .4898 | .4901 | .4904 | .4906 | .4909 | .4911 | .4913 | .4916 |
| 2.4 | .4918 | .4920 | .4922 | .4925 | .4927 | .4929 | .4931 | .4932 | .4934 | .4936 |
| 2.5 | .4938 | .4940 | .4941 | .4943 | .4945 | .4946 | .4948 | .4949 | .4951 | .4952 |
| 2.6 | .4953 | .4955 | .4956 | .4957 | .4959 | .4960 | .4961 | .4962 | .4963 | .4964 |
| 2.7 | .4965 | .4966 | .4967 | .4968 | .4969 | .4970 | .4971 | .4972 | .4973 | .4974 |
| 2.8 | .4974 | .4975 | .4976 | .4977 | .4977 | .4978 | .4979 | .4979 | .4980 | .4981 |
| 2.9 | .4981 | .4982 | .4982 | .4983 | .4984 | .4984 | .4985 | .4985 | .4986 | .4986 |
| 3.0 | .4987 | .4987 | .4987 | .4988 | .4988 | .4989 | .4989 | .4989 | .4990 | .4990 |

Table D.2   Binomial Probabilities. Each tabulated value gives the probability of observing exactly k yeses out of n independent trials with probability p of a yes on a single trial.

| n | k | | | | | p | | | | |
|---|---|------|------|------|------|------|------|------|------|------|
|   |   | .10 | .20 | .30 | .40 | .50 | .60 | .70 | .80 | .90 |
| 1 | 0 | 0.900 | 0.800 | 0.700 | 0.600 | 0.500 | 0.400 | 0.300 | 0.200 | 0.100 |
|   | 1 | 0.100 | 0.200 | 0.300 | 0.400 | 0.500 | 0.600 | 0.700 | 0.800 | 0.900 |
| 2 | 0 | 0.810 | 0.640 | 0.490 | 0.360 | 0.250 | 0.160 | 0.090 | 0.040 | 0.010 |
|   | 1 | 0.180 | 0.320 | 0.420 | 0.480 | 0.500 | 0.480 | 0.420 | 0.320 | 0.180 |
|   | 2 | 0.010 | 0.040 | 0.090 | 0.160 | 0.250 | 0.360 | 0.490 | 0.640 | 0.810 |
| 3 | 0 | 0.729 | 0.512 | 0.343 | 0.216 | 0.125 | 0.064 | 0.027 | 0.008 | 0.001 |
|   | 1 | 0.243 | 0.384 | 0.441 | 0.432 | 0.375 | 0.288 | 0.189 | 0.096 | 0.027 |
|   | 2 | 0.027 | 0.096 | 0.189 | 0.288 | 0.375 | 0.432 | 0.441 | 0.384 | 0.243 |
|   | 3 | 0.001 | 0.008 | 0.027 | 0.064 | 0.125 | 0.216 | 0.343 | 0.512 | 0.729 |
| 4 | 0 | 0.656 | 0.410 | 0.240 | 0.130 | 0.063 | 0.026 | 0.008 | 0.002 | 0.000 |
|   | 1 | 0.292 | 0.410 | 0.412 | 0.346 | 0.250 | 0.154 | 0.076 | 0.026 | 0.004 |
|   | 2 | 0.049 | 0.154 | 0.265 | 0.346 | 0.375 | 0.346 | 0.265 | 0.154 | 0.049 |
|   | 3 | 0.004 | 0.026 | 0.076 | 0.154 | 0.250 | 0.346 | 0.412 | 0.410 | 0.292 |
|   | 4 | 0.000 | 0.002 | 0.008 | 0.026 | 0.063 | 0.130 | 0.240 | 0.410 | 0.656 |
| 5 | 0 | 0.590 | 0.328 | 0.168 | 0.078 | 0.031 | 0.010 | 0.002 | 0.000 | 0.000 |
|   | 1 | 0.328 | 0.410 | 0.360 | 0.259 | 0.156 | 0.077 | 0.028 | 0.006 | 0.000 |
|   | 2 | 0.073 | 0.205 | 0.309 | 0.346 | 0.313 | 0.230 | 0.132 | 0.051 | 0.008 |
|   | 3 | 0.008 | 0.051 | 0.132 | 0.230 | 0.313 | 0.346 | 0.309 | 0.205 | 0.073 |
|   | 4 | 0.000 | 0.006 | 0.028 | 0.077 | 0.156 | 0.259 | 0.360 | 0.410 | 0.328 |
|   | 5 | 0.000 | 0.000 | 0.002 | 0.010 | 0.031 | 0.078 | 0.168 | 0.328 | 0.590 |
| 6 | 0 | 0.531 | 0.262 | 0.118 | 0.047 | 0.016 | 0.004 | 0.001 | 0.000 | 0.000 |
|   | 1 | 0.354 | 0.393 | 0.303 | 0.187 | 0.094 | 0.037 | 0.010 | 0.002 | 0.000 |
|   | 2 | 0.098 | 0.246 | 0.324 | 0.311 | 0.234 | 0.138 | 0.060 | 0.015 | 0.001 |
|   | 3 | 0.015 | 0.082 | 0.185 | 0.276 | 0.313 | 0.276 | 0.185 | 0.082 | 0.015 |
|   | 4 | 0.001 | 0.015 | 0.060 | 0.138 | 0.234 | 0.311 | 0.324 | 0.246 | 0.098 |
|   | 5 | 0.000 | 0.002 | 0.010 | 0.037 | 0.094 | 0.187 | 0.303 | 0.393 | 0.354 |
|   | 6 | 0.000 | 0.000 | 0.001 | 0.004 | 0.016 | 0.047 | 0.118 | 0.262 | 0.531 |
| 7 | 0 | 0.478 | 0.210 | 0.082 | 0.028 | 0.008 | 0.002 | 0.000 | 0.000 | 0.000 |
|   | 1 | 0.372 | 0.367 | 0.247 | 0.131 | 0.055 | 0.017 | 0.004 | 0.000 | 0.000 |
|   | 2 | 0.124 | 0.275 | 0.318 | 0.261 | 0.164 | 0.077 | 0.025 | 0.004 | 0.000 |
|   | 3 | 0.023 | 0.115 | 0.227 | 0.290 | 0.273 | 0.194 | 0.097 | 0.029 | 0.003 |
|   | 4 | 0.003 | 0.029 | 0.097 | 0.194 | 0.273 | 0.290 | 0.227 | 0.115 | 0.023 |
|   | 5 | 0.000 | 0.004 | 0.025 | 0.077 | 0.164 | 0.261 | 0.318 | 0.275 | 0.124 |
|   | 6 | 0.000 | 0.000 | 0.004 | 0.017 | 0.055 | 0.131 | 0.247 | 0.367 | 0.372 |
|   | 7 | 0.000 | 0.000 | 0.000 | 0.002 | 0.008 | 0.028 | 0.082 | 0.210 | 0.478 |
| 8 | 0 | 0.430 | 0.168 | 0.058 | 0.017 | 0.004 | 0.001 | 0.000 | 0.000 | 0.000 |
|   | 1 | 0.383 | 0.336 | 0.198 | 0.090 | 0.031 | 0.008 | 0.001 | 0.000 | 0.000 |
|   | 2 | 0.149 | 0.294 | 0.296 | 0.209 | 0.109 | 0.041 | 0.010 | 0.001 | 0.000 |
|   | 3 | 0.033 | 0.147 | 0.254 | 0.279 | 0.219 | 0.124 | 0.047 | 0.009 | 0.000 |
|   | 4 | 0.005 | 0.046 | 0.136 | 0.232 | 0.273 | 0.232 | 0.136 | 0.046 | 0.005 |
|   | 5 | 0.000 | 0.009 | 0.047 | 0.124 | 0.219 | 0.279 | 0.254 | 0.147 | 0.033 |
|   | 6 | 0.000 | 0.001 | 0.010 | 0.041 | 0.109 | 0.209 | 0.296 | 0.294 | 0.149 |
|   | 7 | 0.000 | 0.000 | 0.001 | 0.008 | 0.031 | 0.090 | 0.198 | 0.336 | 0.383 |
|   | 8 | 0.000 | 0.000 | 0.000 | 0.001 | 0.004 | 0.017 | 0.058 | 0.168 | 0.430 |

Table D.2   (*Continued*)

| n | k | .10 | .20 | .30 | .40 | p .50 | .60 | .70 | .80 | .90 |
|---|---|-----|-----|-----|-----|-----|-----|-----|-----|-----|
| 9 | 0 | 0.387 | 0.134 | 0.040 | 0.010 | 0.002 | 0.000 | 0.000 | 0.000 | 0.000 |
|   | 1 | 0.387 | 0.302 | 0.156 | 0.060 | 0.018 | 0.004 | 0.000 | 0.000 | 0.000 |
|   | 2 | 0.172 | 0.302 | 0.267 | 0.161 | 0.070 | 0.021 | 0.004 | 0.000 | 0.000 |
|   | 3 | 0.045 | 0.176 | 0.267 | 0.251 | 0.164 | 0.074 | 0.021 | 0.003 | 0.000 |
|   | 4 | 0.007 | 0.066 | 0.172 | 0.251 | 0.246 | 0.167 | 0.074 | 0.017 | 0.001 |
|   | 5 | 0.001 | 0.017 | 0.074 | 0.167 | 0.246 | 0.251 | 0.172 | 0.066 | 0.007 |
|   | 6 | 0.000 | 0.003 | 0.021 | 0.074 | 0.164 | 0.251 | 0.267 | 0.176 | 0.045 |
|   | 7 | 0.000 | 0.000 | 0.004 | 0.021 | 0.070 | 0.161 | 0.267 | 0.302 | 0.172 |
|   | 8 | 0.000 | 0.000 | 0.000 | 0.004 | 0.018 | 0.060 | 0.156 | 0.302 | 0.387 |
|   | 9 | 0.000 | 0.000 | 0.000 | 0.000 | 0.002 | 0.010 | 0.040 | 0.134 | 0.387 |
| 10 | 0 | 0.349 | 0.107 | 0.028 | 0.006 | 0.001 | 0.000 | 0.000 | 0.000 | 0.000 |
|   | 1 | 0.387 | 0.268 | 0.121 | 0.040 | 0.010 | 0.002 | 0.000 | 0.000 | 0.000 |
|   | 2 | 0.194 | 0.302 | 0.233 | 0.121 | 0.044 | 0.011 | 0.001 | 0.000 | 0.000 |
|   | 3 | 0.057 | 0.201 | 0.267 | 0.215 | 0.117 | 0.042 | 0.009 | 0.001 | 0.000 |
|   | 4 | 0.011 | 0.088 | 0.200 | 0.251 | 0.205 | 0.111 | 0.037 | 0.006 | 0.000 |
|   | 5 | 0.001 | 0.026 | 0.103 | 0.201 | 0.246 | 0.201 | 0.103 | 0.026 | 0.001 |
|   | 6 | 0.000 | 0.006 | 0.037 | 0.111 | 0.205 | 0.251 | 0.200 | 0.088 | 0.011 |
|   | 7 | 0.000 | 0.001 | 0.009 | 0.042 | 0.117 | 0.215 | 0.267 | 0.201 | 0.057 |
|   | 8 | 0.000 | 0.000 | 0.001 | 0.011 | 0.044 | 0.121 | 0.233 | 0.302 | 0.194 |
|   | 9 | 0.000 | 0.000 | 0.000 | 0.002 | 0.010 | 0.040 | 0.121 | 0.268 | 0.387 |
|   | 10 | 0.000 | 0.000 | 0.000 | 0.000 | 0.001 | 0.006 | 0.028 | 0.107 | 0.349 |
| 11 | 0 | 0.314 | 0.086 | 0.020 | 0.004 | 0.000 | 0.000 | 0.000 | 0.000 | 0.000 |
|   | 1 | 0.384 | 0.236 | 0.093 | 0.027 | 0.005 | 0.001 | 0.000 | 0.000 | 0.000 |
|   | 2 | 0.213 | 0.295 | 0.200 | 0.089 | 0.027 | 0.005 | 0.001 | 0.000 | 0.000 |
|   | 3 | 0.071 | 0.221 | 0.257 | 0.177 | 0.081 | 0.023 | 0.004 | 0.000 | 0.000 |
|   | 4 | 0.016 | 0.111 | 0.220 | 0.236 | 0.161 | 0.070 | 0.017 | 0.002 | 0.000 |
|   | 5 | 0.002 | 0.039 | 0.132 | 0.221 | 0.226 | 0.147 | 0.057 | 0.010 | 0.000 |
|   | 6 | 0.000 | 0.010 | 0.057 | 0.147 | 0.226 | 0.221 | 0.132 | 0.039 | 0.002 |
|   | 7 | 0.000 | 0.002 | 0.017 | 0.070 | 0.161 | 0.236 | 0.220 | 0.111 | 0.016 |
|   | 8 | 0.000 | 0.000 | 0.004 | 0.023 | 0.081 | 0.177 | 0.257 | 0.221 | 0.071 |
|   | 9 | 0.000 | 0.000 | 0.001 | 0.005 | 0.027 | 0.089 | 0.200 | 0.295 | 0.213 |
|   | 10 | 0.000 | 0.000 | 0.000 | 0.001 | 0.005 | 0.027 | 0.093 | 0.236 | 0.384 |
|   | 11 | 0.000 | 0.000 | 0.000 | 0.000 | 0.000 | 0.004 | 0.020 | 0.086 | 0.314 |
| 12 | 0 | 0.282 | 0.069 | 0.014 | 0.002 | 0.000 | 0.000 | 0.000 | 0.000 | 0.000 |
|   | 1 | 0.377 | 0.206 | 0.071 | 0.017 | 0.003 | 0.000 | 0.000 | 0.000 | 0.000 |
|   | 2 | 0.230 | 0.283 | 0.168 | 0.064 | 0.016 | 0.002 | 0.000 | 0.000 | 0.000 |
|   | 3 | 0.085 | 0.236 | 0.240 | 0.142 | 0.054 | 0.012 | 0.001 | 0.000 | 0.000 |
|   | 4 | 0.021 | 0.133 | 0.231 | 0.213 | 0.121 | 0.042 | 0.008 | 0.001 | 0.000 |
|   | 5 | 0.004 | 0.053 | 0.158 | 0.227 | 0.193 | 0.101 | 0.029 | 0.003 | 0.000 |
|   | 6 | 0.000 | 0.016 | 0.079 | 0.177 | 0.226 | 0.177 | 0.079 | 0.016 | 0.000 |
|   | 7 | 0.000 | 0.003 | 0.029 | 0.101 | 0.193 | 0.227 | 0.158 | 0.053 | 0.004 |
|   | 8 | 0.000 | 0.001 | 0.008 | 0.042 | 0.121 | 0.213 | 0.231 | 0.133 | 0.021 |
|   | 9 | 0.000 | 0.000 | 0.001 | 0.012 | 0.054 | 0.142 | 0.240 | 0.236 | 0.085 |
|   | 10 | 0.000 | 0.000 | 0.000 | 0.002 | 0.016 | 0.064 | 0.168 | 0.283 | 0.230 |
|   | 11 | 0.000 | 0.000 | 0.000 | 0.000 | 0.003 | 0.017 | 0.071 | 0.206 | 0.377 |
|   | 12 | 0.000 | 0.000 | 0.000 | 0.000 | 0.000 | 0.002 | 0.014 | 0.069 | 0.282 |

Table D.2   (*Continued*)

| n | k | .10 | .20 | .30 | .40 | .50 | .60 | .70 | .80 | .90 |
|---|---|-----|-----|-----|-----|-----|-----|-----|-----|-----|
| 13 | 0 | 0.254 | 0.055 | 0.010 | 0.001 | 0.000 | 0.000 | 0.000 | 0.000 | 0.000 |
|    | 1 | 0.367 | 0.179 | 0.054 | 0.011 | 0.002 | 0.000 | 0.000 | 0.000 | 0.000 |
|    | 2 | 0.245 | 0.268 | 0.139 | 0.045 | 0.010 | 0.001 | 0.000 | 0.000 | 0.000 |
|    | 3 | 0.100 | 0.246 | 0.218 | 0.111 | 0.035 | 0.006 | 0.001 | 0.000 | 0.000 |
|    | 4 | 0.028 | 0.154 | 0.234 | 0.184 | 0.087 | 0.024 | 0.003 | 0.000 | 0.000 |
|    | 5 | 0.006 | 0.069 | 0.180 | 0.221 | 0.157 | 0.066 | 0.014 | 0.001 | 0.000 |
|    | 6 | 0.001 | 0.023 | 0.103 | 0.197 | 0.209 | 0.131 | 0.044 | 0.006 | 0.000 |
|    | 7 | 0.000 | 0.006 | 0.044 | 0.131 | 0.209 | 0.197 | 0.103 | 0.023 | 0.001 |
|    | 8 | 0.000 | 0.001 | 0.014 | 0.066 | 0.157 | 0.221 | 0.180 | 0.069 | 0.006 |
|    | 9 | 0.000 | 0.000 | 0.003 | 0.024 | 0.087 | 0.184 | 0.234 | 0.154 | 0.028 |
|    | 10 | 0.000 | 0.000 | 0.001 | 0.006 | 0.035 | 0.111 | 0.218 | 0.246 | 0.100 |
|    | 11 | 0.000 | 0.000 | 0.000 | 0.001 | 0.010 | 0.045 | 0.139 | 0.268 | 0.245 |
|    | 12 | 0.000 | 0.000 | 0.000 | 0.000 | 0.002 | 0.011 | 0.054 | 0.179 | 0.367 |
|    | 13 | 0.000 | 0.000 | 0.000 | 0.000 | 0.000 | 0.001 | 0.010 | 0.055 | 0.254 |
| 14 | 0 | 0.229 | 0.044 | 0.007 | 0.001 | 0.000 | 0.000 | 0.000 | 0.000 | 0.000 |
|    | 1 | 0.356 | 0.154 | 0.041 | 0.007 | 0.001 | 0.000 | 0.000 | 0.000 | 0.000 |
|    | 2 | 0.257 | 0.250 | 0.113 | 0.032 | 0.006 | 0.001 | 0.000 | 0.000 | 0.000 |
|    | 3 | 0.114 | 0.250 | 0.194 | 0.085 | 0.022 | 0.003 | 0.000 | 0.000 | 0.000 |
|    | 4 | 0.035 | 0.172 | 0.229 | 0.155 | 0.061 | 0.014 | 0.001 | 0.000 | 0.000 |
|    | 5 | 0.008 | 0.086 | 0.196 | 0.207 | 0.122 | 0.041 | 0.007 | 0.000 | 0.000 |
|    | 6 | 0.001 | 0.032 | 0.126 | 0.207 | 0.183 | 0.092 | 0.023 | 0.002 | 0.000 |
|    | 7 | 0.000 | 0.009 | 0.062 | 0.157 | 0.209 | 0.157 | 0.062 | 0.009 | 0.000 |
|    | 8 | 0.000 | 0.002 | 0.023 | 0.092 | 0.183 | 0.207 | 0.126 | 0.032 | 0.001 |
|    | 9 | 0.000 | 0.000 | 0.007 | 0.041 | 0.122 | 0.207 | 0.196 | 0.086 | 0.008 |
|    | 10 | 0.000 | 0.000 | 0.001 | 0.014 | 0.061 | 0.155 | 0.229 | 0.172 | 0.035 |
|    | 11 | 0.000 | 0.000 | 0.000 | 0.003 | 0.022 | 0.085 | 0.194 | 0.250 | 0.114 |
|    | 12 | 0.000 | 0.000 | 0.000 | 0.001 | 0.006 | 0.032 | 0.113 | 0.250 | 0.257 |
|    | 13 | 0.000 | 0.000 | 0.000 | 0.000 | 0.001 | 0.007 | 0.041 | 0.154 | 0.356 |
|    | 14 | 0.000 | 0.000 | 0.000 | 0.000 | 0.000 | 0.001 | 0.007 | 0.044 | 0.229 |
| 15 | 0 | 0.206 | 0.035 | 0.005 | 0.000 | 0.000 | 0.000 | 0.000 | 0.000 | 0.000 |
|    | 1 | 0.343 | 0.132 | 0.031 | 0.005 | 0.000 | 0.000 | 0.000 | 0.000 | 0.000 |
|    | 2 | 0.267 | 0.231 | 0.092 | 0.022 | 0.003 | 0.000 | 0.000 | 0.000 | 0.000 |
|    | 3 | 0.129 | 0.250 | 0.170 | 0.063 | 0.014 | 0.002 | 0.000 | 0.000 | 0.000 |
|    | 4 | 0.043 | 0.188 | 0.219 | 0.127 | 0.042 | 0.007 | 0.001 | 0.000 | 0.000 |
|    | 5 | 0.010 | 0.103 | 0.206 | 0.186 | 0.092 | 0.024 | 0.003 | 0.000 | 0.000 |
|    | 6 | 0.002 | 0.043 | 0.147 | 0.207 | 0.153 | 0.061 | 0.012 | 0.001 | 0.000 |
|    | 7 | 0.000 | 0.014 | 0.081 | 0.177 | 0.196 | 0.118 | 0.035 | 0.003 | 0.000 |
|    | 8 | 0.000 | 0.003 | 0.035 | 0.118 | 0.196 | 0.177 | 0.081 | 0.014 | 0.000 |
|    | 9 | 0.000 | 0.001 | 0.012 | 0.061 | 0.153 | 0.207 | 0.147 | 0.043 | 0.002 |
|    | 10 | 0.000 | 0.000 | 0.003 | 0.024 | 0.092 | 0.186 | 0.206 | 0.103 | 0.010 |
|    | 11 | 0.000 | 0.000 | 0.001 | 0.007 | 0.042 | 0.127 | 0.219 | 0.188 | 0.043 |
|    | 12 | 0.000 | 0.000 | 0.000 | 0.002 | 0.014 | 0.063 | 0.170 | 0.250 | 0.129 |
|    | 13 | 0.000 | 0.000 | 0.000 | 0.000 | 0.003 | 0.022 | 0.092 | 0.231 | 0.267 |
|    | 14 | 0.000 | 0.000 | 0.000 | 0.000 | 0.000 | 0.005 | 0.031 | 0.132 | 0.343 |
|    | 15 | 0.000 | 0.000 | 0.000 | 0.000 | 0.000 | 0.000 | 0.005 | 0.035 | 0.206 |

Table D.2    (*Continued*)

| n | k | .10 | .20 | .30 | .40 | p .50 | .60 | .70 | .80 | .90 |
|---|---|-----|-----|-----|-----|-----|-----|-----|-----|-----|
| 16 | 0 | 0.185 | 0.028 | 0.003 | 0.000 | 0.000 | 0.000 | 0.000 | 0.000 | 0.000 |
|    | 1 | 0.329 | 0.113 | 0.023 | 0.003 | 0.000 | 0.000 | 0.000 | 0.000 | 0.000 |
|    | 2 | 0.275 | 0.211 | 0.073 | 0.015 | 0.002 | 0.000 | 0.000 | 0.000 | 0.000 |
|    | 3 | 0.142 | 0.246 | 0.146 | 0.047 | 0.009 | 0.001 | 0.000 | 0.000 | 0.000 |
|    | 4 | 0.051 | 0.200 | 0.204 | 0.101 | 0.028 | 0.004 | 0.000 | 0.000 | 0.000 |
|    | 5 | 0.014 | 0.120 | 0.210 | 0.162 | 0.067 | 0.014 | 0.001 | 0.000 | 0.000 |
|    | 6 | 0.003 | 0.055 | 0.165 | 0.198 | 0.122 | 0.039 | 0.006 | 0.000 | 0.000 |
|    | 7 | 0.000 | 0.020 | 0.101 | 0.189 | 0.175 | 0.084 | 0.019 | 0.001 | 0.000 |
|    | 8 | 0.000 | 0.006 | 0.049 | 0.142 | 0.196 | 0.142 | 0.049 | 0.006 | 0.000 |
|    | 9 | 0.000 | 0.001 | 0.019 | 0.084 | 0.175 | 0.189 | 0.101 | 0.020 | 0.000 |
|    | 10 | 0.000 | 0.000 | 0.006 | 0.039 | 0.122 | 0.198 | 0.165 | 0.055 | 0.003 |
|    | 11 | 0.000 | 0.000 | 0.001 | 0.014 | 0.067 | 0.162 | 0.210 | 0.120 | 0.014 |
|    | 12 | 0.000 | 0.000 | 0.000 | 0.004 | 0.028 | 0.101 | 0.204 | 0.200 | 0.051 |
|    | 13 | 0.000 | 0.000 | 0.000 | 0.001 | 0.009 | 0.047 | 0.146 | 0.246 | 0.142 |
|    | 14 | 0.000 | 0.000 | 0.000 | 0.000 | 0.002 | 0.015 | 0.073 | 0.211 | 0.275 |
|    | 15 | 0.000 | 0.000 | 0.000 | 0.000 | 0.000 | 0.003 | 0.023 | 0.113 | 0.329 |
|    | 16 | 0.000 | 0.000 | 0.000 | 0.000 | 0.000 | 0.000 | 0.003 | 0.028 | 0.185 |
| 17 | 0 | 0.167 | 0.023 | 0.002 | 0.000 | 0.000 | 0.000 | 0.000 | 0.000 | 0.000 |
|    | 1 | 0.315 | 0.096 | 0.017 | 0.002 | 0.000 | 0.000 | 0.000 | 0.000 | 0.000 |
|    | 2 | 0.280 | 0.191 | 0.058 | 0.010 | 0.001 | 0.000 | 0.000 | 0.000 | 0.000 |
|    | 3 | 0.156 | 0.239 | 0.125 | 0.034 | 0.005 | 0.000 | 0.000 | 0.000 | 0.000 |
|    | 4 | 0.060 | 0.209 | 0.187 | 0.080 | 0.018 | 0.002 | 0.000 | 0.000 | 0.000 |
|    | 5 | 0.017 | 0.136 | 0.208 | 0.138 | 0.047 | 0.008 | 0.001 | 0.000 | 0.000 |
|    | 6 | 0.004 | 0.068 | 0.178 | 0.184 | 0.094 | 0.024 | 0.003 | 0.000 | 0.000 |
|    | 7 | 0.001 | 0.027 | 0.120 | 0.193 | 0.148 | 0.057 | 0.009 | 0.000 | 0.000 |
|    | 8 | 0.000 | 0.008 | 0.064 | 0.161 | 0.185 | 0.107 | 0.028 | 0.002 | 0.000 |
|    | 9 | 0.000 | 0.002 | 0.028 | 0.107 | 0.185 | 0.161 | 0.064 | 0.008 | 0.000 |
|    | 10 | 0.000 | 0.000 | 0.009 | 0.057 | 0.148 | 0.193 | 0.120 | 0.027 | 0.001 |
|    | 11 | 0.000 | 0.000 | 0.003 | 0.024 | 0.094 | 0.184 | 0.178 | 0.068 | 0.004 |
|    | 12 | 0.000 | 0.000 | 0.001 | 0.008 | 0.047 | 0.138 | 0.208 | 0.136 | 0.017 |
|    | 13 | 0.000 | 0.000 | 0.000 | 0.002 | 0.018 | 0.080 | 0.187 | 0.209 | 0.060 |
|    | 14 | 0.000 | 0.000 | 0.000 | 0.000 | 0.005 | 0.034 | 0.125 | 0.239 | 0.156 |
|    | 15 | 0.000 | 0.000 | 0.000 | 0.000 | 0.001 | 0.010 | 0.058 | 0.191 | 0.280 |
|    | 16 | 0.000 | 0.000 | 0.000 | 0.000 | 0.000 | 0.002 | 0.017 | 0.096 | 0.315 |
|    | 17 | 0.000 | 0.000 | 0.000 | 0.000 | 0.000 | 0.000 | 0.002 | 0.023 | 0.167 |
| 18 | 0 | 0.150 | 0.018 | 0.002 | 0.000 | 0.000 | 0.000 | 0.000 | 0.000 | 0.000 |
|    | 1 | 0.300 | 0.081 | 0.013 | 0.001 | 0.000 | 0.000 | 0.000 | 0.000 | 0.000 |
|    | 2 | 0.284 | 0.172 | 0.046 | 0.007 | 0.001 | 0.000 | 0.000 | 0.000 | 0.000 |
|    | 3 | 0.168 | 0.230 | 0.105 | 0.025 | 0.003 | 0.000 | 0.000 | 0.000 | 0.000 |
|    | 4 | 0.070 | 0.215 | 0.168 | 0.061 | 0.012 | 0.001 | 0.000 | 0.000 | 0.000 |
|    | 5 | 0.022 | 0.151 | 0.202 | 0.115 | 0.033 | 0.004 | 0.000 | 0.000 | 0.000 |
|    | 6 | 0.005 | 0.082 | 0.187 | 0.166 | 0.071 | 0.015 | 0.001 | 0.000 | 0.000 |
|    | 7 | 0.001 | 0.035 | 0.138 | 0.189 | 0.121 | 0.037 | 0.005 | 0.000 | 0.000 |
|    | 8 | 0.000 | 0.012 | 0.081 | 0.173 | 0.167 | 0.077 | 0.015 | 0.001 | 0.000 |
|    | 9 | 0.000 | 0.003 | 0.039 | 0.128 | 0.185 | 0.128 | 0.039 | 0.003 | 0.000 |
|    | 10 | 0.000 | 0.001 | 0.015 | 0.077 | 0.167 | 0.173 | 0.081 | 0.012 | 0.000 |
|    | 11 | 0.000 | 0.000 | 0.005 | 0.037 | 0.121 | 0.189 | 0.138 | 0.035 | 0.001 |
|    | 12 | 0.000 | 0.000 | 0.001 | 0.015 | 0.071 | 0.166 | 0.187 | 0.082 | 0.005 |
|    | 13 | 0.000 | 0.000 | 0.000 | 0.004 | 0.033 | 0.115 | 0.202 | 0.151 | 0.022 |
|    | 14 | 0.000 | 0.000 | 0.000 | 0.001 | 0.012 | 0.061 | 0.168 | 0.215 | 0.070 |
|    | 15 | 0.000 | 0.000 | 0.000 | 0.000 | 0.003 | 0.025 | 0.105 | 0.230 | 0.168 |
|    | 16 | 0.000 | 0.000 | 0.000 | 0.000 | 0.001 | 0.007 | 0.046 | 0.172 | 0.284 |
|    | 17 | 0.000 | 0.000 | 0.000 | 0.000 | 0.000 | 0.001 | 0.013 | 0.081 | 0.300 |
|    | 18 | 0.000 | 0.000 | 0.000 | 0.000 | 0.000 | 0.000 | 0.002 | 0.018 | 0.150 |

Table D.2   (*Continued*)

| n | k | .10 | .20 | .30 | .40 | p .50 | .60 | .70 | .80 | .90 |
|---|---|------|------|------|------|------|------|------|------|------|
| 19 | 0 | 0.135 | 0.014 | 0.001 | 0.000 | 0.000 | 0.000 | 0.000 | 0.000 | 0.000 |
|    | 1 | 0.285 | 0.068 | 0.009 | 0.001 | 0.000 | 0.000 | 0.000 | 0.000 | 0.000 |
|    | 2 | 0.285 | 0.154 | 0.036 | 0.005 | 0.000 | 0.000 | 0.000 | 0.000 | 0.000 |
|    | 3 | 0.180 | 0.218 | 0.087 | 0.017 | 0.002 | 0.000 | 0.000 | 0.000 | 0.000 |
|    | 4 | 0.080 | 0.218 | 0.149 | 0.047 | 0.007 | 0.001 | 0.000 | 0.000 | 0.000 |
|    | 5 | 0.027 | 0.164 | 0.192 | 0.093 | 0.022 | 0.002 | 0.000 | 0.000 | 0.000 |
|    | 6 | 0.007 | 0.095 | 0.192 | 0.145 | 0.052 | 0.008 | 0.001 | 0.000 | 0.000 |
|    | 7 | 0.001 | 0.044 | 0.153 | 0.180 | 0.096 | 0.024 | 0.002 | 0.000 | 0.000 |
|    | 8 | 0.000 | 0.017 | 0.098 | 0.180 | 0.144 | 0.053 | 0.008 | 0.000 | 0.000 |
|    | 9 | 0.000 | 0.005 | 0.051 | 0.146 | 0.176 | 0.098 | 0.022 | 0.001 | 0.000 |
|    | 10 | 0.000 | 0.001 | 0.022 | 0.098 | 0.176 | 0.146 | 0.051 | 0.005 | 0.000 |
|    | 11 | 0.000 | 0.000 | 0.008 | 0.053 | 0.144 | 0.180 | 0.098 | 0.017 | 0.000 |
|    | 12 | 0.000 | 0.000 | 0.002 | 0.024 | 0.096 | 0.180 | 0.153 | 0.044 | 0.001 |
|    | 13 | 0.000 | 0.000 | 0.001 | 0.008 | 0.052 | 0.145 | 0.192 | 0.095 | 0.007 |
|    | 14 | 0.000 | 0.000 | 0.000 | 0.002 | 0.022 | 0.093 | 0.192 | 0.164 | 0.027 |
|    | 15 | 0.000 | 0.000 | 0.000 | 0.001 | 0.007 | 0.047 | 0.149 | 0.218 | 0.080 |
|    | 16 | 0.000 | 0.000 | 0.000 | 0.000 | 0.002 | 0.017 | 0.087 | 0.218 | 0.180 |
|    | 17 | 0.000 | 0.000 | 0.000 | 0.000 | 0.000 | 0.005 | 0.036 | 0.154 | 0.285 |
|    | 18 | 0.000 | 0.000 | 0.000 | 0.000 | 0.000 | 0.001 | 0.009 | 0.068 | 0.285 |
|    | 19 | 0.000 | 0.000 | 0.000 | 0.000 | 0.000 | 0.000 | 0.001 | 0.014 | 0.135 |
| 20 | 0 | 0.122 | 0.012 | 0.001 | 0.000 | 0.000 | 0.000 | 0.000 | 0.000 | 0.000 |
|    | 1 | 0.270 | 0.058 | 0.007 | 0.000 | 0.000 | 0.000 | 0.000 | 0.000 | 0.000 |
|    | 2 | 0.285 | 0.137 | 0.028 | 0.003 | 0.000 | 0.000 | 0.000 | 0.000 | 0.000 |
|    | 3 | 0.190 | 0.205 | 0.072 | 0.012 | 0.001 | 0.000 | 0.000 | 0.000 | 0.000 |
|    | 4 | 0.090 | 0.218 | 0.130 | 0.035 | 0.005 | 0.000 | 0.000 | 0.000 | 0.000 |
|    | 5 | 0.032 | 0.175 | 0.179 | 0.075 | 0.015 | 0.001 | 0.000 | 0.000 | 0.000 |
|    | 6 | 0.009 | 0.109 | 0.192 | 0.124 | 0.037 | 0.005 | 0.000 | 0.000 | 0.000 |
|    | 7 | 0.002 | 0.055 | 0.164 | 0.166 | 0.074 | 0.015 | 0.001 | 0.000 | 0.000 |
|    | 8 | 0.000 | 0.022 | 0.114 | 0.180 | 0.120 | 0.035 | 0.004 | 0.000 | 0.000 |
|    | 9 | 0.000 | 0.007 | 0.065 | 0.160 | 0.160 | 0.071 | 0.012 | 0.000 | 0.000 |
|    | 10 | 0.000 | 0.002 | 0.031 | 0.117 | 0.176 | 0.117 | 0.031 | 0.002 | 0.000 |
|    | 11 | 0.000 | 0.000 | 0.012 | 0.071 | 0.160 | 0.160 | 0.065 | 0.007 | 0.000 |
|    | 12 | 0.000 | 0.000 | 0.004 | 0.035 | 0.120 | 0.180 | 0.114 | 0.022 | 0.000 |
|    | 13 | 0.000 | 0.000 | 0.001 | 0.015 | 0.074 | 0.166 | 0.164 | 0.055 | 0.002 |
|    | 14 | 0.000 | 0.000 | 0.000 | 0.005 | 0.037 | 0.124 | 0.192 | 0.109 | 0.009 |
|    | 15 | 0.000 | 0.000 | 0.000 | 0.001 | 0.015 | 0.075 | 0.179 | 0.175 | 0.032 |
|    | 16 | 0.000 | 0.000 | 0.000 | 0.000 | 0.005 | 0.035 | 0.130 | 0.218 | 0.090 |
|    | 17 | 0.000 | 0.000 | 0.000 | 0.000 | 0.001 | 0.012 | 0.072 | 0.205 | 0.190 |
|    | 18 | 0.000 | 0.000 | 0.000 | 0.000 | 0.000 | 0.003 | 0.028 | 0.137 | 0.285 |
|    | 19 | 0.000 | 0.000 | 0.000 | 0.000 | 0.000 | 0.000 | 0.007 | 0.058 | 0.270 |
|    | 20 | 0.000 | 0.000 | 0.000 | 0.000 | 0.000 | 0.000 | 0.001 | 0.012 | 0.122 |

Table D.3    Student's $t$ Distribution.

Area to right of tabulated $t$ (one-tailed $p$)

| df | .30 | .20 | .10 | .05 | .025 | .01 | .005 |
|---|---|---|---|---|---|---|---|
| 1 | 0.7265 | 1.3764 | 3.0777 | 6.3138 | 12.7062 | 31.8205 | 63.6567 |
| 2 | 0.6172 | 1.0607 | 1.8856 | 2.9200 | 4.3027 | 6.9646 | 9.9248 |
| 3 | 0.5844 | 0.9785 | 1.6377 | 2.3534 | 3.1824 | 4.5407 | 5.8409 |
| 4 | 0.5686 | 0.9410 | 1.5332 | 2.1318 | 2.7764 | 3.7469 | 4.6041 |
| 5 | 0.5594 | 0.9195 | 1.4759 | 2.0150 | 2.5706 | 3.3649 | 4.0321 |
| 6 | 0.5534 | 0.9057 | 1.4398 | 1.9432 | 2.4469 | 3.1427 | 3.7074 |
| 7 | 0.5491 | 0.8960 | 1.4149 | 1.8946 | 2.3646 | 2.9980 | 3.4995 |
| 8 | 0.5459 | 0.8889 | 1.3968 | 1.8595 | 2.3060 | 2.8965 | 3.3554 |
| 9 | 0.5435 | 0.8834 | 1.3830 | 1.8331 | 2.2622 | 2.8214 | 3.2498 |
| 10 | 0.5415 | 0.8791 | 1.3722 | 1.8125 | 2.2281 | 2.7638 | 3.1693 |
| 11 | 0.5399 | 0.8755 | 1.3634 | 1.7959 | 2.2010 | 2.7181 | 3.1058 |
| 12 | 0.5386 | 0.8726 | 1.3562 | 1.7823 | 2.1788 | 2.6810 | 3.0545 |
| 13 | 0.5375 | 0.8702 | 1.3502 | 1.7709 | 2.1604 | 2.6503 | 3.0123 |
| 14 | 0.5366 | 0.8681 | 1.3450 | 1.7613 | 2.1448 | 2.6245 | 2.9768 |
| 15 | 0.5357 | 0.8662 | 1.3406 | 1.7531 | 2.1314 | 2.6025 | 2.9467 |
| 16 | 0.5350 | 0.8647 | 1.3368 | 1.7459 | 2.1199 | 2.5835 | 2.9208 |
| 17 | 0.5344 | 0.8633 | 1.3334 | 1.7396 | 2.1098 | 2.5669 | 2.8982 |
| 18 | 0.5338 | 0.8620 | 1.3304 | 1.7341 | 2.1009 | 2.5524 | 2.8784 |
| 19 | 0.5333 | 0.8610 | 1.3277 | 1.7291 | 2.0930 | 2.5395 | 2.8609 |
| 20 | 0.5329 | 0.8600 | 1.3253 | 1.7247 | 2.0860 | 2.5280 | 2.8453 |
| 21 | 0.5325 | 0.8591 | 1.3232 | 1.7207 | 2.0796 | 2.5176 | 2.8314 |
| 22 | 0.5321 | 0.8583 | 1.3212 | 1.7171 | 2.0739 | 2.5083 | 2.8188 |
| 23 | 0.5317 | 0.8575 | 1.3195 | 1.7139 | 2.0687 | 2.4999 | 2.8073 |
| 24 | 0.5314 | 0.8569 | 1.3178 | 1.7109 | 2.0639 | 2.4922 | 2.7969 |
| 25 | 0.5312 | 0.8562 | 1.3163 | 1.7081 | 2.0595 | 2.4851 | 2.7874 |
| 26 | 0.5309 | 0.8557 | 1.3150 | 1.7056 | 2.0555 | 2.4786 | 2.7787 |
| 27 | 0.5306 | 0.8551 | 1.3137 | 1.7033 | 2.0518 | 2.4727 | 2.7707 |
| 28 | 0.5304 | 0.8546 | 1.3125 | 1.7011 | 2.0484 | 2.4671 | 2.7633 |
| 29 | 0.5302 | 0.8542 | 1.3114 | 1.6991 | 2.0452 | 2.4620 | 2.7564 |
| 30 | 0.5300 | 0.8538 | 1.3104 | 1.6973 | 2.0423 | 2.4573 | 2.7500 |
| 35 | 0.5292 | 0.8520 | 1.3062 | 1.6896 | 2.0301 | 2.4377 | 2.7238 |
| 40 | 0.5286 | 0.8507 | 1.3031 | 1.6839 | 2.0211 | 2.4233 | 2.7045 |
| 50 | 0.5278 | 0.8489 | 1.2987 | 1.6759 | 2.0086 | 2.4033 | 2.6778 |
| 100 | 0.5261 | 0.8452 | 1.2901 | 1.6602 | 1.9840 | 2.3642 | 2.6259 |
| Inf. | 0.5244 | 0.8416 | 1.2816 | 1.6449 | 1.9600 | 2.3263 | 2.5758 |

Table D.4   Critical Values for the Wilcoxon Signed Rank Test. The table shows the values of $W$ (as described in the text) that give one-tailed $p$ values of at most 5% or 1% for sample size $n$.

| | One-tailed p-value | | | | One-tailed p-value | | |
| --- | --- | --- | --- | --- | --- | --- | --- |
| | .05 | .025 | .01 | | .05 | .025 | .01 |
| n less than 5 | — | — | — | n = 28 | 130 | 116 | 101 |
| n =  5 | 0 | — | — | n = 29 | 140 | 126 | 110 |
| n =  6 | 2 | 0 | — | n = 30 | 151 | 137 | 120 |
| n =  7 | 3 | 2 | 0 | n = 31 | 163 | 147 | 130 |
| n =  8 | 5 | 3 | 1 | n = 32 | 175 | 159 | 140 |
| n =  9 | 8 | 5 | 3 | n = 33 | 187 | 170 | 151 |
| n = 10 | 10 | 8 | 5 | n = 34 | 200 | 182 | 162 |
| n = 11 | 13 | 10 | 7 | n = 35 | 213 | 195 | 173 |
| n = 12 | 17 | 13 | 9 | n = 36 | 227 | 208 | 185 |
| n = 13 | 21 | 17 | 12 | n = 37 | 241 | 221 | 198 |
| n = 14 | 25 | 21 | 15 | n = 38 | 256 | 235 | 211 |
| n = 15 | 30 | 25 | 19 | n = 39 | 271 | 249 | 224 |
| n = 16 | 35 | 29 | 23 | n = 40 | 286 | 264 | 238 |
| n = 17 | 41 | 34 | 27 | n = 41 | 302 | 279 | 252 |
| n = 18 | 47 | 40 | 32 | n = 42 | 319 | 294 | 266 |
| n = 19 | 53 | 46 | 37 | n = 43 | 336 | 310 | 281 |
| n = 20 | 60 | 52 | 43 | n = 44 | 353 | 327 | 296 |
| n = 21 | 67 | 58 | 49 | n = 45 | 371 | 343 | 312 |
| n = 22 | 75 | 65 | 55 | n = 46 | 389 | 461 | 328 |
| n = 23 | 83 | 73 | 62 | n = 47 | 407 | 378 | 345 |
| n = 24 | 91 | 81 | 69 | n = 48 | 426 | 396 | 362 |
| n = 25 | 100 | 89 | 76 | n = 49 | 446 | 415 | 379 |
| n = 26 | 110 | 98 | 84 | n = 50 | 466 | 434 | 397 |
| n = 27 | 119 | 107 | 92 | | | | |

*Source*: Adapted from Howell (1982).

Table D.5   Critical Values for the Wilcoxon Rank Sum Test. The tabulated values are the values of *W* (for the smaller sample if the sample sizes are unequal) needed for (at most) the two-tailed *p* value indicated. For unequal sample sizes, the smaller sample is sample 1.

| Size of Sample 2 | 2-tail p | \multicolumn{10}{c}{Size of Sample 1} | | | | | | | | | |
|---|---|---|---|---|---|---|---|---|---|---|---|
| | | \multicolumn{2}{c}{2} | | \multicolumn{2}{c}{3} | | \multicolumn{2}{c}{4} | | \multicolumn{2}{c}{5} | | \multicolumn{2}{c}{6} |
| | | L | U | L | U | L | U | L | U | L | U |
| 4 | .05 | -- | -- | -- | -- | 10 | 26 | | | | |
|   | .02 | -- | -- | -- | -- | -- | -- | | | | |
|   | .01 | -- | -- | -- | -- | -- | -- | | | | |
| 5 | .05 | -- | -- | -- | -- | 11 | 29 | 17 | 38 | | |
|   | .02 | -- | -- | -- | -- | 10 | 30 | 16 | 39 | | |
|   | .01 | -- | -- | -- | -- | -- | -- | 15 | 40 | | |
| 6 | .05 | -- | -- | 7 | 23 | 12 | 32 | 18 | 42 | 26 | 52 |
|   | .02 | -- | -- | -- | -- | 11 | 33 | 17 | 43 | 24 | 54 |
|   | .01 | -- | -- | -- | -- | 10 | 34 | 16 | 44 | 23 | 55 |
| 7 | .05 | -- | -- | 7 | 26 | 13 | 35 | 20 | 45 | 27 | 57 |
|   | .02 | -- | -- | 6 | 27 | 11 | 37 | 18 | 47 | 25 | 59 |
|   | .01 | -- | -- | -- | -- | 10 | 38 | 16 | 49 | 24 | 60 |
| 8 | .05 | 3 | 19 | 8 | 28 | 14 | 38 | 21 | 49 | 29 | 61 |
|   | .02 | -- | -- | 6 | 30 | 12 | 40 | 19 | 51 | 27 | 63 |
|   | .01 | -- | -- | -- | -- | 11 | 41 | 17 | 53 | 25 | 65 |
| 9 | .05 | 3 | 21 | 8 | 31 | 14 | 42 | 22 | 53 | 31 | 65 |
|   | .02 | -- | -- | 7 | 32 | 13 | 43 | 20 | 55 | 28 | 68 |
|   | .01 | -- | -- | 6 | 33 | 11 | 45 | 18 | 57 | 26 | 70 |
| 10 | .05 | 3 | 23 | 9 | 33 | 15 | 45 | 23 | 57 | 32 | 70 |
|   | .02 | -- | -- | 7 | 35 | 13 | 47 | 21 | 59 | 29 | 73 |
|   | .01 | -- | -- | 6 | 36 | 12 | 48 | 19 | 61 | 27 | 75 |
| 11 | .05 | 3 | 25 | 9 | 36 | 16 | 48 | 24 | 61 | 34 | 74 |
|   | .02 | -- | -- | 7 | 38 | 14 | 50 | 22 | 63 | 30 | 78 |
|   | .01 | -- | -- | 6 | 39 | 12 | 52 | 20 | 65 | 28 | 80 |
| 12 | .05 | 4 | 26 | 10 | 38 | 17 | 51 | 26 | 64 | 35 | 79 |
|   | .02 | -- | -- | 8 | 40 | 15 | 53 | 23 | 67 | 32 | 82 |
|   | .01 | -- | -- | 7 | 41 | 13 | 55 | 21 | 69 | 30 | 84 |
| 13 | .05 | 4 | 28 | 10 | 41 | 18 | 54 | 27 | 68 | 37 | 83 |
|   | .02 | 3 | 29 | 8 | 43 | 15 | 57 | 24 | 71 | 33 | 87 |
|   | .01 | -- | -- | 7 | 44 | 13 | 59 | 22 | 73 | 31 | 89 |
| 14 | .05 | 4 | 30 | 11 | 43 | 19 | 57 | 28 | 72 | 38 | 88 |
|   | .02 | 3 | 31 | 8 | 46 | 16 | 60 | 25 | 75 | 34 | 92 |
|   | .01 | -- | -- | 7 | 47 | 14 | 62 | 22 | 78 | 32 | 94 |

Table D.5 *(Continued)* Critical Values for the Wilcoxon Rank Sum Test.

| Size of Sample 2 | 2-tail p | Size of Sample 1 | | | | | | | | | |
|---|---|---|---|---|---|---|---|---|---|---|---|
| | | 2 | | 3 | | 4 | | 5 | | 6 | |
| | | L | U | L | U | L | U | L | U | L | U |
| 15 | .05 | 4 | 32 | 11 | 46 | 20 | 60 | 29 | 76 | 40 | 92 |
| | .02 | 3 | 33 | 9 | 48 | 17 | 63 | 26 | 79 | 36 | 96 |
| | .01 | -- | -- | 7 | 50 | 15 | 65 | 23 | 82 | 33 | 99 |
| 16 | .05 | 4 | 34 | 12 | 48 | 21 | 63 | 30 | 80 | 42 | 96 |
| | .02 | 3 | 35 | 9 | 51 | 17 | 67 | 27 | 83 | 37 | 101 |
| | .01 | -- | -- | 8 | 52 | 15 | 69 | 24 | 86 | 34 | 104 |
| 17 | .05 | 5 | 35 | 12 | 51 | 21 | 67 | 32 | 83 | 43 | 101 |
| | .02 | 3 | 37 | 10 | 53 | 18 | 70 | 28 | 87 | 39 | 105 |
| | .01 | -- | -- | 8 | 55 | 16 | 72 | 25 | 90 | 36 | 108 |
| 18 | .05 | 5 | 37 | 13 | 53 | 22 | 70 | 33 | 87 | 45 | 105 |
| | .02 | 3 | 39 | 10 | 56 | 19 | 73 | 29 | 91 | 40 | 110 |
| | .01 | -- | -- | 8 | 58 | 16 | 76 | 26 | 94 | 37 | 113 |
| 19 | .05 | 5 | 39 | 13 | 56 | 23 | 73 | 34 | 91 | 46 | 110 |
| | .02 | 4 | 40 | 10 | 59 | 19 | 77 | 30 | 95 | 41 | 115 |
| | .01 | 3 | 41 | 9 | 60 | 17 | 79 | 27 | 98 | 38 | 118 |
| 20 | .05 | 5 | 41 | 14 | 58 | 24 | 76 | 35 | 95 | 48 | 114 |
| | .02 | 4 | 42 | 11 | 61 | 20 | 80 | 31 | 99 | 43 | 119 |
| | .01 | 3 | 43 | 9 | 63 | 18 | 82 | 28 | 102 | 39 | 123 |

Table D.5   (*Continued*) Critical Values for the Wilcoxon Rank Sum Test.

| Size of Sample 2 | | Size of Sample 1 | | | | | | | | | |
|---|---|---|---|---|---|---|---|---|---|---|---|
| | | 7 | | 8 | | 9 | | 10 | | 11 | |
| | 2-tail p | L | U | L | U | L | U | L | U | L | U |
| 7 | .05 | 36 | 69 | | | | | | | | |
| | .02 | 34 | 71 | | | | | | | | |
| | .01 | 32 | 73 | | | | | | | | |
| 8 | .05 | 38 | 84 | 49 | 87 | | | | | | |
| | .02 | 35 | 87 | 45 | 91 | | | | | | |
| | .01 | 34 | 88 | 43 | 93 | | | | | | |
| 9 | .05 | 40 | 79 | 51 | 93 | 62 | 109 | | | | |
| | .02 | 37 | 82 | 47 | 97 | 59 | 112 | | | | |
| | .01 | 35 | 84 | 45 | 99 | 56 | 116 | | | | |
| 10 | .05 | 42 | 84 | 53 | 99 | 65 | 115 | 78 | 132 | | |
| | .02 | 39 | 87 | 49 | 103 | 61 | 119 | 74 | 136 | | |
| | .01 | 37 | 89 | 47 | 105 | 58 | 122 | 71 | 139 | | |
| 11 | .05 | 44 | 89 | 55 | 105 | 68 | 121 | 81 | 139 | 96 | 157 |
| | .02 | 40 | 93 | 51 | 109 | 63 | 126 | 77 | 143 | 91 | 162 |
| | .01 | 38 | 95 | 49 | 111 | 61 | 128 | 73 | 147 | 87 | 166 |
| 12 | .05 | 46 | 94 | 58 | 110 | 71 | 128 | 84 | 146 | 99 | 165 |
| | .02 | 42 | 98 | 53 | 115 | 66 | 132 | 79 | 151 | 94 | 170 |
| | .01 | 40 | 100 | 51 | 117 | 63 | 135 | 76 | 154 | 90 | 174 |
| 13 | .05 | 48 | 99 | 60 | 116 | 73 | 134 | 88 | 152 | 103 | 172 |
| | .02 | 44 | 103 | 56 | 120 | 68 | 139 | 82 | 158 | 97 | 178 |
| | .01 | 41 | 106 | 53 | 123 | 65 | 142 | 79 | 161 | 93 | 182 |
| 14 | .05 | 50 | 104 | 62 | 122 | 76 | 140 | 91 | 159 | 106 | 180 |
| | .02 | 45 | 109 | 58 | 126 | 71 | 145 | 85 | 165 | 100 | 186 |
| | .01 | 43 | 111 | 54 | 130 | 67 | 149 | 81 | 169 | 96 | 190 |
| 15 | .05 | 52 | 109 | 65 | 127 | 79 | 146 | 94 | 166 | 110 | 187 |
| | .02 | 47 | 114 | 60 | 132 | 73 | 152 | 88 | 172 | 103 | 194 |
| | .01 | 44 | 117 | 56 | 136 | 69 | 156 | 84 | 176 | 99 | 198 |
| 16 | .05 | 54 | 114 | 67 | 133 | 82 | 152 | 97 | 173 | 113 | 195 |
| | .02 | 49 | 119 | 62 | 138 | 76 | 158 | 91 | 179 | 107 | 201 |
| | .01 | 46 | 122 | 58 | 142 | 72 | 162 | 86 | 184 | 102 | 206 |
| 17 | .05 | 56 | 119 | 70 | 138 | 84 | 159 | 100 | 180 | 117 | 202 |
| | .02 | 51 | 124 | 64 | 144 | 78 | 165 | 93 | 187 | 110 | 209 |
| | .01 | 47 | 128 | 60 | 148 | 74 | 169 | 89 | 191 | 105 | 214 |
| 18 | .05 | 58 | 124 | 72 | 144 | 87 | 163 | 103 | 187 | 121 | 209 |
| | .02 | 52 | 130 | 66 | 150 | 81 | 171 | 96 | 194 | 113 | 217 |
| | .01 | 49 | 133 | 62 | 154 | 76 | 176 | 92 | 198 | 108 | 222 |

Table D.5  (*Continued*) Critical Values for the Wilcoxon Rank Sum Test.

Size of
Sample 2

| | | Size of Sample 1 | | | | | | | | |
|---|---|---|---|---|---|---|---|---|---|---|
| | | 7 | | 8 | | 9 | | 10 | | 11 | |
| | 2-tail p | L | U | L | U | L | U | L | U | L | U |
| 19 | .05 | 60 | 129 | 74 | 150 | 90 | 171 | 107 | 193 | 124 | 217 |
| | .02 | 54 | 135 | 68 | 156 | 83 | 178 | 99 | 201 | 116 | 225 |
| | .01 | 50 | 139 | 64 | 160 | 78 | 183 | 94 | 206 | 111 | 230 |
| 20 | .05 | 62 | 134 | 77 | 155 | 93 | 177 | 110 | 200 | 128 | 224 |
| | .02 | 56 | 140 | 70 | 162 | 85 | 185 | 102 | 208 | 119 | 233 |
| | .01 | 52 | 144 | 66 | 166 | 81 | 189 | 97 | 213 | 114 | 238 |

Size of
Sample 2

| | | Size of Sample 1 | | | | | | | | |
|---|---|---|---|---|---|---|---|---|---|---|
| | | 12 | | 13 | | 14 | | 15 | | 16 | |
| | 2-tail p | L | U | L | U | L | U | L | U | L | U |
| 12 | .05 | 115 | 185 | | | | | | | | |
| | .02 | 109 | 191 | | | | | | | | |
| | .01 | 105 | 195 | | | | | | | | |
| 13 | .05 | 119 | 193 | 136 | 215 | | | | | | |
| | .02 | 113 | 199 | 130 | 221 | | | | | | |
| | .01 | 109 | 203 | 125 | 226 | | | | | | |
| 14 | .05 | 123 | 201 | 141 | 223 | 160 | 246 | | | | |
| | .02 | 116 | 208 | 134 | 230 | 152 | 254 | | | | |
| | .01 | 112 | 212 | 129 | 235 | 147 | 259 | | | | |
| 15 | .05 | 127 | 209 | 145 | 232 | 164 | 256 | 184 | 281 | | |
| | .02 | 120 | 216 | 138 | 239 | 156 | 264 | 176 | 289 | | |
| | .01 | 115 | 221 | 133 | 244 | 151 | 269 | 171 | 294 | | |
| 16 | .05 | 131 | 217 | 150 | 240 | 169 | 265 | 190 | 290 | 211 | 317 |
| | .02 | 124 | 224 | 142 | 248 | 161 | 273 | 181 | 299 | 202 | 326 |
| | .01 | 119 | 229 | 136 | 254 | 155 | 279 | 175 | 205 | 196 | 332 |
| 17 | .05 | 135 | 225 | 154 | 249 | 174 | 274 | 195 | 300 | 217 | 327 |
| | .02 | 127 | 233 | 146 | 257 | 165 | 283 | 186 | 309 | 207 | 337 |
| | .01 | 122 | 238 | 140 | 263 | 159 | 289 | 180 | 315 | 201 | 343 |
| 18 | .05 | 139 | 233 | 158 | 258 | 179 | 283 | 200 | 310 | 222 | 338 |
| | .02 | 131 | 241 | 150 | 266 | 170 | 292 | 190 | 320 | 212 | 348 |
| | .01 | 125 | 247 | 144 | 272 | 163 | 299 | 184 | 326 | 206 | 354 |

Table D.5   (*Continued*) Critical Values for the Wilcoxon Rank Sum Test.

| Size of Sample 2 | 2-tail p | Size of Sample 1 | | | | | | | | | |
|---|---|---|---|---|---|---|---|---|---|---|---|
| | | 12 | | 13 | | 14 | | 15 | | 16 | |
| | | L | U | L | U | L | U | L | U | L | U |
| 19 | .05 | 143 | 241 | 163 | 266 | 183 | 293 | 205 | 320 | 228 | 348 |
| | .02 | 134 | 250 | 154 | 275 | 174 | 302 | 195 | 330 | 218 | 358 |
| | .01 | 129 | 255 | 148 | 281 | 168 | 308 | 189 | 336 | 210 | 366 |
| 20 | .05 | 147 | 249 | 167 | 275 | 188 | 302 | 210 | 330 | 234 | 358 |
| | .02 | 138 | 258 | 158 | 284 | 178 | 312 | 200 | 340 | 223 | 369 |
| | .01 | 132 | 264 | 151 | 291 | 172 | 318 | 193 | 347 | 215 | 377 |

| Size of Sample 2 | 2-tail p | Size of Sample 1 | | | | | | | |
|---|---|---|---|---|---|---|---|---|---|
| | | 17 | | 18 | | 19 | | 20 | |
| | | L | U | L | U | L | U | L | U |
| 17 | .05 | 240 | 355 | | | | | | |
| | .02 | 230 | 365 | | | | | | |
| | .01 | 223 | 372 | | | | | | |
| 18 | .05 | 246 | 366 | 270 | 396 | | | | |
| | .02 | 235 | 377 | 259 | 407 | | | | |
| | .01 | 228 | 384 | 252 | 414 | | | | |
| 19 | .05 | 252 | 377 | 277 | 404 | 303 | 438 | | |
| | .02 | 241 | 388 | 265 | 419 | 291 | 450 | | |
| | .01 | 234 | 395 | 258 | 426 | 283 | 458 | | |
| 20 | .05 | 258 | 388 | 283 | 419 | 309 | 451 | 337 | 483 |
| | .02 | 246 | 400 | 271 | 431 | 297 | 463 | 324 | 496 |
| | .01 | 239 | 407 | 263 | 439 | 289 | 471 | 315 | 505 |

*Source*: Adapted from Howell (1982).

Table D.6    The $F$ Distribution.

| Denominator degrees of freedom | p | 1 | 2 | Numerator degrees of freedom 3 | 4 | 5 | 6 | 7 | 8 | 9 |
|---|---|---|---|---|---|---|---|---|---|---|
| 2 | 0.10 | 8.53 | 9.00 | 9.16 | 9.24 | 9.29 | 9.33 | 9.35 | 9.37 | 9.38 |
|   | 0.05 | 18.51 | 19.00 | 19.16 | 19.25 | 19.30 | 19.33 | 19.35 | 19.37 | 19.38 |
|   | 0.01 | 98.50 | 99.00 | 99.17 | 99.25 | 99.30 | 99.33 | 99.36 | 99.37 | 99.39 |
| 3 | 0.10 | 5.54 | 5.46 | 5.39 | 5.34 | 5.31 | 5.28 | 5.27 | 5.25 | 5.24 |
|   | 0.05 | 10.13 | 9.55 | 9.28 | 9.12 | 9.01 | 8.94 | 8.89 | 8.85 | 8.81 |
|   | 0.01 | 34.12 | 30.82 | 29.46 | 28.71 | 28.24 | 27.91 | 27.67 | 27.49 | 27.35 |
| 4 | 0.10 | 4.54 | 4.32 | 4.19 | 4.11 | 4.05 | 4.01 | 3.98 | 3.95 | 3.94 |
|   | 0.05 | 7.71 | 6.94 | 6.59 | 6.39 | 6.26 | 6.16 | 6.09 | 6.04 | 6.00 |
|   | 0.01 | 21.20 | 18.00 | 16.69 | 15.98 | 15.52 | 15.21 | 14.98 | 14.80 | 14.66 |
| 5 | 0.10 | 4.06 | 3.78 | 3.62 | 3.52 | 3.45 | 3.40 | 3.37 | 3.34 | 3.32 |
|   | 0.05 | 6.61 | 5.79 | 5.41 | 5.19 | 5.05 | 4.95 | 4.88 | 4.82 | 4.77 |
|   | 0.01 | 16.26 | 13.27 | 12.06 | 11.39 | 10.97 | 10.67 | 10.46 | 10.29 | 10.16 |
| 6 | 0.10 | 3.78 | 3.46 | 3.29 | 3.18 | 3.11 | 3.05 | 3.01 | 2.98 | 2.96 |
|   | 0.05 | 5.99 | 5.14 | 4.76 | 4.53 | 4.39 | 4.28 | 4.21 | 4.15 | 4.10 |
|   | 0.01 | 13.75 | 10.92 | 9.78 | 9.15 | 8.75 | 8.47 | 8.26 | 8.10 | 7.98 |
| 7 | 0.10 | 3.59 | 3.26 | 3.07 | 2.96 | 2.88 | 2.83 | 2.78 | 2.75 | 2.72 |
|   | 0.05 | 5.59 | 4.74 | 4.35 | 4.12 | 3.97 | 3.87 | 3.79 | 3.73 | 3.68 |
|   | 0.01 | 12.25 | 9.55 | 8.45 | 7.85 | 7.46 | 7.19 | 6.99 | 6.84 | 6.72 |
| 8 | 0.10 | 3.46 | 3.11 | 2.92 | 2.81 | 2.73 | 2.67 | 2.62 | 2.59 | 2.56 |
|   | 0.05 | 5.32 | 4.46 | 4.07 | 3.84 | 3.69 | 3.58 | 3.50 | 3.44 | 3.39 |
|   | 0.01 | 11.26 | 8.65 | 7.59 | 7.01 | 6.63 | 6.37 | 6.18 | 6.03 | 5.91 |
| 9 | 0.10 | 3.36 | 3.01 | 2.81 | 2.69 | 2.61 | 2.55 | 2.51 | 2.47 | 2.44 |
|   | 0.05 | 5.12 | 4.26 | 3.86 | 3.63 | 3.48 | 3.37 | 3.29 | 3.23 | 3.18 |
|   | 0.01 | 10.56 | 8.02 | 6.99 | 6.42 | 6.06 | 5.80 | 5.61 | 5.47 | 5.35 |
| 10 | 0.10 | 3.29 | 2.92 | 2.73 | 2.61 | 2.52 | 2.46 | 2.41 | 2.38 | 2.35 |
|   | 0.05 | 4.96 | 4.10 | 3.71 | 3.48 | 3.33 | 3.22 | 3.14 | 3.07 | 3.02 |
|   | 0.01 | 10.04 | 7.56 | 6.55 | 5.99 | 5.64 | 5.39 | 5.20 | 5.06 | 4.94 |

Table D.6   (*Continued*)

| Denominator degrees of freedom | p | Numerator degrees of freedom | | | | | | | | |
|---|---|---|---|---|---|---|---|---|---|---|
| | | 10 | 12 | 15 | 20 | 24 | 30 | 40 | 60 | 120 |
| 2 | 0.10 | 9.39 | 9.41 | 9.42 | 9.44 | 9.45 | 9.46 | 9.47 | 9.47 | 9.48 |
| | 0.05 | 19.40 | 19.41 | 19.43 | 19.45 | 19.45 | 19.46 | 19.47 | 19.48 | 19.49 |
| | 0.01 | 99.40 | 99.42 | 99.43 | 99.45 | 99.46 | 99.47 | 99.47 | 99.48 | 99.49 |
| 3 | 0.10 | 5.23 | 5.22 | 5.20 | 5.18 | 5.18 | 5.17 | 5.16 | 5.15 | 5.14 |
| | 0.05 | 8.79 | 8.74 | 8.70 | 8.66 | 8.64 | 8.62 | 8.59 | 8.57 | 8.55 |
| | 0.01 | 27.23 | 27.05 | 26.87 | 26.69 | 26.60 | 26.50 | 26.41 | 26.32 | 26.22 |
| 4 | 0.10 | 3.92 | 3.90 | 3.87 | 3.84 | 3.83 | 3.82 | 3.80 | 3.79 | 3.78 |
| | 0.05 | 5.96 | 5.91 | 5.86 | 5.80 | 5.77 | 5.75 | 5.72 | 5.69 | 5.66 |
| | 0.01 | 14.55 | 14.37 | 14.20 | 14.02 | 13.93 | 13.84 | 13.75 | 13.65 | 13.56 |
| 5 | 0.10 | 3.30 | 3.27 | 3.24 | 3.21 | 3.19 | 3.17 | 3.16 | 3.14 | 3.12 |
| | 0.05 | 4.74 | 4.68 | 4.62 | 4.56 | 4.53 | 4.50 | 4.46 | 4.43 | 4.40 |
| | 0.01 | 10.05 | 9.89 | 9.72 | 9.55 | 9.47 | 9.38 | 9.29 | 9.20 | 9.11 |
| 6 | 0.10 | 2.94 | 2.90 | 2.87 | 2.84 | 2.82 | 2.80 | 2.78 | 2.76 | 2.74 |
| | 0.05 | 4.06 | 4.00 | 3.94 | 3.87 | 3.84 | 3.81 | 3.77 | 3.74 | 3.70 |
| | 0.01 | 7.87 | 7.72 | 7.56 | 7.40 | 7.31 | 7.23 | 7.14 | 7.06 | 6.97 |
| 7 | 0.10 | 2.70 | 2.67 | 2.63 | 2.59 | 2.58 | 2.56 | 2.54 | 2.51 | 2.49 |
| | 0.05 | 3.64 | 3.57 | 3.51 | 3.44 | 3.41 | 3.38 | 3.34 | 3.30 | 3.27 |
| | 0.01 | 6.62 | 6.47 | 6.31 | 6.16 | 6.07 | 5.99 | 5.91 | 5.82 | 5.74 |
| 8 | 0.10 | 2.54 | 2.50 | 2.46 | 2.42 | 2.40 | 2.38 | 2.36 | 2.34 | 2.32 |
| | 0.05 | 3.35 | 3.28 | 3.22 | 3.15 | 3.12 | 3.08 | 3.04 | 3.01 | 2.97 |
| | 0.01 | 5.81 | 5.67 | 5.52 | 5.36 | 5.28 | 5.20 | 5.12 | 5.03 | 4.95 |
| 9 | 0.10 | 2.42 | 2.38 | 2.34 | 2.30 | 2.28 | 2.25 | 2.23 | 2.21 | 2.18 |
| | 0.05 | 3.14 | 3.07 | 3.01 | 2.94 | 2.90 | 2.86 | 2.83 | 2.79 | 2.75 |
| | 0.01 | 5.26 | 5.11 | 4.96 | 4.81 | 4.73 | 4.65 | 4.57 | 4.48 | 4.40 |
| 10 | 0.10 | 2.32 | 2.28 | 2.24 | 2.20 | 2.18 | 2.16 | 2.13 | 2.11 | 2.08 |
| | 0.05 | 2.98 | 2.91 | 2.85 | 2.77 | 2.74 | 2.70 | 2.66 | 2.62 | 2.58 |
| | 0.01 | 4.85 | 4.71 | 4.56 | 4.41 | 4.33 | 4.25 | 4.17 | 4.08 | 4.00 |

Table D.6   (*Continued*) The *F* Distribution.

Denominator
degrees of
freedom                                    Numerator degrees of freedom

|   |      | 1 | 2 | 3 | 4 | 5 | 6 | 7 | 8 | 9 |
|---|------|------|------|------|------|------|------|------|------|------|
|   | p |  |  |  |  |  |  |  |  |  |
| 11 | 0.10 | 3.23 | 2.86 | 2.66 | 2.54 | 2.45 | 2.39 | 2.34 | 2.30 | 2.27 |
|    | 0.05 | 4.84 | 3.98 | 3.59 | 3.36 | 3.20 | 3.09 | 3.01 | 2.95 | 2.90 |
|    | 0.01 | 9.65 | 7.21 | 6.22 | 5.67 | 5.32 | 5.07 | 4.89 | 4.74 | 4.63 |
| 12 | 0.10 | 3.18 | 2.81 | 2.61 | 2.48 | 2.39 | 2.33 | 2.28 | 2.24 | 2.21 |
|    | 0.05 | 4.75 | 3.89 | 3.49 | 3.26 | 3.11 | 3.00 | 2.91 | 2.85 | 2.80 |
|    | 0.01 | 9.33 | 6.93 | 5.95 | 5.41 | 5.06 | 4.82 | 4.64 | 4.50 | 4.39 |
| 13 | 0.10 | 3.14 | 2.76 | 2.56 | 2.43 | 2.35 | 2.28 | 2.23 | 2.20 | 2.16 |
|    | 0.05 | 4.67 | 3.81 | 3.41 | 3.18 | 3.03 | 2.92 | 2.83 | 2.77 | 2.71 |
|    | 0.01 | 9.07 | 6.70 | 5.74 | 5.21 | 4.86 | 4.62 | 4.44 | 4.30 | 4.19 |
| 14 | 0.10 | 3.10 | 2.73 | 2.52 | 2.39 | 2.31 | 2.24 | 2.19 | 2.15 | 2.12 |
|    | 0.05 | 4.60 | 3.74 | 3.34 | 3.11 | 2.96 | 2.85 | 2.76 | 2.70 | 2.65 |
|    | 0.01 | 8.86 | 6.51 | 5.56 | 5.04 | 4.69 | 4.46 | 4.28 | 4.14 | 4.03 |
| 15 | 0.10 | 3.07 | 2.70 | 2.49 | 2.36 | 2.27 | 2.21 | 2.16 | 2.12 | 2.09 |
|    | 0.05 | 4.54 | 3.68 | 3.29 | 3.06 | 2.90 | 2.79 | 2.71 | 2.64 | 2.59 |
|    | 0.01 | 8.68 | 6.36 | 5.42 | 4.89 | 4.56 | 4.32 | 4.14 | 4.00 | 3.89 |
| 16 | 0.10 | 3.05 | 2.67 | 2.46 | 2.33 | 2.24 | 2.18 | 2.13 | 2.09 | 2.06 |
|    | 0.05 | 4.49 | 3.63 | 3.24 | 3.01 | 2.85 | 2.74 | 2.66 | 2.59 | 2.54 |
|    | 0.01 | 8.53 | 6.23 | 5.29 | 4.77 | 4.44 | 4.20 | 4.03 | 3.89 | 3.78 |
| 17 | 0.10 | 3.03 | 2.64 | 2.44 | 2.31 | 2.22 | 2.15 | 2.10 | 2.06 | 2.03 |
|    | 0.05 | 4.45 | 3.59 | 3.20 | 2.96 | 2.81 | 2.70 | 2.61 | 2.55 | 2.49 |
|    | 0.01 | 8.40 | 6.11 | 5.18 | 4.67 | 4.34 | 4.10 | 3.93 | 3.79 | 3.68 |
| 18 | 0.10 | 3.01 | 2.62 | 2.42 | 2.29 | 2.20 | 2.13 | 2.08 | 2.04 | 2.00 |
|    | 0.05 | 4.41 | 3.55 | 3.16 | 2.93 | 2.77 | 2.66 | 2.58 | 2.51 | 2.46 |
|    | 0.01 | 8.29 | 6.01 | 5.09 | 4.58 | 4.25 | 4.01 | 3.84 | 3.71 | 3.60 |
| 19 | 0.10 | 2.99 | 2.61 | 2.40 | 2.27 | 2.18 | 2.11 | 2.06 | 2.02 | 1.98 |
|    | 0.05 | 4.38 | 3.52 | 3.13 | 2.90 | 2.74 | 2.63 | 2.54 | 2.48 | 2.42 |
|    | 0.01 | 8.18 | 5.93 | 5.01 | 4.50 | 4.17 | 3.94 | 3.77 | 3.63 | 3.52 |
| 20 | 0.10 | 2.97 | 2.59 | 2.38 | 2.25 | 2.16 | 2.09 | 2.04 | 2.00 | 1.96 |
|    | 0.05 | 4.35 | 3.49 | 3.10 | 2.87 | 2.71 | 2.60 | 2.51 | 2.45 | 2.39 |
|    | 0.01 | 8.10 | 5.85 | 4.94 | 4.43 | 4.10 | 3.87 | 3.70 | 3.56 | 3.46 |

Table D.6 (*Continued*)

| Denominator degrees of freedom | p | Numerator degrees of freedom | | | | | | | | |
|---|---|---|---|---|---|---|---|---|---|---|
| | | 10 | 12 | 15 | 20 | 24 | 30 | 40 | 60 | 120 |
| 11 | 0.10 | 2.25 | 2.21 | 2.17 | 2.12 | 2.10 | 2.08 | 2.05 | 2.03 | 2.00 |
| | 0.05 | 2.85 | 2.79 | 2.72 | 2.65 | 2.61 | 2.57 | 2.53 | 2.49 | 2.45 |
| | 0.01 | 4.54 | 4.40 | 4.25 | 4.10 | 4.02 | 3.94 | 3.86 | 3.78 | 3.69 |
| 12 | 0.10 | 2.19 | 2.15 | 2.10 | 2.06 | 2.04 | 2.01 | 1.99 | 1.96 | 1.93 |
| | 0.05 | 2.75 | 2.69 | 2.62 | 2.54 | 2.51 | 2.47 | 2.43 | 2.38 | 2.34 |
| | 0.01 | 4.30 | 4.16 | 4.01 | 3.86 | 3.78 | 3.70 | 3.62 | 3.54 | 3.45 |
| 13 | 0.10 | 2.14 | 2.10 | 2.05 | 2.01 | 1.98 | 1.96 | 1.93 | 1.90 | 1.88 |
| | 0.05 | 2.67 | 2.60 | 2.53 | 2.46 | 2.42 | 2.38 | 2.34 | 2.30 | 2.25 |
| | 0.01 | 4.10 | 3.96 | 3.82 | 3.66 | 3.59 | 3.51 | 3.43 | 3.34 | 3.25 |
| 14 | 0.10 | 2.10 | 2.05 | 2.01 | 1.96 | 1.94 | 1.91 | 1.89 | 1.86 | 1.83 |
| | 0.05 | 2.60 | 2.53 | 2.46 | 2.39 | 2.35 | 2.31 | 2.27 | 2.22 | 2.18 |
| | 0.01 | 3.94 | 3.80 | 3.66 | 3.51 | 3.43 | 3.35 | 3.27 | 3.18 | 3.09 |
| 15 | 0.10 | 2.06 | 2.02 | 1.97 | 1.92 | 1.90 | 1.87 | 1.85 | 1.82 | 1.79 |
| | 0.05 | 2.54 | 2.48 | 2.40 | 2.33 | 2.29 | 2.25 | 2.20 | 2.16 | 2.11 |
| | 0.01 | 3.80 | 3.67 | 3.52 | 3.37 | 3.29 | 3.21 | 3.13 | 3.05 | 2.96 |
| 16 | 0.10 | 2.03 | 1.99 | 1.94 | 1.89 | 1.87 | 1.84 | 1.81 | 1.78 | 1.75 |
| | 0.05 | 2.49 | 2.42 | 2.35 | 2.28 | 2.24 | 2.19 | 2.15 | 2.11 | 2.06 |
| | 0.01 | 3.69 | 3.55 | 3.41 | 3.26 | 3.18 | 3.10 | 3.02 | 2.93 | 2.84 |
| 17 | 0.10 | 2.00 | 1.96 | 1.91 | 1.86 | 1.84 | 1.81 | 1.78 | 1.75 | 1.72 |
| | 0.05 | 2.45 | 2.38 | 2.31 | 2.23 | 2.19 | 2.15 | 2.10 | 2.06 | 2.01 |
| | 0.01 | 3.59 | 3.46 | 3.31 | 3.16 | 3.08 | 3.00 | 2.92 | 2.83 | 2.75 |
| 18 | 0.10 | 1.98 | 1.93 | 1.89 | 1.84 | 1.81 | 1.78 | 1.75 | 1.72 | 1.69 |
| | 0.05 | 2.41 | 2.34 | 2.27 | 2.19 | 2.15 | 2.11 | 2.06 | 2.02 | 1.97 |
| | 0.01 | 3.51 | 3.37 | 3.23 | 3.08 | 3.00 | 2.92 | 2.84 | 2.75 | 2.66 |
| 19 | 0.10 | 1.96 | 1.91 | 1.86 | 1.81 | 1.79 | 1.76 | 1.73 | 1.70 | 1.67 |
| | 0.05 | 2.38 | 2.31 | 2.23 | 2.16 | 2.11 | 2.07 | 2.03 | 1.98 | 1.93 |
| | 0.01 | 3.43 | 3.30 | 3.15 | 3.00 | 2.92 | 2.84 | 2.76 | 2.67 | 2.58 |
| 20 | 0.10 | 1.94 | 1.89 | 1.84 | 1.79 | 1.77 | 1.74 | 1.71 | 1.68 | 1.64 |
| | 0.05 | 2.35 | 2.28 | 2.20 | 2.12 | 2.08 | 2.04 | 1.99 | 1.95 | 1.90 |
| | 0.01 | 3.37 | 3.23 | 3.09 | 2.94 | 2.86 | 2.78 | 2.69 | 2.61 | 2.52 |

Table D.6   (*Continued*) The *F* Distribution.

| Denominator degrees of freedom | p | Numerator degrees of freedom | | | | | | | | |
|---|---|---|---|---|---|---|---|---|---|---|
| | | 1 | 2 | 3 | 4 | 5 | 6 | 7 | 8 | 9 |
| 21 | 0.10 | 2.96 | 2.57 | 2.36 | 2.23 | 2.14 | 2.08 | 2.02 | 1.98 | 1.95 |
| | 0.05 | 4.32 | 3.47 | 3.07 | 2.84 | 2.68 | 2.57 | 2.49 | 2.42 | 2.37 |
| | 0.01 | 8.02 | 5.78 | 4.87 | 4.37 | 4.04 | 3.81 | 3.64 | 3.51 | 3.40 |
| 22 | 0.10 | 2.95 | 2.56 | 2.35 | 2.22 | 2.13 | 2.06 | 2.01 | 1.97 | 1.93 |
| | 0.05 | 4.30 | 3.44 | 3.05 | 2.82 | 2.66 | 2.55 | 2.46 | 2.40 | 2.34 |
| | 0.01 | 7.95 | 5.72 | 4.82 | 4.31 | 3.99 | 3.76 | 3.59 | 3.45 | 3.35 |
| 23 | 0.10 | 2.94 | 2.55 | 2.34 | 2.21 | 2.11 | 2.05 | 1.99 | 1.95 | 1.92 |
| | 0.05 | 4.28 | 3.42 | 3.03 | 2.80 | 2.64 | 2.53 | 2.44 | 2.37 | 2.32 |
| | 0.01 | 7.88 | 5.66 | 4.76 | 4.26 | 3.94 | 3.71 | 3.54 | 3.41 | 3.30 |
| 24 | 0.10 | 2.93 | 2.54 | 2.33 | 2.19 | 2.10 | 2.04 | 1.98 | 1.94 | 1.91 |
| | 0.05 | 4.26 | 3.40 | 3.01 | 2.78 | 2.62 | 2.51 | 2.42 | 2.36 | 2.30 |
| | 0.01 | 7.82 | 5.61 | 4.72 | 4.22 | 3.90 | 3.67 | 3.50 | 3.36 | 3.26 |
| 25 | 0.10 | 2.92 | 2.53 | 2.32 | 2.18 | 2.09 | 2.02 | 1.97 | 1.93 | 1.89 |
| | 0.05 | 4.24 | 3.39 | 2.99 | 2.76 | 2.60 | 2.49 | 2.40 | 2.34 | 2.28 |
| | 0.01 | 7.77 | 5.57 | 4.68 | 4.18 | 3.85 | 3.63 | 3.46 | 3.32 | 3.22 |
| 26 | 0.10 | 2.91 | 2.52 | 2.31 | 2.17 | 2.08 | 2.01 | 1.96 | 1.92 | 1.88 |
| | 0.05 | 4.23 | 3.37 | 2.98 | 2.74 | 2.59 | 2.47 | 2.39 | 2.32 | 2.27 |
| | 0.01 | 7.72 | 5.53 | 4.64 | 4.14 | 3.82 | 3.59 | 3.42 | 3.29 | 3.18 |
| 27 | 0.10 | 2.90 | 2.51 | 2.30 | 2.17 | 2.07 | 2.00 | 1.95 | 1.91 | 1.87 |
| | 0.05 | 4.21 | 3.35 | 2.96 | 2.73 | 2.57 | 2.46 | 2.37 | 2.31 | 2.25 |
| | 0.01 | 7.68 | 5.49 | 4.60 | 4.11 | 3.78 | 3.56 | 3.39 | 3.26 | 3.15 |
| 28 | 0.10 | 2.89 | 2.50 | 2.29 | 2.16 | 2.06 | 2.00 | 1.94 | 1.90 | 1.87 |
| | 0.05 | 4.20 | 3.34 | 2.95 | 2.71 | 2.56 | 2.45 | 2.36 | 2.29 | 2.24 |
| | 0.01 | 7.64 | 5.45 | 4.57 | 4.07 | 3.75 | 3.53 | 3.36 | 3.23 | 3.12 |
| 29 | 0.10 | 2.89 | 2.50 | 2.28 | 2.15 | 2.06 | 1.99 | 1.93 | 1.89 | 1.86 |
| | 0.05 | 4.18 | 3.33 | 2.93 | 2.70 | 2.55 | 2.43 | 2.35 | 2.28 | 2.22 |
| | 0.01 | 7.60 | 5.42 | 4.54 | 4.04 | 3.73 | 3.50 | 3.33 | 3.20 | 3.09 |
| 30 | 0.10 | 2.88 | 2.49 | 2.28 | 2.14 | 2.05 | 1.98 | 1.93 | 1.88 | 1.85 |
| | 0.05 | 4.17 | 3.32 | 2.92 | 2.69 | 2.53 | 2.42 | 2.33 | 2.27 | 2.21 |
| | 0.01 | 7.56 | 5.39 | 4.51 | 4.02 | 3.70 | 3.47 | 3.30 | 3.17 | 3.07 |

Table D.6   (*Continued*)

| Denominator degrees of freedom | p | 10 | 12 | 15 | 20 | 24 | 30 | 40 | 60 | 120 |
|---|---|---|---|---|---|---|---|---|---|---|
| | | | | | Numerator degrees of freedom | | | | | |
| 21 | 0.10 | 1.92 | 1.87 | 1.83 | 1.78 | 1.75 | 1.72 | 1.69 | 1.66 | 1.62 |
| | 0.05 | 2.32 | 2.25 | 2.18 | 2.10 | 2.05 | 2.01 | 1.96 | 1.92 | 1.87 |
| | 0.01 | 3.31 | 3.17 | 3.03 | 2.88 | 2.80 | 2.72 | 2.64 | 2.55 | 2.46 |
| 22 | 0.10 | 1.90 | 1.86 | 1.81 | 1.76 | 1.73 | 1.70 | 1.67 | 1.64 | 1.60 |
| | 0.05 | 2.30 | 2.23 | 2.15 | 2.07 | 2.03 | 1.98 | 1.94 | 1.89 | 1.84 |
| | 0.01 | 3.26 | 3.12 | 2.98 | 2.83 | 2.75 | 2.67 | 2.58 | 2.50 | 2.40 |
| 23 | 0.10 | 1.89 | 1.84 | 1.80 | 1.74 | 1.72 | 1.69 | 1.66 | 1.62 | 1.59 |
| | 0.05 | 2.27 | 2.20 | 2.13 | 2.05 | 2.01 | 1.96 | 1.91 | 1.86 | 1.81 |
| | 0.01 | 3.21 | 3.07 | 2.93 | 2.78 | 2.70 | 2.62 | 2.54 | 2.45 | 2.35 |
| 24 | 0.10 | 1.88 | 1.83 | 1.78 | 1.73 | 1.70 | 1.67 | 1.64 | 1.61 | 1.57 |
| | 0.05 | 2.25 | 2.18 | 2.11 | 2.03 | 1.98 | 1.94 | 1.89 | 1.84 | 1.79 |
| | 0.01 | 3.17 | 3.03 | 2.89 | 2.74 | 2.66 | 2.58 | 2.49 | 2.40 | 2.31 |
| 25 | 0.10 | 1.87 | 1.82 | 1.77 | 1.72 | 1.69 | 1.66 | 1.63 | 1.59 | 1.56 |
| | 0.05 | 2.24 | 2.16 | 2.09 | 2.01 | 1.96 | 1.92 | 1.87 | 1.82 | 1.77 |
| | 0.01 | 3.13 | 2.99 | 2.85 | 2.70 | 2.62 | 2.54 | 2.45 | 2.36 | 2.27 |
| 26 | 0.10 | 1.86 | 1.81 | 1.76 | 1.71 | 1.68 | 1.65 | 1.61 | 1.58 | 1.54 |
| | 0.05 | 2.22 | 2.15 | 2.07 | 1.99 | 1.95 | 1.90 | 1.85 | 1.80 | 1.75 |
| | 0.01 | 3.09 | 2.96 | 2.81 | 2.66 | 2.58 | 2.50 | 2.42 | 2.33 | 2.23 |
| 27 | 0.10 | 1.85 | 1.80 | 1.75 | 1.70 | 1.67 | 1.64 | 1.60 | 1.57 | 1.53 |
| | 0.05 | 2.20 | 2.13 | 2.06 | 1.97 | 1.93 | 1.88 | 1.84 | 1.79 | 1.73 |
| | 0.01 | 3.06 | 2.93 | 2.78 | 2.63 | 2.55 | 2.47 | 2.38 | 2.29 | 2.20 |
| 28 | 0.10 | 1.84 | 1.79 | 1.74 | 1.69 | 1.66 | 1.63 | 1.59 | 1.56 | 1.52 |
| | 0.05 | 2.19 | 2.12 | 2.04 | 1.96 | 1.91 | 1.87 | 1.82 | 1.77 | 1.71 |
| | 0.01 | 3.03 | 2.90 | 2.75 | 2.60 | 2.52 | 2.44 | 2.35 | 2.26 | 2.17 |
| 29 | 0.10 | 1.83 | 1.78 | 1.73 | 1.68 | 1.65 | 1.62 | 1.58 | 1.55 | 1.51 |
| | 0.05 | 2.18 | 2.10 | 2.03 | 1.94 | 1.90 | 1.85 | 1.81 | 1.75 | 1.70 |
| | 0.01 | 3.00 | 2.87 | 2.73 | 2.58 | 2.49 | 2.41 | 2.33 | 2.23 | 2.14 |
| 30 | 0.10 | 1.82 | 1.77 | 1.72 | 1.67 | 1.64 | 1.61 | 1.57 | 1.54 | 1.50 |
| | 0.05 | 2.16 | 2.09 | 2.01 | 1.93 | 1.89 | 1.84 | 1.79 | 1.74 | 1.68 |
| | 0.01 | 2.98 | 2.84 | 2.70 | 2.55 | 2.47 | 2.39 | 2.30 | 2.21 | 2.11 |

Table D.6   (*Continued*) The *F* Distribution.

| Denominator degrees of freedom | p | Numerator degrees of freedom | | | | | | | | |
|---|---|---|---|---|---|---|---|---|---|---|
| | | 1 | 2 | 3 | 4 | 5 | 6 | 7 | 8 | 9 |
| 40 | 0.10 | 2.84 | 2.44 | 2.23 | 2.09 | 2.00 | 1.93 | 1.87 | 1.83 | 1.79 |
| | 0.05 | 4.08 | 3.23 | 2.84 | 2.61 | 2.45 | 2.34 | 2.25 | 2.18 | 2.12 |
| | 0.01 | 7.31 | 5.18 | 4.31 | 3.83 | 3.51 | 3.29 | 3.12 | 2.99 | 2.89 |
| 48 | 0.10 | 2.81 | 2.42 | 2.20 | 2.07 | 1.97 | 1.90 | 1.85 | 1.80 | 1.77 |
| | 0.05 | 4.04 | 3.19 | 2.80 | 2.57 | 2.41 | 2.29 | 2.21 | 2.14 | 2.08 |
| | 0.01 | 7.19 | 5.08 | 4.22 | 3.74 | 3.43 | 3.20 | 3.04 | 2.91 | 2.80 |
| 60 | 0.10 | 2.79 | 2.39 | 2.18 | 2.04 | 1.95 | 1.87 | 1.82 | 1.77 | 1.74 |
| | 0.05 | 4.00 | 3.15 | 2.76 | 2.53 | 2.37 | 2.25 | 2.17 | 2.10 | 2.04 |
| | 0.01 | 7.08 | 4.98 | 4.13 | 3.65 | 3.34 | 3.12 | 2.95 | 2.82 | 2.72 |
| 90 | 0.10 | 2.76 | 2.36 | 2.15 | 2.01 | 1.91 | 1.84 | 1.78 | 1.74 | 1.70 |
| | 0.05 | 3.95 | 3.10 | 2.71 | 2.47 | 2.32 | 2.20 | 2.11 | 2.04 | 1.99 |
| | 0.01 | 6.93 | 4.85 | 4.01 | 3.53 | 3.23 | 3.01 | 2.84 | 2.72 | 2.61 |
| 120 | 0.10 | 2.75 | 2.35 | 2.13 | 1.99 | 1.90 | 1.82 | 1.77 | 1.72 | 1.68 |
| | 0.05 | 3.92 | 3.07 | 2.68 | 2.45 | 2.29 | 2.17 | 2.09 | 2.02 | 1.96 |
| | 0.01 | 6.85 | 4.79 | 3.95 | 3.48 | 3.17 | 2.96 | 2.79 | 2.66 | 2.56 |

Table D.6   (*Continued*)

| Denominator degrees of freedom | | Numerator degrees of freedom | | | | | | | | |
|---|---|---|---|---|---|---|---|---|---|---|
| | p | 10 | 12 | 15 | 20 | 24 | 30 | 40 | 60 | 120 |
| 40 | 0.10 | 1.76 | 1.71 | 1.66 | 1.61 | 1.57 | 1.54 | 1.51 | 1.47 | 1.42 |
| | 0.05 | 2.08 | 2.00 | 1.92 | 1.84 | 1.79 | 1.74 | 1.69 | 1.64 | 1.58 |
| | 0.01 | 2.80 | 2.66 | 2.52 | 2.37 | 2.29 | 2.20 | 2.11 | 2.02 | 1.92 |
| 48 | 0.10 | 1.73 | 1.69 | 1.63 | 1.57 | 1.54 | 1.51 | 1.47 | 1.43 | 1.39 |
| | 0.05 | 2.03 | 1.96 | 1.88 | 1.79 | 1.75 | 1.70 | 1.64 | 1.59 | 1.52 |
| | 0.01 | 2.71 | 2.58 | 2.44 | 2.29 | 2.20 | 2.12 | 2.02 | 1.93 | 1.82 |
| 60 | 0.10 | 1.71 | 1.66 | 1.60 | 1.54 | 1.51 | 1.48 | 1.44 | 1.40 | 1.35 |
| | 0.05 | 1.99 | 1.92 | 1.84 | 1.75 | 1.70 | 1.65 | 1.59 | 1.53 | 1.47 |
| | 0.01 | 2.63 | 2.50 | 2.35 | 2.20 | 2.12 | 2.03 | 1.94 | 1.84 | 1.73 |
| 90 | 0.10 | 1.67 | 1.62 | 1.56 | 1.50 | 1.47 | 1.43 | 1.39 | 1.35 | 1.29 |
| | 0.05 | 1.94 | 1.86 | 1.78 | 1.69 | 1.64 | 1.59 | 1.53 | 1.46 | 1.39 |
| | 0.01 | 2.52 | 2.39 | 2.24 | 2.09 | 2.00 | 1.92 | 1.82 | 1.72 | 1.60 |
| 120 | 0.10 | 1.65 | 1.60 | 1.55 | 1.48 | 1.45 | 1.41 | 1.37 | 1.32 | 1.26 |
| | 0.05 | 1.91 | 1.83 | 1.75 | 1.66 | 1.61 | 1.55 | 1.50 | 1.43 | 1.35 |
| | 0.01 | 2.47 | 2.34 | 2.19 | 2.04 | 1.95 | 1.86 | 1.76 | 1.66 | 1.53 |

Table D.7  The Chi-Square Distribution. The tabulated value is the chi-square needed for the indicated $p$ value.

| df | \multicolumn{4}{c}{p} | | | |
|----|-------|-------|-------|-------|
|    | .10   | .05   | .01   | .005  |
| 1   | 2.706   | 3.841   | 6.635   | 7.879   |
| 2   | 4.605   | 5.991   | 9.210   | 10.597  |
| 3   | 6.251   | 7.815   | 11.345  | 12.838  |
| 4   | 7.779   | 9.488   | 13.277  | 14.860  |
| 5   | 9.236   | 11.070  | 15.086  | 16.750  |
| 6   | 10.645  | 12.592  | 16.812  | 18.548  |
| 7   | 12.017  | 14.067  | 18.475  | 20.278  |
| 8   | 13.362  | 15.507  | 20.090  | 21.955  |
| 9   | 14.684  | 16.919  | 21.666  | 23.589  |
| 10  | 15.987  | 18.307  | 23.209  | 25.188  |
| 11  | 17.275  | 19.675  | 24.725  | 26.757  |
| 12  | 18.549  | 21.026  | 26.217  | 28.299  |
| 13  | 19.812  | 22.362  | 27.688  | 29.819  |
| 14  | 21.064  | 23.685  | 29.141  | 31.319  |
| 15  | 22.307  | 24.996  | 30.578  | 32.801  |
| 16  | 23.542  | 26.296  | 32.000  | 34.267  |
| 17  | 24.769  | 27.587  | 33.409  | 35.718  |
| 18  | 25.989  | 28.869  | 34.805  | 37.156  |
| 19  | 27.204  | 30.144  | 36.191  | 38.582  |
| 20  | 28.412  | 31.410  | 37.566  | 39.997  |
| 21  | 29.615  | 32.671  | 38.932  | 41.401  |
| 22  | 30.813  | 33.924  | 40.289  | 42.796  |
| 23  | 32.007  | 35.172  | 41.638  | 44.181  |
| 24  | 33.196  | 36.415  | 42.980  | 45.558  |
| 25  | 34.382  | 37.652  | 44.314  | 46.928  |
| 26  | 35.563  | 38.885  | 45.642  | 48.290  |
| 27  | 36.741  | 40.113  | 46.963  | 49.645  |
| 28  | 37.916  | 41.337  | 48.278  | 50.993  |
| 29  | 39.087  | 42.557  | 49.588  | 52.336  |
| 30  | 40.256  | 43.773  | 50.892  | 53.672  |
| 31  | 41.421  | 44.985  | 52.191  | 55.002  |
| 32  | 42.585  | 46.194  | 53.485  | 56.327  |
| 33  | 43.745  | 47.400  | 54.775  | 57.648  |
| 34  | 44.903  | 48.602  | 56.060  | 58.963  |
| 35  | 46.059  | 49.802  | 57.342  | 60.274  |
| 36  | 47.212  | 50.998  | 58.619  | 61.581  |
| 37  | 48.363  | 52.192  | 59.892  | 62.883  |
| 38  | 49.512  | 53.383  | 61.162  | 64.181  |
| 39  | 50.660  | 54.572  | 62.428  | 65.475  |
| 40  | 51.805  | 55.758  | 63.690  | 66.765  |
| 45  | 57.505  | 61.656  | 69.956  | 73.165  |
| 50  | 63.167  | 67.505  | 76.154  | 79.489  |
| 55  | 68.796  | 73.311  | 82.292  | 85.748  |
| 60  | 74.397  | 79.082  | 88.379  | 91.951  |
| 70  | 85.527  | 90.531  | 100.425 | 104.215 |
| 80  | 96.578  | 101.879 | 112.329 | 116.321 |
| 90  | 107.565 | 113.145 | 124.116 | 128.299 |
| 100 | 118.498 | 124.342 | 135.807 | 140.169 |

Table D.8   Critical Values of $r$, the Correlation Coefficient. Each entry gives the value
of $r$ needed for the indicated $p$ value with the indicated sample size.

|  | One-tail $p$ value | | |
|---|---|---|---|
| $n$ | .05 | .025 | .01 |
| 3 | 0.988 | 0.997 | 1.000 |
| 4 | 0.900 | 0.950 | 0.990 |
| 5 | 0.805 | 0.878 | 0.959 |
| 6 | 0.729 | 0.811 | 0.917 |
| 7 | 0.669 | 0.754 | 0.874 |
| 8 | 0.621 | 0.707 | 0.834 |
| 9 | 0.582 | 0.666 | 0.798 |
| 10 | 0.549 | 0.632 | 0.765 |
| 11 | 0.521 | 0.602 | 0.735 |
| 12 | 0.497 | 0.576 | 0.708 |
| 13 | 0.476 | 0.553 | 0.683 |
| 14 | 0.457 | 0.532 | 0.661 |
| 15 | 0.441 | 0.514 | 0.641 |
| 16 | 0.426 | 0.497 | 0.623 |
| 17 | 0.412 | 0.482 | 0.605 |
| 18 | 0.400 | 0.468 | 0.590 |
| 19 | 0.389 | 0.455 | 0.575 |
| 20 | 0.378 | 0.444 | 0.561 |
| 21 | 0.369 | 0.433 | 0.549 |
| 22 | 0.360 | 0.423 | 0.537 |
| 23 | 0.351 | 0.413 | 0.526 |
| 24 | 0.344 | 0.404 | 0.515 |
| 25 | 0.336 | 0.396 | 0.505 |
| 26 | 0.330 | 0.388 | 0.496 |
| 27 | 0.323 | 0.381 | 0.487 |
| 28 | 0.317 | 0.374 | 0.478 |
| 29 | 0.311 | 0.367 | 0.470 |
| 30 | 0.306 | 0.361 | 0.463 |
| 31 | 0.301 | 0.355 | 0.456 |
| 32 | 0.296 | 0.349 | 0.449 |
| 33 | 0.291 | 0.344 | 0.442 |
| 34 | 0.287 | 0.339 | 0.436 |
| 35 | 0.283 | 0.334 | 0.430 |
| 36 | 0.278 | 0.329 | 0.424 |
| 37 | 0.275 | 0.325 | 0.418 |
| 38 | 0.271 | 0.320 | 0.413 |
| 39 | 0.267 | 0.316 | 0.408 |
| 40 | 0.264 | 0.312 | 0.403 |
| 41 | 0.260 | 0.308 | 0.398 |

Table D.8  (*Continued*)

| n | One-tail $p$ value | | |
|---|---|---|---|
| | .05 | .025 | .01 |
| 42 | 0.257 | 0.304 | 0.393 |
| 43 | 0.254 | 0.301 | 0.389 |
| 44 | 0.251 | 0.297 | 0.384 |
| 45 | 0.248 | 0.294 | 0.380 |
| 46 | 0.245 | 0.291 | 0.376 |
| 47 | 0.243 | 0.288 | 0.372 |
| 48 | 0.240 | 0.284 | 0.368 |
| 49 | 0.238 | 0.282 | 0.365 |
| 50 | 0.235 | 0.279 | 0.361 |
| 60 | 0.214 | 0.254 | 0.330 |
| 70 | 0.198 | 0.235 | 0.306 |
| 80 | 0.185 | 0.220 | 0.286 |
| 90 | 0.174 | 0.207 | 0.270 |
| 100 | 0.165 | 0.197 | 0.256 |

# Index

# Statpal

# Statpal

## A Statistical Package for Microcomputers

PC-DOS version 5.0 for the
IBM PC and compatibles

**BRUCE J. CHALMER**
**DAVID G. WHITMORE**

Statpal Associates
Montpelier, Vermont

**MARCEL DEKKER, INC.**     New York and Basel

# —READ THIS FIRST—

# Implementation-Specific Information for MS-DOS/PC-DOS Statpal Version 5.0

Hardware requirements:
    128K memory, two disk drives; printer optional

Operating system:
    MS-DOS version 2.0/PC-DOS version 2.0 or later

BEFORE STARTING STATPAL FOR THE FIRST TIME, you should do three things:

1.  Read the Copyright notice in the Statpal User's Guide. Removing the Statpal diskette from its envelope constitutes your acceptance of the terms of the Copyright notice. Read it carefully!

2.  Make a working copy of your Statpal diskette, and put the original in a safe place. You might also find it convenient to copy the file COMMAND.COM from your operating system disk to the working Statpal diskette. Instructions for copying a diskette came with your computer.

3.  Read Chapter 1 of the User's Guide.

## Running Statpal from a Diskette

First, follow the instructions that came with your computer to start the operating system. Then, insert the Statpal diskette IN THE ACTIVE DRIVE (generally drive A), and the diskette you will use to hold data in another drive (generally drive B).

Now you are ready to start Statpal. Your operating system will be displaying its prompt, generally indicating the current drive (e.g., "A>"). Simply type:

  STATPAL

and press the Return key (sometimes labeled with a down-and-left arrow). Within a few seconds you should get the Statpal logo, followed by a copyright message with the instruction "Press any key". As soon as you press a key, you should get the Main Menu.

If you don't get the Main Menu, make sure the Statpal diskette is IN THE ACTIVE DRIVE, and try again.

## Running Statpal using a Fixed-disk or a RAM-disk

Statpal is fully compatible with any kind of disk storage or simulated disk storage your operating system is capable of using. The only restriction is that all of the files on your Statpal diskette must reside on the active drive in the active directory.

Further implementation-specific information is available using the HElp facility within Statpal.

# Preface

<u>Welcome to Statpal!</u>

Statpal is a general-purpose statistical package, designed to handle the most common statistical computing needs of both researchers and students. You'll find that Statpal is very easy to learn and use, yet provides the power and sophistication needed by professionals working in the real world.

Unlike some packages, Statpal is designed especially for interactive use on microcomputers. Procedures are set up to take advantage of the strengths of micros (e.g., quick response to commands, and the ability to fill the screen quickly), while avoiding their weaknesses. Prompts are given in clear, concise English, and error messages are informative and useful. With Statpal, you'll be able to concentrate on analyzing your data, rather than your software.

<u>How to use this User's Guide</u>

<u>Before taking the Statpal disk out of its envelope</u>, you should read the copyright notice in this User's Guide.

Then, you should read Chapter 1 thoroughly.  A  quick  skim
through the rest of  the  guide  is also a good idea at this
point, to give you an idea of how  the  guide  is organized.
Once you have started to use Statpal,  Chapters 2 and 3 will
come in handy for reference.

      In the beginning of this manual you'll find information
specific  to  the particular implementation of  Statpal  you
ordered.   You  will need this information  to  get  Statpal
started  on  your   computer.   Follow   the   instructions
carefully.

      If you are  upgrading  from Statpal version 4.0, you'll
be pleased with  the  many new features of version 5.0. The
on-line  HElp  utility  offers  a  special  category   of
information on new features, including information on how to
convert a version 4.0 Statpal system file for use by version
5.0.

      We welcome your  suggestions  and  comments on Statpal.
Please let us know what features you would  like  to  see in
future releases.  Write  to  us at the address shown on the
copyright page.

                                                    B.J.C.
                                                    D.G.W.

# Contents

# 1

# Getting Started with Statpal

## 1.1 INTRODUCTION

Statpal is a powerful, versatile tool for statistical analysis on your microcomputer. Using Statpal, you can enter data, summarize sets of data numerically and graphically, and perform a wide variety of standard statistical tests. Statpal combines the simplicity of menu-driven structure with sophisticated features such as variable names and labels, missing value specification, case selection, accurate computational algorithms, on-line help, and informative messages in clear, straightforward English.

The best way to learn to use Statpal is by using it. You'll find all you need to know to get started with Statpal in the remainder of this chapter. After you have read this chapter, you'll be ready to use any Statpal procedure. The remaining chapters in this User's Guide provide detailed information on all Statpal procedures.

## 1.2 IMPLEMENTATION-SPECIFIC FEATURES OF STATPAL

Statpal is available for a variety of machines and operating systems, and certain information necessarily

varies from one machine or operating system to another.   One
particularly critical example of a feature that varies among
implementations is the procedure for starting Statpal on
your computer.


## 1.2.1 Implementation-specific Documentation

Implementation-specific features of Statpal are
documented in two ways.  First, you'll find a special sheet
of instructions specific to your implementation of Statpal
inserted in this User's Guide.  This sheet will detail the
specific hardware and software needed by Statpal on your
computer, as well as the specific commands needed to run
Statpal.   Second, a special category of on-line help is
provided for each implementation of Statpal, giving
information about commands or limitations specific to that
implementation.  We'll refer to these two forms of
implementation-specific information as necessary throughout
this User's Guide.


## 1.2.2 Special Keys

Another thing that varies from computer to computer is
the nomenclature used for special keys.  Different computers
use different terms for special keys on the keyboard.   The
only two special keys used by Statpal are the "Return" key
and the "Delete" key.   The Return key is the key used to
signal the end of a line of input to your computer.  If you
don't have a key marked "Return" on your computer, this key
might be labeled "Enter" or "CR" (for carriage return), or
simply have a picture suggesting a return/line feed
sequence.  The Delete key is the key used to correct typing
errors.  On your computer, this key might be labeled with
the words "Rubout" or "Del", or with a picture of a left
arrow.

We'll use the terms Return and Delete in this User's
Guide to indicate these special keys.

## 1.2.3 Printed Output From Statpal

All Statpal procedures that produce statistical output offer you the explicit option of printing the results on your printer. In addition, some computers have a special key supported by the operating system that causes the printing of whatever happens to be on the screen at the moment.

You may also use the PRinter utility (on the Main Menu) to instruct Statpal to direct "printed" output to a file instead of to your printer. This allows you to incorporate Statpal output in documents using any standard word processing package.

Statpal output is formatted for a line width of 80 columns.

## 1.3 COMMUNICATING WITH STATPAL

### 1.3.1 Menus

Statpal is "menu-driven"--that is, you select what you want to do from a menu of options at any given point. When you first start Statpal, you are presented with the "Main Menu". To make your choice, simply type the first two letters of the menu item, and press Return. You need not type the entire name as shown on the Menu. The letters needed for each menu item are highlighted on your screen (using bold and/or capital letters, depending on your implementation). If a Statpal menu requires only one letter, Statpal will respond immediately as soon as you type the letter of your choice--that is, you do not need to press Return.

## 1.3.2 Responding to Statpal Prompts

Once you have made your selection from the Main Menu, Statpal will ask you for more information as necessary to carry out the procedure you selected. Some information you supply by selecting from a menu; other information you supply by typing a response to a prompt.

In some cases, pressing Return without anything else in response to a prompt will cause Statpal to use a "default" response. Statpal tells you when there is a default response for a particular prompt.

## 1.3.3 Error Messages

If you instruct Statpal to do something impossible (such as reading data from a non-existent file), Statpal will beep and give you an infomative error message explaining what went wrong. Then Statpal will reprompt you for the correct information.

In addition, Statpal will inform you if something goes wrong in the course of a statistical procedure (for example, trying to calculate a standard deviation with only one valid case), or if something might go wrong (such as severe multicollinearity in a regression problem). In some situations Statpal will give you a warning and ask if you wish to proceed.

"Fatal" errors are errors which cause your computer to abort execution of Statpal. Statpal is designed to "trap" errors--that is, to handle errors without aborting the program, so that you should generally not experience fatal errors. However, there are unusual circumstances which might arise, depending on your computer, which could lead to fatal errors.

If a fatal error occurs, you can simply start Statpal again, and you'll generally find that you can get back to where you were quite easily.

## 1.4 ORGANIZING DATA FOR ANALYSIS USING STATPAL

### 1.4.1 Statpal System Files

To use Statpal to analyze a set of data, the first thing you need to do is create a "Statpal system file". A Statpal system file is a special file containing information about the data set (such as the number of cases in the set, the names of the variables in the set, etc.), along with the data values themselves. All Statpal statistical procedures require a Statpal system file. Statpal system files are created using the CReate facility described below.

Once you have created a Statpal system file for a particular data set, you can use it as often as you wish--that is, you need not make a new one every time you use Statpal. In addition, you can use SYstem file utilities to change and update Statpal system files in a variety of ways.

### 1.4.2 Cases and Variables

A data set to be analyzed using Statpal consists of a set of "cases", with each case having one or more "variables". For example, if you have a set of questionnaire responses from 100 people, with each respondent answering 30 questions, your data set contains 100 cases, each with 30 variables.

### 1.4.3 Coding the Data

Data to be analyzed using Statpal must be numeric--that is, no letters are allowed. So, for example, if you have a variable indicating the sex of each respondent, you should code the information as, say, 1 and 2 rather than M and F. You may use both integers (whole numbers) and reals (numbers with a decimal point), and both positive and negative numbers, for any variable. You may also use scientific notation (for example, 3E-5 for .00003).

A Statpal system file stores data internally as a two-digit exponent and an eleven-digit mantissa. Numbers can be positive or negative. The range of allowable numbers (in absolute value) is 1.0E-38 through 1.0E+38 (i.e., ten to the minus 38 through ten to the plus 38). However, it is strongly recommended that variables with very large or very small numbers (say, outside the range 1.0E-10 to 1.0E+10) be coded in different units so as to make the numbers less extreme.

Before you start to use Statpal, you should have your data organized by case, coded as numbers. As we'll note below, in creating your Statpal system file you may either enter the data directly from the keyboard, with Statpal prompting you for the data for each variable for each case, or you may instruct Statpal to read the data from a disk file in either fixed or free format. It is even possible to enter some of the data from keyboard, and then add other data from a disk file (or vice versa). The section on the ADd utility (one of the SYstem file utilities) describes how to add data to an existing Statpal system file.

## 1.5 CREATING A STATPAL SYSTEM FILE: THE CREATE FACILITY

### 1.5.1 Accessing the CReate Facility

The CReate facility is accessed from the Main Menu. The first thing Statpal asks you for is a name to use for the Statpal system file about to be created. If you specify the name of a file that already exists, Statpal prints a warning and asks you if you wish to overwrite the existing file. Be sure that is what you really want to do before you allow Statpal to overwrite an existing file, because anything in the existing file will be lost when the file is overwritten.

Using the extension .STP for Statpal system files is a good convention, to help you distinguish between Statpal

system files and other files. You should not try to change
a Statpal system file using any program except Statpal.
Using a special extension for Statpal system files will help
you avoid accidental attempts to edit a Statpal system file
with another program.

When you have entered the file name, Statpal asks
whether you will be entering the data from a file (i.e., a
disk file which you have already created outside of Statpal)
or from the keyboard. Your choice here will depend on which
way is more convenient. If you do not already have your
data in a disk file, of course, you will select the keyboard
option here. You'll find Statpal offers a very convenient
way of prompting for data entry, easily handled by people
with very little computer experience.

However, if you have a good text editor and are adept
at its use, you might prefer to put your data into a disk
file before you run Statpal. The format of a data file to
be read by Statpal may be either "fixed" or "free". These
terms are discussed below.

Still a third option for putting data into a Statpal
system file is to enter no data at all at this point, but
rather to make use of Statpal's system file utilities to
generate dummy cases and then enter data using the EDit
utility. These utilities are described in Chapter 3. If you
choose this method, you should tell Statpal at this point
that you will be entering data from keyboard; later, when
Statpal prompts you for data, you can get out of CReate
without entering data (see below).


1.5.2 Variable Names, Labels and Missing Values

Once you have specified how you will be entering data,
Statpal asks you to specify three pieces of information
about each variable:

   1.  A brief (8 characters or fewer) variable name, by
       which you will refer to the variable when you want to
       use it in a procedure--for example, AGE, QUESTN37,

APLUSB, VAR1, etc. You may type variable names in upper-case, lower-case, or any combination of the two (assuming your computer can do so). Statpal always changes variable names to upper-case before it does anything with them, to avoid potential confusion. Certain words and symbols are reserved for use by Statpal. If you try to give a variable name consisting of a reserved word or including a reserved symbol, Statpal will give you an error message and request a different name.

2.  An optional variable label (40 characters or fewer) describing the variable, which will be displayed by procedures using the variable. For example, if question 17 on your questionnaire asks "How many children do you have?", you might wish to use the wording of the question as the label. The variable label may include any displayable characters.

3.  An optional "missing value", which is used by Statpal to flag cases for which the information on the variable was not available (for example, if a respondent gave no response to a questionnaire item). Cases with the missing value for a variable are excluded from computations involving that variable. If no missing value is specified for a variable, the default value is -999999.

These three pieces of information (variable name, variable label, and missing value) are stored for each variable as part of the Statpal system file. You may change them using the REvise utility (accessed through the SYstem file utilities menu).

Note that the CReate facility supplies a default variable name of the form VARx, where x is a number--e.g., VAR37. If you wish to use the default variable name for a variable, just press Return when Statpal asks you to specify the variable name.

Note also that Statpal will keep prompting for variable names, labels and missing values until one of two events occurs: (1) you tell Statpal you are finished (by typing !!!, read "three bangs"), or (2) you reach the maximum number of variables allowed. The maximum number of variables allowed is an implementation-specific feature, and is documented in the on-line help. (Incidentally, don't worry about remembering special sequences such as !!!, because Statpal always tells you about them when you might need them.)

If you forget to type !!!, and find that you have accidentally specified more variables than you intended, you can get rid of the extra variable(s) using the DElete utility (accessed through the SYstem file utilities menu). You should do this before entering any data. (See below for information on how to get out of CReate without entering any data.)

In addition to the variables you name, Statpal always supplies one additional variable with the special name CASE#. (Statpal will not let you name another variable CASE#.) The variable CASE# contains for each case the sequence number of the case in the Statpal system file, (i.e., 1 for the first case, 2 for the second case, etc.). You may use CASE# in any Statpal procedure; however, Statpal will not allow you to change CASE# in any way (as, for example, using a transformation). Statpal supplies the label "Case number" for the variable CASE#.

There is, of course, no missing value for CASE#, since if a case is present at all it has a valid CASE#. Note that the value of CASE# for a given case is always the same, regardless of any missing data or case selection involved in a given procedure. If a case is the 17th case in the system file, its CASE# is 17.


## 1.5.3 Entering Data from the Keyboard

If you have chosen to enter the data from the keyboard, Statpal will now prompt you for the data. Any of the following forms may be used to enter data:

1.  Integers, with or without a sign; for example: -117

2.  Reals, with or without a sign; for example: 117.3298

3.  Scientific notation; for example: 1.17E4 (which means 1.17 times 10 to the 4th power, or 11700).

Press Return after each entry. Be careful not to use commas in your entries; for example, the number one thousand seven hundred must be entered as 1700 (rather than 1,700).

To indicate missing values, you may either enter the missing value yourself, or simply press Return, in which case the missing value you set previously for the variable will be assigned to the case. If you use the latter method, Statpal will inform you that the value has been set to the missing value.

If you make a mistake in entering a value before you press Return, you may use the Delete key to backtrack one character at a time. If you notice that you have made a mistake in entering a value, but have already pressed Return, you should make a note of which case and variable contains the error. You'll be able to fix it using the EDit utility (accessed through the SYstem file utilities menu).

When you have finished entering data, type !!!. Statpal will report how many cases with how many variables have been entered, and send you back to the Main Menu.

To get out of CReate without entering any data (if, for example, you need to fix your variable specifications), simply type !!! when Statpal asks you for the data for the first case. The Statpal system file will then be saved with no cases.

Note that it is not necessary to enter all of your data in one session; just enter as much as you wish. You can easily use the ADd cases utility (accessed through the SYstem file utilities menu) to enter more data at a later time. It is also possible to add more variables for the same cases using the ADd utility. Still another way to combine data from several sources is through the use of the MErge utility (also accessed through the SYstem file utilities menu).

## 1.5.4 Entering Data from a Disk File

If you chose to have Statpal read your data from a disk file, Statpal will now ask you to indicate whether the data file was in a fixed or free format.

A fixed format is one in which the data for each variable occupies a fixed row and column location in the file. A free format is one in which the data for successive variables are separated by at least one space. (Note that fixed and free formats are not mutually exclusive; it is possible to have a file with a fixed format that can also be read using a free format.)

Let's consider these terms in the context of an example. (We'll be using this example throughout this User's Guide.) Suppose you have data for 50 patients involved in a study of the effects of four different drugs on blood pressure. For each patient, you have five variables: (1) the patient's age in years (called AGE); (2) the sex of the patient (called SEX), coded as 1 for males and 2 for females; (3) the drug received by the patient (called DRUG), coded as 1, 2, 3 or 4; (4) the patient's systolic blood pressure (called SYSTOL); and (5) the patient's diastolic blood pressure (called DIASTOL).

## 1.5.4.1 Fixed-format Data Files

The key feature of a fixed-format data file is that the data lines for any case case have the same format as the data lines for every other case; that is, the value for a given variable occupies the same spaces for every case. A fixed-format data file for our example could look like this:

```
64   1   1    126   107
36   1   1    112    92
30   2   2
45   2   1    120    98
    [etc.]
```

Thus, the first patient was a 64-year-old male, receiving drug 1, with systolic pressure 126 and diastolic pressure 107. Note that the data for variables SYSTOL and DIASTOL were missing for the third patient. With fixed-format data files, if all columns specified for a particular variable are left blank in the data file for a case, that case will be assigned the missing value currently in effect for the variable. So, with fixed format data files, you may leave blanks in the file to indicate missing information.

In this example, the data for each patient fit easily on a single record (i.e., line) of the data file. In general, however, it is permissible to use as many lines of data as you wish for each case. For fixed-format data files, Statpal will ask you how many records there are per case (with a default of 1), and will then prompt you for the location of each variable. The location of a variable means the particular columns (i.e., spaces) the variable occupies in the data file. In our example, AGE occupies columns 1 through 3 (including the leading space), SEX occupies columns 4 through 7, etc. (The example Statpal session at the end of this section shows how the locations are specified.) If we were using a data file with, say, three records of data per case, we would also need to specify which of the three records each variable is on. Statpal will ask you to specify a record number only if you have indicated that there are two or more records per case.

## 1.5.4.2 Free-format Data Files

A free-format data file is one in which successive data values for each case are separated by at least one space. (Note to users of Statpal version 4.0: values can no longer

be separated by commas in Statpal version 5.0 and later.)  A
free-format data file for our example data could  look  like
this:


64 1 1 126 107
36 1 1 112 92
30 2 2 -999999 -999999
45 2 1 120 98
    [etc.]


     Note that the missing value for the  third  patient  is
entered  explicitly.   In a free-format data  file,  missing
values must be explicitly entered; that is, you cannot leave
blanks for missing data in a free-format data file.  This is
because Statpal  has no way of distinguishing between blanks
used  to  separate variables, and blanks  used  to  indicate
missing data for a variable.

     With free-format data files, every case should begin on
a new line in the data file (although you may have  as  many
lines as you need for each case).  The data need not line up
in  fixed  columns (as for free formats).  So, if  you  have
specified a free format, Statpal does not ask you to specify
column locations.


1.5.4.3 Specifying the Data File

     Next, Statpal  asks  you  for  the  name  of  the  file
containing the data, and proceeds to read the  data from the
file.  If something goes wrong as  Statpal  is  reading  the
data, Statpal will save everything read up to  that point in
the Statpal system file.  You can then fix  any problems and
try again.  If the problem was in the data file, you can use
the DElete utility (from the SYstem file utilities menu)  to
delete all  cases  in  the Statpal system file, fix the data
file using your text editor, and then use the ADd utility to
read the data correctly.

     To get out of the CReate facility without  reading  any
data, press Return (without specifying  a  file  name)  when

Statpal asks for the name of the data file. Statpal will save the system file, with its variable specifications, as a Statpal system file with no data.

When Statpal has successfully read the data, it reports the number of cases and variables in the new system file, and sends you back to the Main Menu.

## 1.5.5 The CReate Facility: Example Statpal Session

---

Note:   Underlined text is what the user types.   The symbol <RET> shows when
        the Return key is pressed.

---

Example of creating a Statpal system file from a fixed-format data file:

What should the new Statpal system file be named?   (Press Return to stop.)
   ===> B:EXAMPLE.STP<RET>

Do you wish to input data from a File or Keyboard? (S to Stop) ===> F

Enter specifications for variable # 1 ...
     Variable name (default is VAR1 - !!! to stop):   AGE<RET>
     Variable label (up to 40 characters):   Patient's age in years<RET>
     Missing value (default is -999999):<RET>

Enter specifications for variable # 2 ...
     Variable name (default is VAR2 - !!! to stop):   SEX<RET>
     Variable label (up to 40 characters):   1=male, 2=female<RET>
     Missing value (default is -999999):<RET>

Enter specifications for variable # 3 ...
     Variable name (default is VAR3 - !!! to stop):   DRUG<RET>
     Variable label (up to 40 characters):   1=A, 2=B, 3=C, 4=D<RET>
     Missing value (default is -999999):<RET>

Enter specifications for variable # 4 ...
     Variable name (default is VAR4 - !!! to stop):   SYSTOL<RET>
     Variable label (up to 40 characters):   Systolic blood pressure<RET>
     Missing value (default is -999999):<RET>

Enter specifications for variable # 5 ...
     Variable name (default is VAR5 - !!! to stop):   DIASTOL<RET>
     Variable label (up to 40 characters):   Diastolic blood pressure<RET>
     Missing value (default is -999999):<RET>

Enter specifications for variable # 6 ...
    Variable name (default is VAR6 - !!! to stop):  !!!<RET>

A total of 5 variables will be read.
Is your data file in a FIxed format or a FRee format?  (Press Return to stop.)


    ===> FI<RET>

How many lines in the data file does each case have?  (Default is 1,
    !!! to cancel procedure) ===> <RET>

Enter starting and ending column for each variable...

Variable:  AGE
    Starting column ===> 1<RET>
    Ending column    ===> 3<RET>

Variable:  SEX
    Starting column ===> 4<RET>
    Ending column    ===> 7<RET>

Variable:  DRUG
    Starting column ===> 8<RET>
    Ending column    ===> 11<RET>

Variable:  SYSTOL
    Starting column ===> 12<RET>
    Ending column    ===> 18<RET>

Variable:  DIASTOL
    Starting column ===> 19<RET>
    Ending column    ===> 23<RET>

Name of the raw data file:
Enter file name (!!! to cancel) ===> B:EXAMPLE.DAT<RET>

Processing data for case ____  ...
Statpal system file B:EXAMPLE.STP has 50 cases with 5 variables.

Press any key...  <RET>

_____


Example of creating a Statpal system file from keyboard:

What should the new Statpal system file be named?  (Press Return to stop.)
    ===> B:EXAMPLE2.STP<RET>

Do you wish to input data from a File or Keyboard? (S to stop) ===> K

```
Enter specifications for variable # 1 ...
    Variable name (default is VAR1 - !!! to stop):  AGE<RET>
    Variable label (up to 40 characters):  Patient's age in years<RET>




    Missing value (default is -999999):<RET>

Enter specifications for variable # 2 ...
    Variable name (default is VAR2 - !!! to stop):  SEX<RET>
    Variable label (up to 40 characters):  1=male, 2=female<RET>
    Missing value (default is -999999):<RET>

Enter specifications for variable # 3 ...
    Variable name (default is VAR3 - !!! to stop):  DRUG<RET>
    Variable label (up to 40 characters):  1=A, 2=B, 3=C, 4=D<RET>
    Missing value (default is -999999):<RET>

Enter specifications for variable # 4 ...
    Variable name (default is VAR4 - !!! to stop):  !!!<RET>

Enter data for each case, one variable at a time.  Type !!! to stop.

Case number  1

    Variable:  AGE        ===> 64<RET>
    Variable:  SEX        ===> 1<RET>
    Variable:  DRUG       ===> 1<RET>

Case number  2

    Variable:  AGE        ===> <RET>
         (Set to missing value of -999999)
    Variable:  SEX        ===> 1<RET>
    Variable:  DRUG       ===> 2<RET>

    [etc.]

Case number  51

    Variable:  AGE        ===> !!!<RET>

Statpal system file B:EXAMPLE2.STP has 50 cases with 3 variables.

Press any key... <RET>
```

## 1.6 CHOOSING A PROCEDURE

### 1.6.1 Checking the Data

Once you have created a Statpal system file, you are ready to choose a procedure from the Main Menu. One of the first you might want to use is the LIst procedure, which lets you look at the data for whichever variables and cases you wish. Listing the data for all variables for the first few and last few cases in a Statpal system file is a good way to check that the data have been read properly. You might also want to use the EDit utility (on the SYstem file utilities menu) to check the data and make any necessary changes.

### 1.6.2 The Current Statpal System File

In any Statpal session, once you have done something with a particular Statpal system file (either created one or used an existing one in some procedure), Statpal considers that file to be the "current" file, and assumes that you want to continue using it for additional procedures. If you wish to use a different Statpal system file in the same session, you can use the SPecify utility to specify a different Statpal system file as current. For convenience, the SPecify utility is accessible from both the Main Menu and the SYstem file utilities menu.

Some system file utilities (such as MErge) will always ask you to specify the file or files to be used, regardless of which file (if any) is current. In general, if there is the possibility of ambiguity, Statpal will ask you to specify which system file to use.

### 1.6.3 Getting a List of Variables in the Current Statpal System File

All Statpal statistical procedures, and many of the utilities, ask you to specify which variables are to be

used. Whenever you are asked to specify an existing
variable, you may type $$$ to get a list of the variables,
as well as a report of the number of cases and variables, in
the current Statpal system file. (Again, you do not need to
remember special sequences such as $$$, since Statpal will
always tell you when you can use them.)

## 1.7 WHERE TO GO FROM HERE

### 1.7.1 Trying It Out

Now you are ready to try out Statpal. Follow the
instructions on the printed sheet of implementation-specific
information to get Statpal running on your computer. Then,
give it a try. Remember that Statpal is very forgiving of
mistakes; if you're not sure what something does, try it!
Just be sure that you know what is being asked for whenever
you type a file name, so that you don't inadvertently
overwrite a file you really wanted.

### 1.7.2 Garbage In, Garbage Out

Statpal is a powerful tool to help you perform
statistical analyses. However, there are some things
Statpal (or any computer program, so far) cannot do.

First, Statpal cannot make sure the data you entered
are correct. Errors in coding or data entry can have an
astonishing effect on the results of an analysis. There is
little point in performing a series of analyses, only to
find that errors in the data render the results
meaningless. It is vital to verify, at both the coding
stage and the data entry stage, that the data are correct.
You'll find the LIst procedure and EDit utility particularly
helpful for this. Another good practice is to use the
DEscriptive statistics procedure, paying particular
attention to the minimum and maximum values of each

variable. Some data entry problems (such as putting a number in the wrong columns in a fixed-format file, or misspecifying the column locations) can manifest themselves as numbers off by a power of 10, and these often turn up as the minimum or maximum value for a variable.

Second, Statpal has no way of knowing whether you are using the right procedure. Picking the wrong procedure--say, a grouped t-test when you have a paired design--can seriously inflate the probability of an incorrect conclusion, be it a type I or a type II error. Even more common is a kind of error usually referred to as type III: giving the right answer to the wrong question! You might find that your data are correct, all necessary assumptions fulfilled, and all procedures performed correctly, and still have no information relevant to what you want to find out, because the design of the study or the data collection procedures used were inappropriate.

Finally, Statpal cannot make sense out of your results. You'll find output from Statpal is generally clear and informative, but you must supply the interpretation.

# 2

# Statpal Statistical Procedures

This chapter describes the statistical procedures available in Statpal. Chapter 3 covers utilities for maintaining and changing Statpal system files, and transformations to create new variables.

All Statpal statistical procedures work with the current Statpal system file; if there is no current Statpal system file, the first thing each procedure does is ask you to specify one.

The descriptions below assume that you have already created a Statpal system file, and that you have read Chapter 1. The descriptions are arranged in alphabetical order as the procedures appear on the Main Menu. For each procedure, four categories of information are given: (1) an overview of the procedure; (2) details about the specific prompts used by the procedure; (3) information about the computational algorithms used by the procedure, with references as appropriate; and (4) an example of part of a Statpal session using the procedure.

No attempt has been made to give an introductory explanation of the statistical concepts embodied in these procedures. For that, you'll need to consult a statistics text.

## 2.1 ANALYSIS OF VARIANCE

### 2.1.1 ANalysis of Variance: Overview

Statpal's analysis of variance procedure performs an analysis of variance and covariance for a factorial design with one dependent variable, up to five factors, and as many covariates as workspace allows. Interaction terms may be included or excluded in any combination. All factors must be crossed (not nested).

Each factor in the design is represented in the data by a grouping variable, with each level of the factor represented by a code. We'll explain the idea in the context of the example introduced in Chapter 1. Suppose you are studying drug effects on systolic blood pressure. You have four groups of subjects, with the subjects in each group receiving a different drug. Both male and female subjects are included in each drug group, and you are also interested in determining if there are any sex differences in the effects of the drugs.

In terms of the Statpal system file for this example, SYSTOL is the dependent variable, and DRUG and SEX are the grouping variables. Recall that the four drug types are coded as 1, 2, 3, and 4, and the two sexes as 1 and 2. (Any coding scheme is acceptable to Statpal; grouping variables need not have integer values, although they usually do.) If, in addition, you wish to incorporate each subject's age as a covariate, you would make use of the variable AGE.

### 2.1.2 ANalysis of Variance: Prompts

Statpal first asks you to enter the variables to be used in the analysis. Enter the dependent variable first, followed by the grouping variable(s), followed by the covariates (if any). If you enter more than one variable after the dependent variable, Statpal asks you how many of those should be considered covariates. If you simply press

Return, Statpal assumes all variables (after the dependent
variable) should be considered grouping variables.

Next, Statpal asks you to specify the codes that define
the levels of each factor (as defined by each grouping
variable). You must specify all codes you wish to include
in the analysis; thus, if you wish to exclude cases with
certain codes on a grouping variable--say, you wish to
exclude subjects with drug type 2 from the analysis--simply
leave out code 2 when you specify the levels for the
grouping variable DRUG. You must specify at least two codes
for each grouping variable.

If you have more than one grouping variable, Statpal
will now ask you whether each possible interaction term
should be included in the analysis. (In our example,
Statpal would ask if we wish to include the SEX by DRUG
interaction.) Simply press Y (yes) or N (no) to indicate
your choice. Note that interaction terms can require a
substantial amount of memory space to process; if Statpal
reports that it cannot analyze a particularly complex design
because of lack of memory space, try omitting some
interaction terms. Note also that the more complex the
design, the longer the analysis will take.

Statpal will now process the data for each case. Any
case with a missing value for any of the variables involved
in the procedure, and any case with a value on a grouping
variable not specified as one of the codes defining a level,
is excluded from the computations.

Next, Statpal will inform you how many sweep operations
(a kind of matrix operation) are required to analyze the
data, and will keep you posted on how many it has
completed. By watching the number of sweeps completed, you
can judge how long the analysis will take. A fairly simple
analysis will require only a few seconds; however, a large
design with several interaction terms can require several
minutes to perform all the necessary computations.

When all sweep operations have been completed, Statpal
displays the analysis of variance table. You are then asked

to choose from among the commands Newmodel, Print, Help, and
Stop.  Newmodel goes back to the beginning of  the  ANalysis
of variance procedure,  prompting  you  for variables; Print
prints  the  current  analysis  of  variance  table  on  the
printer; Help explains  these  options; and Stop returns you
to the Main Menu.


## 2.1.3 ANalysis of Variance: Computational Algorithms

    Statpal  first  converts  the specified  model  into  a
regression  problem,  and  then  uses  the  sweep  operator
(Goodnight, 1982)  to  enter  (and  remove)  the appropriate
terms into the regression model to find the appropriate sums
of squares.  The coding scheme used for  main  effect  terms
uses 1, 0 and -1 (Neter and Wasserman, 1974, page 633).

    The method used for calculating sums of squares is  the
regression method.  All terms are added to the model;  then,
the terms corresponding  to  each covariate, main effect, or
interaction are removed (and  re-entered),  with  the sum of
squares  for  the  effect  calculated  as  the  increase  in
residual sum of squares resulting from the  removal  of  the
terms.  That  is,  all  sums of squares are adjusted for all
other terms in the model.


## 2.1.4 ANalysis of Variance: Example Statpal Session

Variables to be used:  Enter the dependent variable, followed by the grouping
    variables, followed by covariates (if any)...
    (Return to stop, $$$ to see list of available variables, !!! to cancel
    procedure, ALL to use all variables.)
    ===> SYSTOL<RET>
    ===> SEX<RET>
    ===> DRUG<RET>
    ===> AGE<RET>
    ===> <RET>

Enter the number of covariates you specified (default is 0) ===> 1<RET>

Enter the codes for levels of the grouping variable(s)...

Codes for variable SEX (Press Return to stop entering codes,
   !!!to go back to Main Menu...)
   Level 1 ===> 1<RET>
   Level 2 ===> 2<RET>
   Level 3 ===> <RET>

Codes for variable DRUG (Press Return to stop entering codes,
   !!! to go back to Main Menu...)
   Level 1 ===> 1<RET>
   Level 2 ===> 2<RET>
   Level 3 ===> 3<RET>
   Level 4 ===> 4<RET>
   Level 5 ===> <RET>

Interaction terms:  Type Y to indicate that the term should be included...
   SEX by DRUG ===> Y

Processing data for case ___ of 50...

Performing sweep number ___ of 22...

    Statpal - Analysis of Variance  -   7/14/85  16:19 - File: EXAMPLE.STP

Dependent variable: SYSTOL      Systolic blood pressure
Source              SS        DF      MS          F       Prob

| Source | SS | DF | MS | F | Prob |
|---|---|---|---|---|---|
| Covariates | | | | | |
| AGE | 33.5128 | 1 | 33.5128 | 0.2368 | 0.6293 |
| Main Effects | | | | | |
| SEX | 2.8206 | 1 | 2.8206 | 0.0199 | 0.8885 |
| DRUG | 131.7227 | 3 | 43.9076 | 0.3102 | 0.8178 |
| Interactions | | | | | |
| SEX by DRUG | | | | | |
| ====> | 128.2707 | 3 | 42.7569 | 0.3021 | 0.8237 |
| Residual | 5378.0289 | 38 | 141.5271 | | |
| Total | 5676.4681 | 46 | 123.4015 | | |

Valid cases:  47   Missing cases:  3

Command?  (Newmodel, Print, Help, Stop) ===> S

## 2.2 BREAKDOWN

### 2.2.1 BReakdown: Overview

Statpal's BReakdown procedure displays summary statistics for a selected variable (the dependent variable), calculated separately for each group of cases defined by one or two grouping variables. For example, you could find out the mean systolic pressure of the subjects in each of the four groups defined by drug type, or in each of the two groups defined by sex, or in each of the eight groups defined jointly by drug type and sex. In terms of the BReakdown procedure, SYSTOL would be the dependent variable, and DRUG and SEX would be the grouping variables.

In BReakdown, you do not need to tell Statpal the codes used to define groups. Statpal figures them out as it reads the data. For a breakdown involving one grouping variable, a maximum of 400 distinct groups is allowed. For a breakdown involving two grouping variables, the maximum number of categories allowed for each of the two grouping variables is 20.

### 2.2.2 BReakdown: Prompts

Statpal first asks you to specify the dependent and grouping variables, in that order. Statpal then processes the data. Any case with a missing value on the dependent variable or the grouping variable(s) is excluded from the computations.

If you specified a single grouping variable, Statpal displays the mean, standard deviation, minimum, and maximum of the dependent variable, along with the number of cases, for each subgroup as defined by the grouping variable. It also displays these same statistics for the entire sample.

If you specified two grouping variables, Statpal displays a two-way table with the first grouping variable defining rows and the second grouping variable defining columns. In each cell of the table, Statpal initially displays the mean of the dependent variable for the cases in that cell. You are then given the option of redisplaying the table to show standard deviations, numbers of observations, or the means again, as many times as you wish. If the entire table is too big to fit on one screen, Statpal tells you how many screenfuls are required, and lets you move to the next screenful or the previous screenful as you wish.

If it is impossible to calculate a mean in a cell of the table because the cell has no cases, Statpal displays "(n=0)" in the cell instead of the mean. Similarly, if it is impossible to calculate the standard deviation because there are fewer than 2 cases in the cell, Statpal displays "(n<2)" in the cell instead of the standard deviation.

## 2.2.3 BReakdown: Computational Algorithms

BReakdown uses the same provisional accumulation algorithms as DEscriptive statistics to calculate means and standard deviations for each group. The Super-Shellsort algorithm (Barron and Diehr, 1983) is used to sort the values found for the grouping variables. Standard deviations are calculated using the unbiased sample variance formula (i.e., using n-1 rather than n).

## 2.2.4 BReakdown: Example Statpal Session

Enter the dependent variable (i.e., the variable to be broken down) followed
   by one or two grouping variables (i.e., the variables that define groups).
   (Return to stop, $$$ to see list of available variables, !!! to cancel
   procedure.)
   ===> SYSTOL<RET>
   ===> SEX<RET>
   ===> DRUG<RET>

Processing data for case ___ of 50 ...

```
          Statpal - Breakdown   -   7/14/85  9:29 - File: EXAMPLE.STP
Dependent variable:   SYSTOL     Systolic blood pressure
  broken down by
  Row variable:       SEX        1=male, 2=female
  Column variable:    DRUG       1=A, 2=B, 3=C, 4=D

Mean of SYSTOL
              1      2      3      4  Total
      1  115.20 120.38 119.33 119.50 118.88
      2  118.00 120.00 116.00 124.83 119.78

  Total  116.60 120.21 117.67 122.17 119.31

Valid cases: 48   Missing cases: 2

Command? (Mean, Std.dev., Observations, Print, Quit) ===> S
```

```
          Statpal - Breakdown   -   7/14/85  9:29 - File: EXAMPLE.STP
Dependent variable:   SYSTOL     Systolic blood pressure
  broken down by
  Row variable:       SEX        1=male, 2=female
  Column variable:    DRUG       1=A, 2=B, 3=C, 4=D

S.D. of SYSTOL
              1      2      3      4  Total
      1  16.037 12.501 13.216 16.196 13.535
      2  6.9282 6.1319 8.9889 14.091 9.6104

  Total  11.740 9.9319 10.916 14.739 11.704

Valid cases: 48   Missing cases: 2

Command? (Mean, Std.dev., Observations, Print, Quit) ===> O
```

```
          Statpal - Breakdown   -   7/14/85  9:29 - File: EXAMPLE.STP
Dependent variable:   SYSTOL     Systolic blood pressure
  broken down by
  Row variable:       SEX        1=male, 2=female
  Column variable:    DRUG       1=A, 2=B, 3=C, 4=D

Number of cases
              1      2      3      4  Total
      1       5      8      6      6     25
      2       5      6      6      6     23

  Total      10     14     12     12     48

Valid cases: 48   Missing cases: 2

Command? (Mean, Std.dev., Observations, Print, Quit) ===> Q
Do you wish to do another breakdown? (Y/N) Y
```

Enter the dependent variable (i.e., the variable to be broken down) followed
  by one or two grouping variables (i.e., the variables that define groups).
  (Return to stop, $$$ to see list of available variables, !!! to cancel
  procedure.)
    ===> SYSTOL<RET>
    ===> DRUG<RET>
    ===> <RET>

Processing data for case ___ of 50 ...

              Statpal - Breakdown   -  7/14/85  16:22 - File: EXAMPLE.STP
Dependent variable: SYSTOL        Systolic blood pressure
broken down by:     DRUG          1=A, 2=B, 3=C, 4=D

                        Mean      Std.Dev.     Minimum     Maximum      N

DRUG    =      1      116.6000    11.7398     92.0000    134.0000     10
DRUG    =      2      120.2143     9.9319    104.0000    144.0000     14
DRUG    =      3      117.6667    10.9157    101.0000    140.0000     12
DRUG    =      4      122.1667    14.7391     96.0000    142.0000     12

All cases             119.3125    11.7041     92.0000    144.0000     48

Missing cases: 2

Press C to continue, P to print... C


## 2.3 CORRELATION


### 2.3.1 COrrelation: Overview

    COrrelation produces a matrix of Pearson product-moment
correlation coefficients, along with a one-tail significance
level (derived from the t-distribution) for each
coefficient. You may specify up to 20 variables to include
in the matrix.


### 2.3.2 COrrelation: Prompts

    When you choose the COrrelation procedure, Statpal asks
you to specify a list of up to 20 variables for which a
correlation matrix will be computed.

Once you have specified the variables, Statpal asks you how it should handle missing data. You have two choices: (1) listwise deletion of missing cases, and (2) pairwise deletion of missing cases.

If you choose listwise deletion, a case will be excluded from the computations of all the correlation coefficients in the entire matrix if it has a missing value on any of the variables in the matrix. Thus, with listwise deletion, a case must have complete data on all the variables in the matrix to be included in the calculation of any part of it.

If you choose pairwise deletion, a case will be included in the computation of the correlation coefficient for any two variables for which the case has valid data. Thus, with pairwise deletion it is possible that the coefficients in the matrix might not all be based on the same cases.

In general, listwise deletion takes less time.

When you have made your choice, Statpal processes the data and displays the results in the form of a square matrix. For each cell of the matrix, three numbers are given: (1) the correlation coefficient, (2) the number of cases used in the calculation, and (3) the one-tailed significance level, from a t-test of the null hypothesis that the population correlation coefficient equals zero.

You are then given the option of redisplaying or printing the data, specifying variables for a new matrix, or returning to the Main Menu.


## 2.3.3 COrrelation: Computational Algorithms

Each correlation coefficient is computed using a provisional accumulation algorithm similar to that used by the DEscriptive statistics procedure. For each coefficient in the matrix, a provisional mean and sum of squared deviations is accumulated for each of the two variables involved, along with a provisional sum of cross-product deviations. For listwise deletion of missing cases, provisional means and sums of squares need only be accumulated once for each variable (rather than once for each coefficient), thus yielding a considerable saving in computation time.

The one-tailed significance level is derived from the t-distribution via the relation t= r times the square root of [(n-2)/(1-r squared)], where r is the correlation coefficient and n is the number of cases used in its computation.

## 2.3.4 COrrelation: Example Statpal Session

```
Enter up to 20 variables to be included in the correlation matrix.
  (Return to stop, $$$ to see list of available variables, !!! to cancel
  procedure, ALL to use all variables.)
      ===> SYSTOL<RET>
      ===> DIASTOL<RET>
      ===> AGE<RET>
      ===> SEX<RET>
      ===> <RET>

How should missing data be handled?  (Listwise, Pairwise, Help, Stop) ===> L

Processing data for case ___ of 50 ...

          Statpal - Correlation  -  7/14/85 16:24 - File: EXAMPLE.STP
Each cell shows the correlation, sample size, and one-tailed significance.

          SYSTOL   DIASTOL  AGE      SEX

SYSTOL    1.0000   0.9568   0.0771   -0.0099
          n:  47   n:  47   n:  47   n:  47
          p:0.000  p:0.000  p:0.304  p:0.474

DIASTOL   0.9568   1.0000   0.1226   -0.1088
          n:  47   n:  47   n:  47   n:  47
          p:0.000  p:0.000  p:0.206  p:0.234

AGE       0.0771   0.1226   1.0000   0.0860
          n:  47   n:  47   n:  47   n:  47
          p:0.304  p:0.206  p:0.000  p:0.283

SEX       -0.0099  -0.1088  0.0860   1.0000
          n:  47   n:  47   n:  47   n:  47
          p:0.474  p:0.234  p:0.283  p:0.000
Listwise deletion of missing data in effect.
  Valid cases: 47   Missing cases: 3

Press any key... <RET>
Command? (New matrix, Redisplay, Stop) ===> S
```

## 2.4 DESCRIPTIVE STATISTICS

### 2.4.1 DEscriptive Statistics: Overview

Statpal's DEscriptive statistics procedure calculates the mean, standard deviation, standard error, minimum, maximum, and range for as many variables as you wish.

### 2.4.2 DEscriptive Statistics: Prompts

When you choose DEscriptive statistics, Statpal asks you for the list of variables for which you wish to see statistics. It then processes the data and displays the results. Once the data have been displayed (and optionally printed), Statpal returns to the Main Menu.

Missing data are handled individually for each variable in the list; that is, a case is included in the computation of statistics for any variable for which it has non-missing data.

### 2.4.3 DEscriptive Statistics: Computational Algorithms

The mean and standard deviation are calculated using a provisional accumulation algorithm. As the data for each case are processed, Statpal accumulates a provisional mean and sum of squared deviations for each variable. The standard deviation is calculated using the unbiased form of the variance (i.e., using n-1 rather than n). The standard error equals the standard deviation divided by the square root of the sample size.

### 2.4.4 DEscriptive Statistics: Example Statpal Session

```
Enter the variables for which statistics are desired.
  (Return to stop, $$$ to see list of available variables, !!! to cancel
  procedure, ALL to use all variables.)
  ===> ALL<RET>
```

Processing data for case ___ of 50 ...

   Statpal - Descriptive Statistics  -  7/14/85  16:25 - File: EXAMPLE.STP

Statistics for variable AGE       Patient's age in years
Mean:      39.8333    Std. Dev.:    12.9833    Std. Error:    1.8740
Range:         46     Minimum:          18     Maximum:          64
   Valid cases: 48  Missing cases: 2

Statistics for variable SEX       1=male, 2=female
Mean:       1.5000    Std. Dev.:     0.5051    Std. Error:    0.0714
Range:          1     Minimum:           1     Maximum:           2
   Valid cases: 50  Missing cases: 0

Statistics for variable DRUG      1=A, 2=B, 3=C, 4=D
Mean:       2.5400    Std. Dev.:     1.0730    Std. Error:    0.1518
Range:          3     Minimum:           1     Maximum:           4
   Valid cases: 50  Missing cases: 0

Statistics for variable SYSTOL    Systolic blood pressure
Mean:     119.3125    Std. Dev.:    11.7041    Std. Error:    1.6893
Range:         52     Minimum:          92     Maximum:         144
   Valid cases: 48  Missing cases: 2

Statistics for variable DIASTOL   Diastolic blood pressure
Mean:      99.0625    Std. Dev.:    12.9457    Std. Error:    1.8686
Range:         53     Minimum:          74     Maximum:         127
   Valid cases: 48  Missing cases: 2

Press C to continue, P to print... C

## 2.5 HISTOGRAM AND FREQUENCY DISTRIBUTION

### 2.5.1 HIstogram and Frequency Distribution: Overview

    The HIstogram and frequency distribution procedure
allows you to examine the frequency distribution of a
selected variable both graphically and numerically. For the
selected variable, Statpal displays the number and
proportion of cases with each score (or range of scores,
depending on the number of distinct scores in the data),
along with a histogram. You may redisplay the results as
many times as you wish, specifying different category ranges
each time.

2.5.2 HIstogram and Frequency Distribution: Prompts

After you specify the variable for which you wish to see a histogram and frequency distribution, Statpal processes the data and displays a histogram and frequency distribution that fit on a single screen. If the number of distinct scores in the data is twelve or fewer, the initial display will show individual scores (rather than score ranges).

Score ranges are indicated using intervals of the form [lower, upper) --that is, the interval includes the lower bound, but not the upper bound. (Statpal uses this notation on the displayed histogram to remind you of the form of the interval.) The last interval, however, is of the form [lower, upper] --that is, both lower and upper bounds are included.

Statpal then gives you the option of redisplaying the histogram and frequency distribution with different category bounds, starting again with a new variable, or returning to the main menu. Redisplaying the results does not require reprocessing the data and requires very little computation time, so you can feel free to do so until you get the most useful display.

If you elect to redisplay the results, Statpal will give you a choice between two different methods of specifying category bounds. The choices are (1) "Lower bound and increment"--i.e., to specify the lower bound of the first category and the increment to subsequent categories, or (2) "Number of categories"--i.e., to specify a specific number of categories, and let Statpal calculate the category bounds for you. (Exception: if the number of distinct scores in the data is greater than 288, as explained below, you are not given choice 1.) Once you have made your choice, you are prompted for the necessary information, and Statpal proceeds to redisplay the histogram.

Provided the number of distinct scores in the data is 288 or fewer, Statpal is capable of displaying frequencies

for individual scores (rather than score ranges). To make
this happen, choose the Redisplay option, specify "Number of
caterogies" as the method, and give the number of distinct
scores as the number of categories. Don't worry if you
don't know the number of distinct scores in advance; Statpal
will tell you.

If the number of distinct scores was greater than 288,
Statpal groups the data as they are processed into 288 equal
score ranges. If you wish to look at a frequency
distribution with categories of equal width, you should
specify a number of categories that divides 288 evenly.
(The number 288 was chosen precisely because it has so many
divisors, including 2, 3, 4, 6, 8, 12, 16, 18, 24, and 36.)
Otherwise, the last category will have a different width
from the other categories. Statpal displays a message
suggesting this when the situation arises.

If you are working with a variable with more than 288
distinct scores in the data, but you nevertheless want to
obtain the frequency for each individual score, you can use
the SElect cases utility (before you get into the HIstogram
procedure) to break up the range of scores into chunks, each
with 288 or fewer distinct scores represented in the data.


2.5.3 HIstogram and Frequency Distribution: Computational
Algorithms

As Statpal processes the data, it keeps track of how
many times each distinct score occurs until it reaches 288
distinct scores. If a 289th distinct score is observed,
Statpal reads through the remainder of the data to find the
minimum and maximum scores, and then defines 288 score
ranges of equal width. It then counts up the number of
cases in each score range, which entails rereading the data
from the point at which the 289th score was encountered.

When this process is completed, Statpal displays the
results, and, if requested, redisplays them as requested.
Since the information on the (up to) 288 scores or score
ranges retained in memory, it is not necessary to reprocess
the data in order to redisplay the histogram.

## 2.5.4 HIstogram and Frequency Distribution: Example Statpal Session

Enter the variable you wish to use.
   (Return to stop, $$$ to see list of available variables, !!! to cancel
   procedure.)
   ===> SYSTOL<RET>

Processing data for case ___ of 50 ...


                Statpal - Histogram  -  7/14/85  16:26 - File: EXAMPLE.STP
For variable: SYSTOL     Systolic blood pressure
     Lower       Upper        Frequency
     bound       bound      N     prop.

[   92.0000    96.3333 )    2     0.042   **
[   96.3333   100.6667 )    1     0.021   *
[  100.6667   105.0000 )    3     0.063   ***
[  105.0000   109.3333 )    2     0.042   **
[  109.3333   113.6667 )    6     0.125   ******
[  113.6667   118.0000 )    7     0.146   *******
[  118.0000   122.3333 )    7     0.146   *******
[  122.3333   126.6667 )   10     0.208   *********
[  126.6667   131.0000 )    1     0.021   *
[  131.0000   135.3333 )    4     0.083   ****
[  135.3333   139.6667 )    2     0.042   **
[  139.6667   144.0000 ]    3     0.063   ***

Valid cases: 48   Missing cases: 2

Command? (New var, Print, Redisplay with different categories, Stop) ===> R

Specify? (Help, Lower bound and increment, Number of categories, Stop) ===> L

Enter the lower bound you wish to specify for the first category.
   Note:  The range of the data is from 92 to 144
   (Press Return to use minimum, !!! to stop) ===> 100<RET>

Enter the increment to succeeding categories.  (Default is 5.3333
   Press Return to use default, !!! to stop.) ===> 10<RET>

```
        Statpal - Histogram   -   7/14/85  16:29 - File: EXAMPLE:STP
For variable: SYSTOL      Systolic blood pressure
   Lower       Upper       Frequency
   bound       bound       N     prop.

 Below first category       2    0.042  **
[ 100.0000    110.0000 )    6    0.125  ******
[ 110.0000    120.0000 )   14    0.292  **************
[ 120.0000    130.0000 )   17    0.354  *****************
[ 130.0000    140.0000 )    6    0.125  ******
[ 140.0000    150.0000 ]    3    0.063  ***

Valid cases: 48   Missing cases: 2

Command? (New var, Print, Redisplay with different categories, Stop) ===> S
```

## 2.6 LIST DATA

### 2.6.1 LIst Data: Overview

The LIst data procedure provides two different ways of listing data for specified cases and variables: "Plain" and "Fancy". The Plain option lets you list the data with a simple heading indicating the order in which variables are listed. The Fancy option allows you to specify column widths, numbers of decimals to report, and column headings, and allows you to specify the number of lines to output to a printed page. Both options provide for both screen and printer (or print file) output.

### 2.6.2 LIst Data: Prompts

Statpal first asks you to choose a list option from the choices Fancy, Plain, Help, and Stop. (Help provides a brief description of the Fancy and Plain options, and Stop returns you to the Main Menu.)

Statpal next asks you to specify the variables to be listed. The order in which you specify the variables will be the order in which they are listed. Statpal next asks

you to specify the starting case number, with a default of
1, and then asks whether you wish to list the data on the
screen or the printer.

If you selected the Plain option, Statpal will now list
the data in a simple format, with a heading at the top of
the list indicating the order in which variables are
listed. The special variable CASE# (indicating the case
number in the file) is always listed as the first variable.
Missing values are listed followed by "(M)".

If you selected the Fancy option, Statpal will take you
through the set of variables you requested, asking you for
some optional specifications. For each variable, Statpal
first asks you to give a column width, with a default of 10.
If your specifications would result in a page greater than
80 columns wide, Statpal warns you of this fact and requests
confirmation before proceeding. Depending on your response,
Statpal may then ask you to specify the number of decimals
to be listed, and a column heading (with the default heading
being the variable name). A response of !!! to any of
these prompts will cancel the procedure. Once you have
given specifications for all variables, Statpal will produce
the list on either the screen or the printer (or print
file), depending on what you chose above. Once the list has
been produced you are given the option of producing another
list with the same specifications on the other device (i.e.,
on the printer if you have just displayed it on the screen,
or vice versa). Before sending a list to the printer (or
print file), Statpal will ask you to specify the number of
lines on a page. The formfeed character (ASCII code 12) is
used to go to a new page.

A Fancy list includes a heading for each variable at
the top of each page of output. The headings are arranged
so as to fit in the column width. For columns less than 8
spaces wide, the heading will consist of the variable name
printed vertically, centered in the column.

Whichever method you choose, you can interrupt the
listing in progress by pressing and holding any key
(preferably the space bar).

## 2.6.3 LIst Data: Example Statpal Session

```
List Option? (Fancy, Plain, Help, Stop) ===> P
Enter the variables to be listed.
  (Return to stop, $$$ to see list of available variables, !!! to cancel
  procedure, ALL to use all variables.)
   ===> ALL<RET>

Enter starting case number (default is 1; !!! to stop) ===> <RET>
Do you wish to list data on the Screen or Printer (N for neither) ===> S


            Statpal - List Data   -   7/19/85  15:17 - File: EXAMPLE.STP
Variables for each case will be listed as follows:
CASE#      AGE        SEX         DRUG        SYSTOL       DIASTOL
Data in Statpal system file EXAMPLE.STP:

    1          64          1           1           126          107
    2          36          1           1           112           92
    3          30          2           1           114           88
    4          45          2           1           120           98
    5          35          1           1           134          117
    6          61          2           1           124          110
    7          31          2           1           108           85
    8          24          1           1           112           94
    9     -999999(M)       1           1            92           74
   10          58          2           1           124          106

   [etc.]

   48          30          1           4           117           95
   49          21          1           4           126          113
   50          27          2           4           132          110

Press any key... C

Do you wish to do more listing? (Y/N) Y

List Option? (Fancy, Plain, Help, Stop) ===> F
Enter the variables to be listed.
  (Return to stop, $$$ to see list of available variables, !!! to cancel
  procedure, ALL to use all variables.)
   ===> ALL<RET>

Enter starting case number (default is 1; !!! to stop) ===> <RET>
```

Do you wish to list data on the Screen or Printer (N for neither) ===> <u>S</u>
Specify column width for each variable (!!! to cancel):

Variable  AGE
    Column width (default 10) ===> <u>3</u><RET>

Variable  SEX
    Column width (default 10) ===> <u>3</u><RET>

Variable  DRUG
    Column width (default 10) ===> <u>12</u><RET>
    Number of decimals (default 4) ===> <u>0</u><RET>
    Column heading (12 characters max; default is DRUG    ):
       ===> <u>Drug Type</u><RET>

Variable  SYSTOL
    Column width (default 10) ===> <u>13</u><RET>
    Number of decimals (default 4) ===> <u>0</u><RET>
    Column heading (13 characters max; default is SYSTOL   ):
       ===> <u>Systolic BP</u>

Variable  DIASTOL
    Column width (default 10) ===> <u>13</u><RET>
    Number of decimals (default 4) ===> <u>0</u><RET>
    Column heading (13 characters max; default is DIASTOL  ):
       ===> <u>Diastolic BP</u>


            Statpal - List Data   -   7/19/85  15:31 - File: EXAMPLE.STP


                  D
              A  S  R
              G  E  U
CASE#         E  X  G  Systolic BP  Diastolic BP

       1    64  1  1         126          107
       2    36  1  1         112           92
       3    30  2  1         114           88
       4    45  2  1         120           98
       5    35  1  1         134          117
       6    61  2  1         124          110
       7    31  2  1         108           85
       8    24  1  1         112           94
       9  ****  1  1          92           74
      10    58  2  1         124          106
      11    37  1  2         123           99
      12    46  1  2         111           90
      13    60  2  2         118           92
      14    21  2  2*************************

```
15   42 1 2           114             98
16   42 2 2           121             93
17   45 2 2           114            100
18   52 1 2           104             89

[etc.]

48   30 1 4           117             95
49   21 1 4           126            113
50   27 2 4           132            110
```

Do you wish to print this listing? (Y/N) Y
How many lines do you wish to put on a page?  (Default 55; 999 for "infinite"
  page length; !!! to cancel) ===> <RET>

Do you wish to display this listing on the screen? (Y/N) N
Do you wish to do more listing? N


## 2.7 NONPARAMETRIC TESTS


### 2.7.1 NOnparametric Tests: Overview

Statpal offers the following nonparametric procedures:

- Friedman test

- Kendall's tau (measure of association)

- Kruskal-Wallis nonparametric ANOVA

- Wilcoxon Rank Sum test

- Sign test

- Spearman's rho (rank correlation coefficient)

- Wilcoxon Signed Rank test

The procedures are accessed through a special
Nonparametric Tests menu presented by Statpal when you
choose NOnparametric tests from the Main Menu.

There is an implementation-specific limit to the number
of cases Statpal can use in carrying out all nonparametric

tests except the Friedman test and the Sign test. See the on-line help regarding implementation-specific features for information on this limit. If the limit is exceeded, Statpal gives you the option of either performing the procedure using only the cases processed up to the point at which the limit was reached, or aborting the procedure. Cases with missing values on any of the variables involved in a procedure, and cases not meeting a SElect criterion, do not count toward the limit.


## 2.7.2 NOnparametric Tests: Prompts, Algorithms, and Examples

Each item on the Nonparametric Tests menu leads to its own series of prompts. We shall discuss each item in turn.


## 2.7.2.1 FR - Friedman Test for Correlated Samples

Statpal asks you to specify two or more variables to be compared using the Friedman test (Siegel, 1956). Statpal then processes the data and reports the results. The significance test is based on the chi-square distribution.


```
Friedman Test:  Example Statpal Session

Enter the variables to be compared.
  (Return to stop, $$$ to see list of available variables, !!! to cancel
  procedure, ALL to use all variables.)
  ===> SYSTOL<RET>
  ===> DIASTOL<RET>
  ===> <RET>

Processing data for case ___ of 50 ...

        Statpal - Friedman Test   -   7/14/85  8:51 - File: EXAMPLE.STP

Average rank     Variable

       2         SYSTOL     Systolic blood pressure
       1         DIASTOL    Diastolic blood pressure

Chi-square = 48.0000 with 1 df...Significance = 0.0000

Valid cases:  48  Missing cases:  2
```

## 2.7.2.2 KE - Kendall's Tau

Statpal asks for two variables for which you wish to calculate Kendall's tau, and then processes the data and reports the results. The one-tailed significance level is based on a Normal approximation (see Siegel, 1956). Chalmer (Understanding Statistics, Chapter 11) gives a description of Kendall's tau.


Kendall's Tau:  Example Statpal Session

Enter the two variables for which Kendall's tau is to be computed.
  (Return to stop, $$$ to see list of available variables, !!! to cancel
  procedure.)
    ===> SYSTOL<RET>
    ===> DIASTOL<RET>

Processing data for case ___ of 50 ...
Sorting...
Evaluating concordance (press and hold any key to interrupt) for cases:
    ___ and ___ of 48...

          Statpal - Kendall's Tau   -   7/14/85  13:05 - File: EXAMPLE.STP

Between SYSTOL      Systolic blood pressure
    and DIASTOL     Diastolic blood pressure

Kendall's tau =  0.8312   One-tail significance = 0.0000

Valid cases:  48  Missing cases:  2


## 2.7.2.3 KW - Kruskal-Wallis Nonparametric ANOVA

Statpal asks for one dependent variable and one grouping variable. Groups are defined by individual (non-missing) values on the grouping variable (e.g., a grouping variable for which the values 1, 4 and 5 occur in the data defines three groups).

Statpal then processes the data and reports the results, both unadjusted and adjusted for ties. The significance level is based on the Chi-square distribution. See Chalmer (Understanding Statistics, Chapter 15) and Siegel (1956) for descriptions of this test.

Kruskal-Wallis Test:   Example Statpal Session

Enter the dependent variable, followed by the grouping variable.
  (Return to stop, $$$ to see list of available variables, !!! to cancel
  procedure.)
  ===> SYSTOL<RET>
  ===> DRUG<RET>

Processing data for case ___ of 50 ...

        Statpal - Kruskal-Wallis Test   -   7/14/85  13:05 - File: EXAMPLE.STP
Dependent variable: SYSTOL       Systolic blood pressure
Grouping variable:  DRUG         1=A, 2=B, 3=C, 4=D

                          Average Rank        n

DRUG    =        1        22.1500          10
DRUG    =        2        24.6071          14
DRUG    =        3        .22.3750         12
DRUG    =        4        28.4583          12

Kruskal-Wallis H statistic (unadjusted)        =   1.5183  Significance =  0.6773

Kruskal-Wallis H statistic adjusted for ties = 1.5215  Significance =  0.6773

Valid cases:  48  Missing cases:  2


## 2.7.2.4 RS - Wilcoxon Rank Sum Test

    Statpal asks for one dependent variable and one
grouping variable. Then, Statpal asks you to specify the
value of the grouping variable that defines group 1. All
cases with the specified value on the grouping variable will
be assigned to group 1; all other cases will be assigned to
group 2.

    Statpal then processes the data and reports the
results. The test statistic, W, is computed with a
correction for ties (Siegel, 1956). If the numbers of cases
in both groups are 10 or fewer, the significance level comes
from a table look-up, and Statpal reports whether the
p-value is greater than .05, .05 or less, or .01 or less.
If the number of cases in either group is greater than 10,
the significance level comes from a Normal approximation
(Siegel, 1956). In either case, the significance level is
two-tailed.

Wilcoxon Rank Sum Test:  Example Statpal Session

Enter the dependent variable, followed by the grouping variable.
  (Return to stop, $$$ to see list of available variables, !!! to cancel
  procedure.)
  ===> SYSTOL<RET>
  ===> SEX<RET>

Enter the code for variable SEX that defines group 1.  (Default is 1.
  Group 2 will consist of all cases not in group 1.  Enter !!! to stop.)
  ===> <RET>

Processing data for case ___ of 50 ...

    Statpal - Wilcoxon Rank Sum Test  -   7/14/85  13:06 - File: EXAMPLE.STP
Dependent variable: SYSTOL     Systolic blood pressure
Grouping variable:  SEX        1=male, 2=female

Group 1:  SEX       equals 1
    Rank sum =      598   n =      25

Group 2:  SEX       does not equal 1
    Rank sum =      578   n =      23

Test statistic W =      578  Sig. (two-tail, using Normal approx.) =  0.7648

Valid cases: 48  Missing cases:  2

Press C to continue, P to print... C


## 2.7.2.5 SI - Sign Test

    Statpal asks you to specify the  two  variables  to  be
compared using the Sign test, and processes the  data.  The
results  reported  are  the  number of  cases  with  positive,
negative, and  zero  differences,  and  the  two-tailed
significance of the difference, where the null hypothesis is
that the  probability of a positive (or negative) difference
for any one case is .5.  Zero differences are ignored in the
computations.  If  the  number  of  cases  with  non-zero
difference is 25 or fewer, both the exact significance level
(using the Binomial distribution)  and a Normal approxmation
are reported; otherwise,  only  the  Normal approximation is
used.  See Chalmer (Understanding Statistics, Chapter 8) for
a description of the Sign test.

Sign Test:  Example Statpal Session

Enter the two variables to be compared.
  (Return to stop, $$$ to see list of available variables, !!! to cancel
  procedure.)
  ===> SYSTOL<RET>
  ===> DIASTOL<RET>

Processing data for case ___ of 50 ...

        Statpal - Sign Test   -   7/14/85  13:06 - File: EXAMPLE.STP
Between SYSTOL      Systolic blood pressure
    and DIASTOL     Diastolic blood pressure

Number of cases with positive difference (SYSTOL   minus DIASTOL  > 0): 48

Number of cases with negative difference (SYSTOL   minus DIASTOL  < 0): 0

Two-tailed significance (using Normal approximation) =  0.0000

Valid cases:  48  Missing cases:  2

Press C to continue, P to print... C


## 2.7.2.6 SP - Spearman's Rho (Rank Correlation Coefficient)

Statpal asks you to specify the two variables for which
Spearman's rho is to be computed, and then processes the
data and reports the results. The one-tailed significance
level is based on the t-distribution (Understanding Statis-
tics, Chapter 11).

Spearman's Rho:  Example Statpal Session

Enter the two variables for which Spearman's rho is to be computed.
  (Return to stop, $$$ to see list of available variables, !!! to cancel
  procedure.)
  ===> SYSTOL<RET>
  ===> DIASTOL<RET>

Processing data for case ___ of 50 ...

        Statpal - Spearman's Rho   -   7/14/85  13:07 - File: EXAMPLE.STP

Between SYSTOL      Systolic blood pressure
    and DIASTOL     Diastolic blood pressure

Spearman's rho =  0.9484   One-tail significance = 0.0000

Valid cases:  48  Missing cases:  2

Press C to continue, P to print... <u>C</u>


## 2.7.2.7 SR - Wilcoxon Signed Rank Test

Statpal asks you to specify the two variables to be compared using the Wilcoxon Signed Rank test, and then processes the data. The reported results include the sums of the ranks of the positive and negative differences, the number of cases with zero difference, and a two-tailed significance level. If the number of cases with non-zero difference was 50 or fewer, the significance level comes from a table look-up; otherwise, the significance level comes from a Normal approximation (Understanding Statistics, Chapter 8).


Wilcoxon Signed Rank Test:  Example Statpal Session

Enter the two variables to be compared.
  (Return to stop, $$$ to see list of available variables, !!! to cancel
  procedure.)
  ===> <u>SYSTOL</u><RET>
  ===> <u>DIASTOL</u><RET>

Processing data for case ___ of 50 ...

  Statpal - Wilcoxon Signed Rank Test   -   7/14/85  13:07 - File: EXAMPLE.STP
Between SYSTOL      Systolic blood pressure
    and DIASTOL     Diastolic blood pressure

Cases with positive difference (SYSTOL   - DIASTOL > 0): 48
  Sum of ranks =    1176   Mean rank = 24.5000

Cases with negative difference (SYSTOL   - DIASTOL < 0): 0
  Sum of ranks =       0

Two-tailed significance from table < .05.
Two-tailed significance using Normal approximation =  0.0000

Valid cases:  48  Missing cases:  2
Press C to continue, P to print... <u>C</u>

## 2.8 REGRESSION

### 2.8.1 REgression: Overview

Statpal's REgression procedure allows you to build a
linear regression model, adding and removing independent
variables and displaying intermediate results
interactively. At any point in the procedure you may
instruct Statpal to compute residuals and predicted values
for the current model, saving either or both as new
variables in the current Statpal system file.

### 2.8.2 REgression: Prompts

Statpal asks you to specify a list of variables to be
included in the regression procedure. The first variable in
the list is treated as the dependent variable, with the
other variables independent. The maximum number of
variables allowed is implementation-specific, and is
documented in the on-line help category for
implementation-specific features.

Next, Statpal processes the data. Any case with a
missing value on any of the variables included in the
procedure is excluded from the computations.

If you specified a total of two variables (one
dependent variable and one independent variable), Statpal
will immediately display regression results for the linear
regression model using the independent variable. Statpal
will then prompt you for a command as described below.

If you specified more than two variables (i.e., more
than one independent variable), Statpal will NOT immediately
display results. Rather, Statpal waits for you to build a
model, using the commands described below to add independent
variables to the model. Thus, when you are given the first
command prompt after Statpal has processed the data, there
are not yet any independent variables in the model. If you
display results at this point, Statpal will use the "means"

model--that is, regression results for the model with an
intercept only.  (The coefficient for the intercept will be
simply the mean of the dependent variable.)

The command prompt offers you the choices ADD, DISplay,
HELp, NEWmodel, OUTs, PRInt, REMove, RESids, or STOp. The
commands have the following functions:

- ADD: Statpal recaps which variables are currrently in
  and out of the model, and asks you which one you wish
  to add.

- DISplay: Statpal displays information about the current
  model. For the intercept and each independent
  variable, statistics displayed include: the regression
  coefficient for the intercept and for each independent
  variable in the equation, labeled "B"; the standard
  error of B; a t score for B (B divided by its standard
  error); and a one-tailed p-value labeled "Prob",
  derived from the t score. For the model overall,
  Statpal reports the analysis of variance table,
  R-squared, and R-squared adjusted for degrees of
  freedom.

- HELp: Statpal displays a brief description of each
  command.

- NEWmodel: Statpal goes back to the original prompt for
  the regression variable list, allowing you to specify a
  new model. This command has the same effect as
  starting the REgression procedure over.

- OUTs: Statpal displays information about the variables
  not included in the model. Information displayed
  includes the regression coefficient for each variable
  if it were the next variable added to the model,
  labeled "B-in"; the standard error of B-in; the
  tolerance of each variable (equal to 1 minus the
  squared multiple correlation of the variable with the
  independent variables already in the model); the t
  score for B-in (equal to B-in divided by its standard
  error); and a one-tailed probability derived from the t
  score.

- PRInt: Statpal asks whether you wish to print results
  for the Current model (i.e., the results you get on the
  screen when you use the DISplay option) or Outs (i.e.,
  the results you get on the screen when you use the OUTs
  option), and then sends the requested output to the
  printer (or print file).

- REMove: Same as ADD, except that Statpal asks you which
  variable you wish to remove from the model. (If there
  are no independent variables in the model, as is the
  case at the beginning of the process if you have more
  than one independent variable, Statpal will not offer
  you this option.)

- RESids: Statpal asks you whether you wish to compute
  residuals or predicted values, and prompts you for
  specifications for the variable to hold the resulting
  values. You may name either a new variable or an
  existing variable; if you name an existing variable,
  Statpal will request confirmation. Statpal then
  computes the selected values (residuals or predicted
  values), based on the current model, for each case.
  You may calculate residuals and predicted values
  whenever you wish in the REgression procedure, and then
  use other Statpal procedures (such as SCatterplot and
  HIstogram) to examine them. Any case excluded from the
  computations because of missing values will be assigned
  the missing value for the variable of residuals or
  predicted values. Similarly, if case selection is in
  effect, Statpal will set the residual or predicted
  value to missing for any case not meeting the selection
  criterion.

- STOp: Statpal returns to the Main Menu.


2.8.3 REgression: Computational Algorithms

The REgression procedure uses the sweep operator
(Goodnight, 1982) to add and remove variables from the
equation and obtain the appropriate sums of squares. The
original matrix to be swept is built using a one-pass

provisional accumulation algorithm to produce corrected sums
of squares and cross-products (Kennedy and Gentle, 1980,
pp. 322-323).

R-squared adjusted for df is computed as R-squared
times (N-k-1) divided by (N-1), where N is the sample size
and k is the number of independent variables included in the
model (see Neter and Wasserman, 1974).

## 2.8.4 REgression: Example Statpal Session

```
Variables to be used:  Enter the dependent variable, followed by independent
  variables.
  (Return to stop, $$$ to see list of available variables, !!! to cancel
  procedure, ALL to use all variables.)
    ===> SYSTOL<RET>
    ===> AGE<RET>
    ===> DIASTOL<RET>
    ===> SEX<RET>
    ===> <RET>

Processing data for case ___ of 50 ...

Command? (ADD,DISplay,HELp,NEWmodel,OUTs,PRInt,RESids,STOp) ===> ADD<RET>

DEPENDENT VARIABLE:  SYSTOL
  No independent variables are in the model.
  Independent variables not in the model:  AGE DIASTOL SEX
  Which variable do you wish to add to the model (press Return to stop)? DIASTOL

Command? (ADD,DISplay,HELp,NEWmodel,OUTs,PRInt,REMove,RESids,STOp) ===> DIS<RET>


        Statpal - Regression  -  7/14/85  13:10 - File: EXAMPLE.STP


Dependent variable:  SYSTOL
  Independent variables in the model:
    DIASTOL
  Independent variables not in the model:
    AGE      SEX
For variables not in the model:

Variable       B-in      Std Error  Tolerance  t Score     2-tail Sig.

AGE          -0.0353      0.0379     0.9850    -0.9324      0.3563
SEX           2.0950      0.9162     0.9882     2.2867      0.0228

Command? (ADD,DISplay,HELp,NEWmodel,OUTs,PRInt,REMove,RESids,STOp) ===> RES<RET>
```

Do you wish to save Residuals or Predicted values (N for neither)? R
Enter specifications for new variable of residuals...
  Variable name (Return to stop, $$$ to see list):  RESIDS<RET>
  Label:  Using DIASTOL to predict SYSTOL<RET>
  Missing value (default is -999999): <RET>

Creating scratch file...

Processing data for case ___ of 50 ...

Command? (ADD,DISplay,HELp,NEWmodel,OUTs,PRInt,REMove,RESids,STOp) ===> STO<RET>


          Statpal - Regression   -   7/14/85  13:10 - File: EXAMPLE.STP


Dependent variable:  SYSTOL
  Independent variables in the model:
     DIASTOL
  Independent variables not in the model:
     AGE      SEX

Variable          B          Std Error    t Score    2-tail Sig.

Intercept      35.4856        3.8519       9.2124      0.0000
DIASTOL         0.8475        0.0384      22.0825      0.0000

Valid cases:  47   Missing cases:  3


Analysis of Variance

Source            SS          DF      MS         F       Sig.
Regression     5196.8907       1   5196.8907   487.6379 0.0000
Residual        479.5773      45     10.6573

Total          5676.4681      46    123.4015

R-squared =  0.9155
R-squared adjusted for DF =  0.9136

Command? (ADD,DISplay,HELp,NEWmodel,OUTs,PRInt,REMove,RESids,STOp) ===> OUT<RET>


## 2.9 SCATTERPLOT


### 2.9.1 SCatterplot: Overview

     Statpal's SCatterplot  procedure  displays  a bivariate
scatterplot, with each case represented by a  point  on  the
plot.

## 2.9.2 SCatterplot: Prompts

Statpal asks you to specify the two variables to be used for the plot, with the first variable defining the vertical (Y) axis and the second the horizontal (X) axis. Next, Statpal processes the data and displays the plot.

Individual cases are represented by a single point. If two cases must be plotted at the same screen position, a colon (:) is displayed. If more than two cases must be plotted at the same screen position, the appropriate digit is used up through 9; for more than 9, the letter M (for "many") is used.

After the scatterplot has been displayed (and optionally printed), Statpal gives you the option of doing another scatterplot or returning to the Main Menu.

## 2.9.3 SCatterplot: Computational Algorithms

Statpal takes two passes through the data to do a scatterplot. On the first pass, Statpal finds the minimum and maximum values for each of the two variables; Statpal then scales the plot and displays the axes on the screen. On the second pass, Statpal assigns each case to the appropriate screen position, keeping count of the number of cases plotted at each screen position. Once all cases have been displayed on the screen, Statpal has the counts for each position in memory, and therefore does not need to reprocess the data if you wish to print the scatterplot on the printer.

## 2.9.4 SCatterplot: Example Statpal Session

Enter the vertical variable, then the horizontal variable.
  (Return to stop, $$$ to see list of available variables, !!! to cancel
  procedure.)
    ===> DIASTOL<RET>
    ===> SYSTOL<RET>

Processing data for case ___ of 50 ...

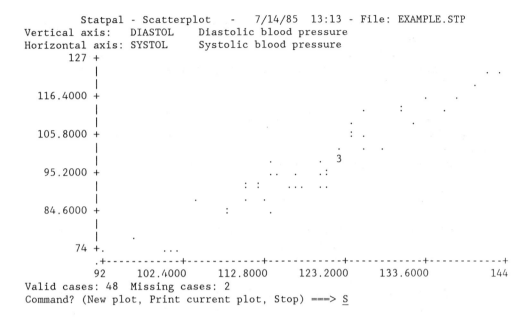

```
          Statpal - Scatterplot  -   7/14/85  13:13 - File: EXAMPLE.STP
Vertical axis:   DIASTOL     Diastolic blood pressure
Horizontal axis: SYSTOL      Systolic blood pressure
     127 +                                                        . .
         |                                                      .
         |                                                  .       .
 116.4000 +                                           .          .
         |                                          .      :      .
         |                                        .
 105.8000 +                                      : .
         |                                       .   .
         |                             .       . 3
  95.2000 +                           .. . .    .:
         |                     : :    ...   ..
         |                  .        .
  84.6000 +              :         .
         |
         |     .
     74 +.         ...
         .+------------+------------+------------+------------+-----------+
          92    102.4000     112.8000     123.2000     133.6000     144
Valid cases: 48  Missing cases: 2
Command? (New plot, Print current plot, Stop) ===> S
```

## 2.10 TABLE (CROSSTABULATE TWO VARIABLES)

### 2.10.1 TAble: Overview

Statpal's TAble procedure gives you a crosstabulation of two variables, showing the number of cases with each possible combination of values on the two variables. You can also display the data as percents of rows, columns, or total sample size, display expected values, and calculate the Chi-square statistic and its significance level. If you wish, you may instruct Statpal to merge several categories into one (on either or both variables) and redisplay the results, without the necessity of reprocessing the data.

### 2.10.2 TAble: Prompts

Statpal asks you to specify two variables, with the first variable defining the rows of the table and the second

defining the columns. Statpal then processes the data and
displays the number of cases with each possible combination
of values on the two variables. If more than one screenful
is required to display the complete table, Statpal reports
the number of displays required.

Statpal then prompts you for a command from the list
CHisq, EXp, HElp, MErge, OBs, PCts, PRint, and STop, plus
NExt and LAst if two or more displays are required for the
complete table. The commands are as follows:

- CHisq: Statpal calculates the Chi-square statistic for
  the table and its significance level. If any of the
  cells of the table have expected counts less than five,
  Statpal tells you how many (both the number and the
  percent of the total number of cells in the table).

- EXp: Statpal displays the table (or current display of
  the table, if two or more are required), showing the
  expected count in each cell assuming independence with
  fixed marginals.

- HElp: Statpal gives a brief description of each
  command.

- MErge: Statpal lists the row and column values, and
  asks you to specify whether you wish to merge row
  categories or column categories. Then, Statpal asks
  you to specify the values you wish to merge to form a
  new category, and prompts you for a new value to use
  for the merged category. The default new value is the
  first old value (e.g., if you are merging categories
  coded 3, 5, and 8, the default new value for the merged
  category will be 3). You may merge categories as many
  times as you wish during the TAble procedure. Note,
  however, that once merged, values cannot be unmerged
  within the TAble procedure. The merging does not
  affect the data in your Statpal system file; you can
  always go back to the beginning of the TAble procedure
  if you want to see the unmerged categories again, or
  merge them in a different way.

- OBs: Statpal displays the observed number of cases in each category. This is what Statpal displays for the first table; you need only use this command if you have previously used PCts or EXp during the TAble procedure.

- PCts: Statpal asks you whether you wish to display the observed counts as percentages of Rows, Columns, or Total sample size. Then Statpal redisplays the table (or current display of the table, if two or more are required) as per your request.

- PRint: Statpal prints the most recently displayed results on the printer (or print file). This will either be the current table (or current display of the table, if two or more are required), or the chi-square results.

- STop: Statpal leaves the TAble procedure and returns to the Main Menu.

- NExt and LAst: If two or more screenfuls are required to display the entire table, Statpal displays the next or previous (last) display. Statpal cycles through the table by incrementing rows first; when all rows have been displayed for a given set of columns, Statpal goes on to the next set of columns.

Whichever form of data you have selected (observed counts, percentages, or expected counts), Statpal always indicates in the upper left corner of the table what it is displaying.

There is an implementation-specific limit to the number of distinct values allowed for each value. See the on-line help category of implementation-specific features for more information. If this limit presents a problem, you can use the REcode transformation to collapse the data into fewer categories before running TAble.

## 2.10.3 TAble: Computational Algorithms

The expected count for the cell in row i and column j is computed as Ri times Cj divided by N, where Ri is the total number of cases in row i, Cj is the total number of cases in column j, and N is the total number of cases in the entire table. Chi-square is calculated as the sum over all cells in the table of [(O minus E) squared] divided by E, where O and E are the observed and expected counts in each cell, respectively. The degrees of freedom for the significance level is computed as (r minus 1) times (c minus 1), where r and c are the number of rows and columns in the table, respectively.

See Chalmer (Understanding Statistics, Chapter 15) for a description of the chi-square test.

## 2.10.4 TAble: Example Statpal Session

Enter the two variables to be cross-tabulated: row first, then column variable.
  (Return to stop, $$$ to see list of available variables, !!! to cancel
  procedure.)
    ===> DRUG<RET>
    ===> SEX<RET>

Processing data for case ___ of 50 ...


            Statpal - Crosstabulation   -   7/14/85  13:16 - File: EXAMPLE.STP
Row variable:    DRUG        1=A, 2=B, 3=C, 4=D
Column variable: SEX         1=male, 2=female

Observed Frequencies
                  1      2   Total
        1         5      5     10
        2         8      7     15
        3         6      7     13
        4         6      6     12

   Total         25     25     50

Valid cases: 50    Missing cases: 0

Command? (HElp, PCts, OBs, EXp, MErge, CHisq, PRint, STop) ===> PC<RET>

Display the frequencies as percent of Row, Column, or Total n? ===> R

```
        Statpal - Crosstabulation  -  7/14/85  13:16 - File: EXAMPLE.STP
Row variable:    DRUG        1=A, 2=B, 3=C, 4=D
Column variable: SEX         1=male, 2=female

Percent of Row
              1     2   Total
     1     50.00 50.00 100.00
     2     53.33 46.67 100.00
     3     46.15 53.85 100.00
     4     50.00 50.00 100.00

  Total   50.00 50.00 100.00

Valid cases: 50   Missing cases: 0


Command? (HElp, PCts, OBs, EXp, MErge, CHisq, PRint, STop) ===> CH<RET>

For entire table, Chi-square = 0.1436 with 3 df...Significance = 0.986

Command? (HElp, PCts, OBs, EXp, MErge, CHisq, PRint, STop) ===> ME<RET>


Current categories:
For row variable DRUG:  1  2  3  4
For column variable SEX:  1  2

Note:  Once merged, categories cannot be separated within this procedure.

Merge? (Row, Column, Help, Stop) ===> R

Enter categories to be merged (press Return to stop):
   ===> 1<RET>
   ===> 2<RET>
   ===> <RET>

Enter code for merged category (default is 1, or !!! to abort) ===> <RET>

Current categories:
For row variable DRUG:  1  3  4
For column variable SEX:  1  2

Note:  Once merged, categories cannot be separated within this procedure.

Merge? (Row, Column, Help, Stop) ===> S

Command? (HElp, PCts, OBs, EXp, MErge, CHisq, PRint, STop) ===> OB<RET>
```

```
        Statpal - Crosstabulation  -   7/14/85  13:58 - File: EXAMPLE.STP
Row variable:    DRUG        1=A, 2=B, 3=C, 4=D
Column variable: SEX         1=male, 2=female

Observed Frequencies
                 1     2  Total
        1       13    12    25
        3        6     7    13
        4        6     6    12

  Total         25    25    50
```

Valid cases: 50   Missing cases: 0


Command? (HElp, PCts, OBs, EXp, MErge, CHisq, PRint, STop) ===> <u>ST</u><RET>


## 2.11 TTEST


### 2.11.1 TTest: Overview

Statpal's TTest procedure allows you to perform either a grouped (i.e., independent samples) or a paired t-test. For a grouped t-test, you specify a dependent variable and a grouping variable, and Statpal compares the means of the two groups as defined by the grouping variable. For a paired t-test, you specify two variables, and Statpal tests the mean difference score to determine if it is significantly different from zero.

Note that you can use the paired t-test procedure to perform a one-sample t-test--that is, a test of the null hypothesis that the mean of a single population is different from some specified constant. To do this, you need only use the TRansformation utility to define a variable with a constant value (specifically, the constant specified in the null hypothesis). You can then perform a paired t-test comparing that variable with the variable of interest, and you will have performed a one-sample t-test.

2.11.2 TTest: Prompts

Statpal first asks you whether you wish to do a grouped
or a paired t-test, which you indicate by pressing either G
or P.


2.11.2.1 Grouped T-test

If you are doing a grouped t-test, Statpal asks you to
specify the dependent variable followed by the grouping
variable. The grouping variable is used to define the two
independent samples to be compared in terms of the dependent
variable. Statpal next asks you to specify the value for
the grouping variable that defines Group 1. All cases with a
different value on the grouping variable will be assigned to
Group 2.

Once you have specified the value that defines Group 1,
Statpal processes the data and reports the results. These
include the means, standard deviations, and sample sizes for
the two groups, the difference between the two means (Group
1 mean minus Group 2 mean), an F-test comparing the
variances of the two groups, and the t-test performed with
both pooled and separate variance estimates, together with a
two-tailed significance level for each method.

Statpal then gives you the opportunity to print or
redisplay the results, perform another t-test, or return to
the Main Menu.


2.11.2.2 Paired T-test

If you are doing a paired t-test, Statpal asks you to
specify the two variables to be compared, and then processes
the data and reports the results. These include the means
and standard deviations for the two variables, the mean,
standard deviation and number of the difference scores
(computed as first variable minus second variable), the
Pearson correlation coefficient between the two variables

and its one-tailed significance, and the t-score for the
mean difference score with its two-tailed significance.

As with a grouped t-test, Statpal then allows you to
print the results, perform another t-test, or return to the
Main Menu.


## 2.11.3 TTest: Computational Algorithms

For both types of t-test, Statpal uses the same
provisional accumulation algorithm referred to in the
DEscriptive statistics procedure for the computation of
means and standard deviations in each group (for a grouped
t-test) or for each variable and the difference score (for a
paired t-test).

For the grouped t-test, Statpal reports two different
t-tests: one using a pooled variance estimate in the
computation of the standard error of the difference between
means, and the other using separate variance estimates. The
formulas are given in Chalmer (Understanding Statistics,
Chapter 9). For the test using separate variance estimates,
the number of degrees of freedom is calculated using the
formula given in Afifi and Azen (1979), truncated to the next
lower integer.

## 2.11.4 TTest: Example Statpal Session

Do you wish to do a grouped (G) or paired (P) t-test? (S to stop) ===> G̲

Enter the dependent variable, followed by the grouping variable.
   (Return to stop, $$$ to see list of available variables, !!! to cancel
   procedure.)
   ===> SYSTOL<RET>
   ===> SEX<RET>

Enter the code for variable SEX that defines group 1.  (Group 2 will
   consist of all cases not in group 1.  Press Return to stop.)
   ===> 1̲<RET>

Processing data for case ___ of 50 ...

```
        Statpal - Grouped T-test   -   7/14/85  14:04 - File: EXAMPLE.STP
Dependent variable: SYSTOL      Systolic blood pressure
Grouping variable:  SEX         1=male, 2=female
                             Mean      Std.Dev.    Minimum    Maximum     N

SEX       equals 1
                           118.8800    13.5349    92.0000   144.0000    25
SEX       does not equal 1
                           119.7826     9.6104   100.0000   138.0000    23
```

F-test for equality of variances:
  F score: 1.9835 with 24 and 22 df...Two-tail significance = 0.1110

Mean of group 1 minus mean of group 2: -0.9026

Pooled variance estimate t-test:
  Standard error of difference: 3.4156
  t score: -0.2643 with 46 df...Two-tail significance = 0.7934

Separate variance estimate t-test:
  Standard error of difference: 3.3680
  t score: -0.2680 with 43 df...Two-tail significance = 0.7906

Valid cases: 48  Missing cases: 2

Command? (Print, Redisplay, Stop) ===> S

Do you wish to do another t-test? (Y/N) Y


Do you wish to do a grouped (G) or paired (P) t-test? (S to stop) ===> P

Enter the two variables to be compared.
  (Return to stop, $$$ to see list of available variables, !!! to cancel
  procedure.)
    ===> SYSTOL<RET>
    ===> DIASTOL<RET>

Processing data for case ___ of 50 ...

```
        Statpal - Paired T-test   -   7/14/85  14:05 - File: EXAMPLE.STP
Variable 1: SYSTOL      Systolic blood pressure
Variable 2: DIASTOL     Diastolic blood pressure
                    Mean      Std.Dev.   Minimum    Maximum     N

    SYSTOL        119.3125    11.7041    92.0000    144.0000
    DIASTOL        99.0625    12.9457    74.0000    127.0000
    Difference     20.2500     3.7274    13.0000     28.0000     48
```

Correlation coefficient:
  Pearson's r =  0.9592    One-tail significance = 0.0000

Standard error of mean difference:  0.5380

t score:  37.6390 with 47 df...Two-tail significance = 0.0000

Valid cases: 48  Missing cases: 2

Command? (Print, Redisplay, Stop) ===> S

Do you wish to do another t-test? N

Case selection remains in effect for the duration of the Statpal session, unless you either turn it off explicitly (also done using the SElect utility) or do something that changes which Statpal system file is the "current" file. Case selection is an attribute not of the Statpal system file, but rather of the session. When you leave Statpal and start again, case selection will no longer be in effect.

Every Statpal procedure that makes use of case selection informs you conspicuously when case selection is in effect. The message includes a statement of the selection criterion.

Case selection is respected by all statistical procedures (i.e., those documented in Chapter 2). In REgression, if you request that residuals or predicted values be calculated, cases not meeting the selection criterion are assigned the missing value for the variable of residuals or predicted values.

In the TRansformation utility, if case selection is in effect, cases not meeting the selection criterion will be left <u>unchanged</u> by transformation commands. For newly created variables, all cases start out with missing values prior to the execution of the transformation commands. Thus, cases not meeting the selection criterion will be left with missing values on newly created variables.

Other Statpal utilities handle case selection as follows:

- BAckup utility: Case selection is respected (with a warning issued). Only cases meeting the selection criterion will be saved in the backup file.

- DElete utility: Case selection is ignored. If you are deleting cases, the case numbers refer to the sequential position of cases in the file, regardless of any selection that may be in effect.

- EDit utility: Case selection is ignored. All cases are included in the procedure; you may change a case regardless of any selection criterion.

# 3
# Statpal Utilities

This chapter describes the items on the Main Menu other than the CReate facility (described in Chapter 1) and statistical procedures (described in Chapter 2). It is assumed that you have read Chapter 1.


## 3.1 HELP (THE ON-LINE HELP UTILITY)

The HElp utility is self-explanatory; try it out!


## 3.2 PRINTER CONTROL UTILITY

### 3.2.1 PRinter Control: Description

The PRinter control utility allows you to redirect what Statpal calls "printed" output to go to a file (which we'll call a "print file") instead. Thus, the results from any statistical procedure can be saved in a file which you can subsequently edit (e.g., for publication) using any standard word processing software.

When you select the PRinter control utility on the Main Menu, Statpal will ask you to specify either (1) the name of the print file, if you want output to go into a file, or (2) the special device name LST: (be sure to include the colon), if you want output to go directly to your printer. Statpal will then confirm your response. (Note: Statpal uses standard PC-DOS/MS-DOS device names. Feel free to use any legal file name for a print file, but be careful with device names. You should not attempt to use a device other than LST: unless you are familiar with device names and their effects.)

The output destination you specify will remain in effect during the Statpal session until you change it again. If you are sending output to an existing print file, all output will be appended to the file. This applies even to the first output generated during a session; that is, if you specify an already existing file as a print file, Statpal will append to it, not overwrite it. Since all output indicates the date and time it was generated, you can easily keep track of multiple analyses in the same print file.

### 3.2.2 PRinter Control: Example Statpal Session

```
Enter name of print file, or LST: (including the colon) to use the printer.
  (Press Return to leave the printer destination as LST:.)
    ===> EXAMPLE.TXT<RET>
Printed output will be appended to file EXAMPLE.TXT.
Press any key... C
```

### 3.3 SELECT CASES UTILITY

#### 3.3.1 SElect Cases: Description

The SElect cases utility allows you to perform statistical procedures (and listing of cases) for a subset of the cases in a Statpal system file. Using the Select cases utility, you specify a criterion that must be before a case is included in subsequent procedures. you have specified the selection criterion, Statpal each case against the criterion before including it i subsequent statistical procedure.

A criterion for case selection contains three ele a variable name, a relational operator, and a value. example, suppose you have a variable called AGE, an wish to include in subsequent procedures only those for which AGE is greater than 30. Your selection cri would then specify the variable name AGE, the rela operator "greater than", and the value 30.

Statpal prompts you for the three elements criterion one at a time, asking first for the variabl then for the relational operator (selected from a men two-letter abbreviations), and finally for the value.

Although you may not specify criteria involvi than one variable or relational operator, you can acc the same thing by creating a new variable (usi TRansformation utility) that has one value for the c want selected, and another value for the cases y excluded.

For example, suppose you wish to select cases t both AGE less than 30 and SEX equal to 1, and exc others. Using the TRansformation utility, you firs a new variable--say, AGESEX--that has a value of 1 f with both AGE less than 30 and SEX equal to 1, and ( other cases. Transformation statements that will are:

```
AGESEX = 0
IF (AGE LT 30 AND SEX EQ 1) AGESEX=1
```

If you then use SElect, specifying "AGESEX equ your criterion, you will be accomplishing what you to do: select only those cases for which both AGE than 30 and SEX equals 1.

- SOrt cases utility: Case selection is respected (with a warning issued). Only cases meeting the selection criterion will be saved in the sorted file.

- WRite data utility: Case selection is respected (with a warning issued). Only cases meeting the selection criterion will be written out to the text or DIF file.

## 3.3.2 SElect Cases: Example Statpal Session

Enter variable to use for case selection (press Return to go back to Main Menu,
  !!! to cancel case selection, $$$ for list of variables) ===> AGE<RET>

Enter a relational operator:
            LT (less than)          LE (less than or equal to)
            GT (greater than)       GE (greater than or equal to)
            EQ (equal to)           NE (not equal to)
                          ST (stop--cancel case selection)

                    What is your choice? ===> GT<RET>

Enter the value to be used for case selection:
      AGE       GT       ===> 30<RET>

Case selection is now in effect.  Press any key... C

## 3.4 SPECIFY A STATPAL SYSTEM FILE

   The SPecify utility allows you to change the Statpal system file you are working on during a Statpal session. As discussed in Chapter 1, Statpal assumes you wish to continue to work with the current Statpal system file unless you tell it otherwise. The SPecify utility is the means for telling Statpal that you wish to change what it considers to be the current file.

   When you choose the SPecify utility, Statpal asks you for the name of the Statpal system file you wish to use. When you have named a valid Statpal system file, Statpal returns to the Main Menu.

If there is no current file, every statistical procedure or utility that requires an existing Statpal system file will automatically ask you to specify one. Thus, you need not use the SPecify utility unless you wish to change from one current file to another during the same Statpal session.

For convenience, you can access the SPecify utility from either the Main Menu or the SYstem file utilities menu.

## 3.5 STOP UTILITY

The STop utility gets you out of Statpal, returning you to your computer's operating system.

## 3.6 SYSTEM FILE UTILITIES

Choosing the SYstem file utilities item on the Main Menu sends you to the SYstem file utilities menu. The SYstem file utilities menu includes a number of utilities for handling Statpal system files. Each item will be descibed in turn.

### 3.6.1 ADd Cases or Variables Utility

### 3.6.1.1 ADd: Description

The ADd utility allows you to add either new cases or new variables from raw data to an existing Statpal system file. You can add the raw data either from keyboard or from a disk file.

The ADd utility is designed for situations in which you have additional data that matches the form of the existing system file. That is, if you are adding cases, the ADd utility will look for data for all existing variables; if you are adding variables, the ADd utility will look for data for all existing cases. In many circumstances, things are more complicated. If so, you need to use the MErge utility (documented later in this chapter), rather than the ADd utility.

When you select ADd, Statpal asks you whether you wish to add cases or variables. Then, Statpal asks you whether you wish to add data from keyboard or from a disk file.

If you are adding cases, the prompts from this point on are the same as those for the CReate facility. Refer to Chapter 1 for a description of how Statpal reads data from keyboard or disk file.

If you are adding variables, Statpal will first ask you for specifications (name, label, and missing value) for each variable to be added, and will then continue as in the CReate facility. The example below shows how this looks in the context of the ADd utility.


## 3.6.1.2 ADd: Example Statpal Session

```
Do you wish to add new Cases or new Variables? (H for help, S to stop) ===> V
Do you wish to input data from a File or Keyboard? (S to stop) ===> F
Enter specifications for the variables to be added...

  Variable name (Return to stop, $$$ to see list):  SYSTOL2<RET>
  Variable label:  Systolic BP at time 2<RET>
  Missing value (default is -999999): -1<RET>

  Variable name (Return to stop, $$$ to see list):  <RET>

Creating scratch file...

Is your data file in a FIxed format or a FRee format?  (Press Return to stop).
  ===> FR<RET>
A free format will be used.

Name of the raw data file:
Enter file name (!!! to cancel) ===> TIME2.DAT<RET>
```

Processing data for case ___ ...

Statpal system file EXAMPLE.STP has 50 cases with 6 variables.
Press any key... <RET>


## 3.6.2 BAckup Utility

The BAckup utility allows you to make a backup copy of
the current Statpal system file. (The file considered to be
"current" remains unchanged.)

A useful feature of the BAckup utility is that it
respects case selection. Thus, the BAckup utility allows
you to make a Statpal system file containing only those
cases that meet the case selection criterion, without
affecting the current Statpal system file.

If your intention is to make an exact copy of the
current Statpal system file, be sure to turn off case
selection (using the SElect utility on the Main Menu) before
you use BAckup. As a safeguard, Statpal will warn you if
case selection is in effect, allowing you to cancel the
BAckup if you wish.


## 3.6.3 COnvert a DIF File Utility

The Data Interchange Format (DIF) is a protocol
originally designed by Software Arts, Inc. to facilitate
the exchange of data among programs (especially spreadsheet
programs). Statpal is capable of both reading and writing
DIF files. The COnvert utility is used to create a Statpal
system file from a DIF file; the WRite utility (documented
later in this chapter) can be used to create a DIF file from
a Statpal system file.

As described in Chapter 1, a Statpal system file
contains two kinds of information: (1) variable
specifications (name, label, and missing value) and (2)
data. Statpal is capable of recovering both kinds of
information from a DIF file, provided the program that
produced the DIF file is capable of handling them. We'll
cover data first, and then variable specifications.

Data: In DIF terminology, Statpal's cases are called "tuples", and variables are called "vectors". In most spreadsheet applications, each column of the spreadsheet would constitute a vector (i.e., a variable), with each row constituting a tuple (i.e., a case). Thus, if you wish to move data between a spreadsheet program and Statpal, you would arrange your spreadsheet data with one row per case, with each column containing data for a variable. (This is the same way Statpal's EDit utility represents data.)

Variable specifications:

- Variable names and labels: Whether or not Statpal can recover information on names and labels from the DIF file depends on what the program producing the DIF file is capable of doing. If names and labels are to be recovered, Statpal requires that the DIF file have a "LABEL" header item (in DIF terminology) for each variable. The content of the label must be a string consisting of the variable name (padded on the right with blanks, if necessary, to bring it out to eight spaces), followed by a single space, followed by the variable label (40 spaces maximum). If Statpal does not find any LABEL header items (presumably because the program that created the DIF file could not produce them), Statpal will use default variable names of the form VAR1, VAR2, etc., and labels will be null. You can then use Statpal's REvise utility to change names and labels.

- Missing values: Statpal requires that the first tuple (i.e., row) of data consist of the missing values for each variable. Thus, before instructing your spreadsheet program to prepare a DIF file from a spreadsheet of data, insert a row of missing values as the first row.

Since Statpal does not use non-numeric data, you should not attempt to recover such data from a DIF file. The best practice is to remove non-numeric data from your spreadsheet before instructing the program to save the DIF file.

Since programs that make use of the DIF format vary greatly in how they use it, it is not possible to cover all possibilities. You may need to experiment somewhat to determine what works. If you find the DIF file produced by another program will not work with Statpal, you might find it easier to have the program write out an ASCII file containing data values only, and then use Statpal's CReate facility to read it.


## 3.6.4 DElete Cases or Variables Utility


### 3.6.4.1 DElete Cases or Variables: Description

The DElete utility allows you to delete a specified range of cases, or a specified list of variables, from an existing Statpal system file.

Statpal first asks you whether you wish to delete cases or variables, and then asks you to name the new Statpal system file to be created.

If you are deleting cases, Statpal asks you to specify a range of case numbers to be deleted, by specifying a starting and ending case number. If you wish to delete only one case, you may name the same case as the starting and ending number.

Note that is is permissible to delete <u>all</u> cases, by specifying 1 as the starting case number and the number of cases in the file as the ending case number. (Statpal will request confirmation if you do this.) This will make a Statpal system file containing all of the variable specifications, but no data. You might wish to do this if there were extensive errors in the raw data file from which the file was created; after you delete all cases, you can make the necessary corrections in the original raw data file (outside of Statpal), and then use the ADd utility to read back in the raw data without having to redo the variable specifications.

If you are deleting variables, Statpal asks you to specify a list of variables to be deleted. Deleting all variables has the effect of deleting the entire file. Statpal will request confirmation if you wish to do this.

The DElete utility deletes data from the current Statpal system file. It is often prudent to use the BAckup utility to make a backup copy of the current Statpal system file before using DElete.

## 3.6.4.2 DElete Cases or Variables: Example Statpal Session

Do you wish to delete Cases or Variables? (H for help, S to stop) ===> V

Enter the variables to be deleted.
  (Return to stop, $$$ to see list of available variables, !!! to cancel
  procedure, ALL to use all variables.)
   ===> SYSTOL<RET>
   ===> DIASTOL<RET>
   ===> <RET>

Creating scratch file...
Transferring variable specifications...
Processing data for case ___ of 50 ...

Statpal system file B:EXAMPLE.STP has 50 cases with 3 variables..
Press any key... <RET>

## 3.6.5 EDit Utility

The EDit utility is a screen-oriented data editor that allows you to view and change the data in a Statpal system file. It is largely self-explanatory, using the built-in Help command, and the best way to learn to use the EDit utility is to try it. This section gives a brief summary of how EDit operates.

Statpal first asks you to specify the variables to be used, and then creates a scratch file. Statpal will then display a screenful of data. The first column is always the special variable CASE# (i.e., the sequence number of each case in the file); Statpal will not allow any changes to this variable.  Subsequent columns display the data for each specified variable.

Near the bottom of the screen is a highlighted  command line showing your options at any point.  At the beginning of the EDit session, the  command  line  includes the following commands:

- Find Case (C): allows you to move to  a particular case (for the current variable).

- Find Variable (V): allows you to move to  a  particular variable (for the current case).

- Insert (I): puts you in Insert mode for the highlighted data point.  In Insert mode, you can enter the  desired number (followed by  Return),  the letter "M" to assign the missing value, or simply press Return to leave  the data point unchanged.  In  addition, you can enter !!! to  cancel  all changes made during  the  current  EDit session, leaving the original system file untouched.

- Stop (S): ends  the  current EDit session, updating the current Statpal system file.

- Help (H): summarizes EDit commands.

You can also use special keys (such as the arrow keys); to find  out  how  these  work  in  your  implementation  of Statpal, try them.


3.6.6 GEnerate Dummy Cases


3.6.6.1 GEnerate Dummy Cases: Description

The GEnerate utility allows you to add cases consisting entirely of missing values to  the  current  Statpal  system file.  You can then use the TRansformation or EDit utilities to insert the data.  For example, you might wish to generate some dummy cases, and then use the TRansformation utility to compute random numbers for one or more variables.

The operation of the GEnerate utility is simple:
Statpal asks you how many cases to generate, and then
proceeds to add them to the current Statpal system file
(adding them after any existing cases). For each case, each
variable will be assigned the missing value in effect for
that variable.


## 3.6.7 MErge Statpal System Files

The MErge utility is a very powerful tool for combining
data in two Statpal system files. Using MErge, you can do
any of the following:

-   Combine two Statpal system files containing data for
    different cases. This is called <u>concatenating</u> files,
    because in the resulting Statpal system file data for
    cases from the second file are concatenated to (i.e.,
    tacked on the end of) data for cases from the first
    file. The two source files would generally have the
    same variables, but Statpal can handle the situation
    even if they do not.

-   Combine two Statpal system files containing data for
    the same cases, but with (mostly) different variables.
    This is called <u>matching</u> files, because data for
    corresponding cases in the two source files are matched
    case by case to form the resulting Statpal system
    file.

-   Combine two Statpal system files that contain data for
    some of the same cases, but also some different cases,
    producing a resulting file that contains all cases
    represented in either source file. This method is
    called <u>matching with key variables</u>, because Statpal
    uses up to four "key" variables (which you specify) to
    identify which cases should be matched with which.

When you enter the MErge utility, Statpal will ask you
to specify the two files to be merged, and will then ask you
to choose either Concatenate or Match. If you choose Match,
Statpal will then ask you to specify key variables, if any.

Regardless of which method you choose, Statpal determines the order of variables in the resulting file in the same way. The order of variables in the new file is as follows: (1) all variables that appear in in the first file (regardless of whether they also appear in the second file), and (2) variables that appear in the second file only. Thus, the set of variables in the new file will be the union of the two sets of variables in the source files. In general terms, the number of variables in the result file will be v1 plus v2 minus vboth, where vk is the number of variables in file k and vboth is the number of variables common to the two files.

It is not necessary that variables common to the two files be in the same order in the two files. Thus, for example, if file 1 contains variables V1, V2, V3, and V6 (in that order), while file 2 contains variables V4, V2, and V5 (in that order), the resulting file will contain variables V1, V2, V3, V6, V4, and V5 (in that order).

We'll consider each method in turn. Note that the MErge utility requires that both source files be Statpal system files. If you are starting with raw data, you need to use the CReate facility to create Statpal system files before using MErge. (Another option that may be appropriate is the ADd utility, which can add to an existing Statpal system file starting with raw data, provided the raw data file has the necessary structure.)

## 3.6.7.1 Concatenating Files

This method is used to combine data from two existing Statpal system files when the two files contain entirely different cases. When you instruct Statpal to concatenate the files, Statpal will proceed as follows:

1.  Statpal will determine the order in which the variables will appear in the new file, and transfer the variable specifications to the new file.

2.  Statpal will transfer the data for cases from the first source file to the new file. Any variables in

the new file not present in the first source file
(i.e., variables that appear only in the second source
file) will be assigned missing values.

3. Finally, Statpal will transfer the data for cases from
   the second source file, appending these to the cases
   already transferred from the first source file.
   Again, any variables in the new file not present in
   the second source file (i.e., variables that appear
   only in the first source file) will be assigned
   missing values.

The result of this procedure is a Statpal system file
containing n1 plus n2 cases, where nk is the number of cases
in the kth file.

3.6.7.2 Matching Cases (Without Key Variables)

As noted above, when you select the Match type of
merge, Statpal will ask you to specify key variables, if
any. If you press Return at that point without specifying
any key variables, Statpal will perform a one-for-one match
between the cases in the two source files. The procedure is
as follows:

1. Statpal will determine the order in which variables
   will appear in the new file, and transfer variable
   specifications to the new file. A warning will be
   given if a variable occurs in both files (see step 2
   below).

2. Statpal will transfer the data from both source files
   to the new file. The first case in the new file will
   contain the data from the first case in each of the
   two files. If a variable occurs in both files, the
   value from the _first_ file will be used for that
   variable, unless that value is the missing value for
   the variable, in which case the value from the second
   file will be used.

3. Statpal will continue to transfer the data from both
   files until all data have been transferred. If one of

the source files is longer than the other, Statpal
will still transfer the data, supplying missing values
as needed for the variables not found.

The result of this procedure is a Statpal system file
with either n1 or n2 cases, whichever is larger, where nk is
the number of cases in file k.


### 3.6.7.3 Matching Files With Key Variables

If you do specify one or more key variables (you may
specify up to four), Statpal will use those variables to
identify which cases in the two files should be matched.
Both source files must contain all key variables; however,
the key variables need not be in the any particular order or
any particular position in either file. The special
variable CASE# cannot be used as a key variable. In the
rare circumstance where this might be desirable, you should
use the TRansformation utility to compute another variable
equal to CASE#, and then specify that variable as a key
variable instead of CASE#.

In order to work properly, <u>Statpal requires that both
source files be sorted in ascending order on all key
variables.</u> This is critical! If a file has not been
sorted, you should use the SOrt utility to sort it,
specifying an ascending sort on the variables you intend to
use as key variables (in the same order).

Statpal proceeds as follows:

1.  Statpal will determine the order in which variables
    will appear in the new file, and transfer variable
    specifications to the new file. A warning will be
    given if any variable other than a key variable occurs
    in both files (see step 2 below).

2.  Statpal then begins the data transfer by looking at
    the key variables for the first case in source file 1
    and the first case in source file 2. If all key
    variables match up, Statpal combines the data into a
    single case (exactly as above when matching without

key variables), writes it into the result file, and
proceeds to the next case from each source file.  If
the key variables do not match up, Statpal writes a
case with data from the file with the smaller values
on the key variables (filling in missing values as
needed), reads another case from that same file, and
checks again for a match.

3.  The process continues until all cases have been read
    from both files.

Note that each case in either source file can have only
one match.  If a set of two or more cases in file 1 (say)
have the same values on the key variables as a single case
in file 2, only the first of the set will be matched up; the
others will be treated as unmatched data (and will therefore
have missing data supplied for the variables that come from
file 2).

The result of this procedure is a Statpal system file
with n1 plus n2 minus nboth cases, where nk is the number of
cases in file k and nboth is the number of cases in each
source file having matching cases in the other source file.


3.6.8 REvise Variable Name, Label, or Missing Value

The REvise utility lets you change variable names,
variable labels, and missing values in an existing Statpal
system file.

When you choose REvise, Statpal asks you to specify the
variable for which you wish to make revisions.  When you
have given your choice, Statpal lists the current variable
name, variable label, and missing value, and prompts you for
a new variable name, variable label, and missing value.  If
you wish to leave the entity as is, press Return without
typing anything.

Statpal will not allow you to rename a variable to have
the same name as an existing variable, or to have the
special variable name CASE#.  However, you may give any

variable label you wish (up to 40 characters), and you may specify any number for a missing value.

Be careful about changing the missing value, since the data values themselves are not changed, but only the value Statpal considers to represent missing data. If, for example, you had entered data for a variable using the default code of -999999 as the code to represent missing information, and you then use REvise to change the missing value to some other number (say, -99), Statpal will treat the values of -999999 in the data as valid values. If this is what you want, fine; otherwise, it's best not to change the missing value once the data are in the system file.


3.6.9 SOrt Cases Utility

The SOrt utility allows you to reorder the cases in an existing Statpal system file according to the values of up to four variables. You may specify either an ascending or descending sort on each variable (independently of the others).

When you enter the SOrt utility, Statpal asks you to specify the variables on which you wish to sort. The order in which you specify the variables determines the precedence of each variable in doing the sort. For example, suppose you have a system file containing (among others) the variables SEX (coded 1 or 2), AGE (in years), and DRUG (coded 1, 2, 3 or 4). If you specify an ascending sort on SEX, AGE, and DRUG, in that order, the resulting file will be ordered as follows: (1) all cases with SEX equal to 1, ordered by AGE, cases with the same age being ordered by DRUG; (2) all cases with SEX equal to 2, ordered by AGE, cases with the same age being ordered by DRUG.

In some situations you might wish to retain information about the previous ordering of the cases in the file. You cannot rely on the special variable CASE# to do this, because CASE# is always equal to the sequence number of the case in the file, and the reordering will therefore change the values of CASE# to reflect the new order. The solution

is use the TRansformation utility before using SOrt to
create a new variable--say, OLDCASE--equal to CASE#. Then,
after you do the sort, the variable OLDCASE for each case
will still be equal to the sequence number of that case
previous to the sort.

There is an implementation-specific limit to the number
of cases that can be sorted. This limit can be found using
the on-line HElp facility. (This limit is the same as the
limit for certain nonparametric test procedures.)

If you have a file with more cases than the SOrt
utility can deal with, you can still accomplish the sort by
using the TRansformation, SElect and BAckup utilities to
divide the file into smaller files. To do this: (1) use the
TRansformation utility to create a variable that divides the
cases into groups according to ranges of values of the sort
variables; (2) use the SElect utility to specify a criterion
that selects one group; (3) use the BAckup utility to save a
system file including only those cases that pass the
selection criterion; (4) repeat steps 2 and 3 until a file
has been created for each of the groups created in step 1.
Then each file can be sorted, and the files concatenated
using the MErge utility.


3.6.10 WRite Utility

The WRite utility allows you to write data from a
Statpal system file into a disk file, creating either a Text
file or a DIF (Data Interchange Format) file. Case
selection (as defined by the SElect cases utility) is
respected by WRite, allowing you to write out data for a
specified subset of cases.

When you enter the WRite utility, Statpal asks you to
indicate which kind of file (Text or DIF) you wish to
write. (Information on DIF files is found in the
description of the COnvert utility elsewhere in this
chapter.) Statpal then asks you to specify the name of the
output file to be written, the variables to be written out,
and the number of decimals to write in the case of
non-integers.

### 3.6.10.1 Writing a Text File

In the text file format, Statpal will write out the data in a format amenable for use as either a fixed-format or a free-format data file. Each case will always start on a new line in the output file. Within a case, the value for each variable will take up a fixed number of spaces (Statpal will inform you how many it used), with a maximum of 80 spaces used on a line. Statpal will use as many lines as necessary for each case (and will inform you how many were necessary).

### 3.6.10.2 Writing a DIF File

In the DIF format, Statpal will give variable names and labels as "LABEL" header items (as described in connection with the COnvert utility). The first "tuple" of data will consist of a case with all missing values. Subsequent tuples will be the actual cases. Thus, if you use WRite to write a DIF file and then read it in using a spreadsheet program, the first row of the spreadsheet will indicate the missing values in effect for all variables.

## 3.7 TRANSFORMATIONS

### 3.7.1 TRansformation Utility: Description

The TRansformation utility allows you to compute new variables or recompute existing variables using a variety of operators and functions. You instruct Statpal what transformations you wish to perform by giving a set of transformation statements. Up to 20 statements can be given at a time, and you can have Statpal save your statements in a file for later use.

When you enter the TRansformation utility, Statpal first asks you to give specifications for any new variables you might wish to add to the current Statpal system file.

If you specify one or more new variables at this point, Statpal will add them to the system file with all missing values. The new variables are then available (as are old variables) to be used in transformation statements.

Next, Statpal will prompt you for up to 20 transformation statements. Two basic types of statement can be used: (1) assignments and (2) conditional statements. The statements will be executed in the order you specify them; what you are doing is writing a program for Statpal to use in processing the data. Statements must not exceed 80 characters in length.


## 3.7.1.1 Assignments

Assignments are statements of the form variable = expression, where "expression" indicates how "variable" is to be computed. Here are some examples of assignments:

```
PI = 3.14159
YEARS = MONTHS / 12
X = (-B + SQRT(B^2 - 4*A*C))/(2*A)
RANDOMIQ = NRAND(10) + 100
CASEPLS2 = CASE# + 2
```

The first example statement above would set the variable PI equal to 3.14159. The second would compute the variable YEARS equal to the variable MONTHS divided by 12. The third shows a more complicated expression involving the variables X, A, B and C, along with the function SQRT. The fourth shows the use of the NRAND function to generate Normal random numbers. The last shows the use of the special variable CASE# in a transformation statement. Statpal would not allow any statement in which CASE# appears on the left of the equals sign, because CASE# cannot be changed in any way.

The expression part of an assignment can include the following operators and functions:

Arithmetic operators:
   +    plus       -    minus
   *    times     /    divided by

   ^    exponent
(e.g., VARA^VARB means VARA to the VARB power)

   MOD  modulus
(e.g., VARA MOD VARB means the remainder of VARA
  divided by VARB)

Functions:

All functions take a single argument (called arg below) in parentheses; arg can be any valid expression, including constants, variables, or functions.

| | |
|---|---|
| ABS | Absolute value |
| ARCSIN | Arcsine |
| ARCTAN | Arctangent |
| COS | Cosine |
| IRAND | Random integer between 1 and arg |
| LAG | Value for previous case for arg |
| LN | Natural logarithm (base e) |
| LOG | Common logarithm (base 10) |
| NRAND | Normal random number: mean 0, sd arg |
| ROUND | Round to nearest integer |
| SIN | Sine |
| SQRT | Square root |
| TRUNC | Truncate to integer |
| URAND | Uniform random number between 0 and arg |

In giving an expression, you can (and often should) use parentheses to indicate the order in which computations will be performed. For example, if VARA equals 3 and VARB equals 2, the expression VARA*(VARA+VARB) evaluates to 3 times (3 plus 2), or 15, while the expression (VARA*VARA)+VARB evaluates to (3 times 3) plus 2, or 11.

If you do not give parentheses, Statpal will evaluate expressions in the following order: functions ^ MOD * / +

-.  Thus,  for  example,  the  expression  VARA $^\wedge$  VARB  *
SQRT(VARC*VARD) + VARE  will  be  evaluated  as  ((VARA$^\wedge$VARB) *
SQRT(VARC*VARD)) + VARE.


## 3.7.1.2 Conditional statements

Conditional  statements  have  the  form  IF  (logical
expression) assignment.  That is, a conditional statement is
an assignment preceded by the word IF followed  by a logical
expression  in  parentheses.   Here  are  some  examples  of
conditional statements:

    IF (VARA EQ 1) VARB = 17
    IF (AGE GT 30 AND AGE LT 50) AGEGROUP = 2
    IF ((B$^\wedge$2-4*A*C) LE 0) ROOTTYPE = -1
    IF (DENOM) FRAC = NUM/DENOM

In  the  first  example,  the  variable  VARB  will  be
assigned the value 17 for those  cases  for  which  VARA  is
equal to 1. In the second example, AGEGROUP will be set to 2
for cases with  AGE  greater  than  30 and less than 50. The
third  example  illustrates  the  use  of  an  arithmetic
expression  within  the  logical  expression;  any  valid
arithmetic  expression (including functions) can be used  in
this way.

The last example shows  a  special  property of logical
expressions  in  Statpal:  you  can  use  any  arithmetic
expression  as  a  logical  expression,  because  Statpal
considers the number zero to  be  equivalent  to  a  logical
"false", and any other number to be equivalent  to a logical
"true".  Thus,  in the example, the assignment that computes
the variable FRAC will be  carried  out only if the variable
DENOM is non-zero.

You can  also  use  logical  expressions  as arithmetic
expressions, because Statpal will evaluate a "false" logical
expression  as  the  number  zero,  and  a  "true"  logical
expression as the number 1. For example, the statement YOUNG
= (AGE LE 30) would assign 1 to the variable YOUNG for cases
with AGE less than or  equal to 30, and 0 for cases with AGE
greater than 30.

The following logical operators are available for use in logical expressions:

    EQ    equals
    LE    less than or equal to
    LT    less than
    GE    greater than or equal to
    GT    greater than
    NE    not equal to
    NOT   negation
    AND   and (true if both operands are true)
    OR    or (true if either operand is true)

3.7.1.3 Missing Values in Transformation Statements

In assignment statements, Statpal will assign a missing value to the variable on the left of the equals sign if the expression on the right of the equals sign cannot be computed. There are two general reasons why the expression might not be computable: either (1) the expression involves one or more variables with missing values, or (2) the expression involves an impossible operation, such as dividing by zero. Either way, Statpal will set the result equal to the missing value for the variable being computed.

In conditional statements, Statpal will only carry out the assignment part of the statement if the logical expression is evaluated as "true". If the logical expression is evaluated as "missing" (for either of the two reasons mentioned above), Statpal will not carry out the assignment part. Thus, missing values or impossible computations in the logical expression of a conditional statement will leave the target variable of the assignment part unchanged, just as if the logical expression evaluated to "false".

3.7.1.4 The Command Prompt

Once you have entered your transformation program (including up to 20 statements), Statpal will display a

listing of the statements and then give a command prompt
with the following choices:

- Cancel: Cancels the transformation and returns to the
  Main Menu.

- Delete: Lets you delete a statement.

- Help: Displays these options.

- Insert: Lets you insert one or more additional
  statements into your program.

- Okay: Tells Statpal that the statements are okay as is
  and initiates processing of the data.

- Replace: Lets you replace a statement.

- Save: Saves the current program in a file; the default
  file name is TRANSCOM.TXT. You may subsequently edit
  this file outside of Statpal, if you wish. Commands
  saved in a file can be read in using the READ command
  (documented below).

3.7.1.5 Reading Statements from a File

In addition to entering your transformation program
from the keyboard, Statpal allows you to read transformation
statements from a file. You can even mix both methods,
reading some commands from a file and entering additional
commands from the keyboard. This is accomplished using the
special keywords READ and PAUSE.

To instruct Statpal to read commands from a file, enter
the following command when Statpal is prompting you for a
transformation statement:

READ filename

"Filename" is the name of the file containing your
transformation statements, one statement per line. If

"filename" is omitted, Statpal will look for a file called TRANSCOM.TXT (the same default file name as that used by the Save command documented above).

You may use the READ command at any point in your program, and Statpal will add the commands read from the file to the current program at that point. Thus, you can have a library of commonly used transformations, and can combine them as necessary. However, you may not include a READ command in a file to be read by Statpal (i.e., READ instructions cannot be nested).

In some situations you might wish to have Statpal read statements from a file up to a point, then pause for keyboard commands, and then resume reading from the same file. To do this, put the word PAUSE in the file on a line by itself (using a text editor outside Statpal). When Statpal encounters PAUSE, it stops reading from the file and returns control to the keyboard. You can then enter additional statements. If you then wish to resume reading from the same file, give the command READ again, without specifying a file name, and Statpal will start reading from the line following the PAUSE command.


## 3.7.2 TRansformation: Example Statpal Session


Enter specifications for new variables to be created, if any...

    Variable name (Return to stop, $$$ to see list): AGEGROUP<RET>
    Variable label: Age group: 1=under 30, 2=30-49, 3=50 up<RET>
    Missing value (default -999999): <RET>

    Variable name (Return to stop, $$$ to see list): <RET>

Enter transformation statements (press Return to stop)

1: AGEGROUP = 1
2: IF (AGE GE 30 AND AGE LE 49) AGEGROUP = 2
3: IF (AGE GT 49) AGEGROUP = 3
4: <RET>

Statements have been understood as follows:

1: AGEGROUP=1
2: IF(AGE GE 30 AND AGE LE 49)AGEGROUP=2
3: IF(AGE GT 49)AGEGROUP=3

Command? (Cancel, Delete, Help, Insert, Okay, Replace, Save) ===> O
Processing data for case ___ of 50...

Do you wish to do more transformation? (Y/N) N

# References

Afifi, A. A. and Azen, S. P. (1979). _Statistical Analysis: A Computer Oriented Approach_, 2d Ed. New York: Academic Press.

Barron, T. and Diehr, G. (1983). Sorting algorithms for microcomputers. _Byte_, May 1983, 482-490.

Box, G. E. P. and Muller, M. E. (1958). A note on the generation of random normal deviates. _Annals of Mathematical Statistics_, 29, 610-611.

Goodnight, J. H. (1982). A tutorial on the SWEEP operator. _American Statistician_, 33(3), 149-158.

Kennedy, W. J. and Gentle, J. E. (1980). _Statistical Computing_. New York: Marcel Dekker, Inc.

Neter, J. and Wasserman, W. (1974). _Applied Linear Statistical Models_. Homewood, IL: Richard D. Irwin, Inc.

Siegel, S. (1956). _Nonparametric Statistics for the Behavioral Sciences_. New York: McGraw-Hill.

# Index